ACTING BODIES AND SOCIAL NETWORKS

A Bridge between Technology and Working Memory

Edited and Introduced by

Bianca Maria Pirani and Ivan Varga

University Press of America,® Inc.
Lanham · Boulder · New York · Toronto · Plymouth, UK

University Press of America,® Inc.
4501 Forbes Boulevard
Suite 200
Lanham, Maryland 20706
UPA Acquisitions Department (301) 459-3366

Estover Road
Plymouth PL6 7PY
United Kingdom

Printed in the United States of America
British Library Cataloging in Publication Information Available

Library of Congress Control Number: 2009939316
ISBN: 978-0-7618-4997-1 (paperback : alk. paper)
eISBN: 978-0-7618-4998-8

™ The paper used in this publication meets the minimum requirements of American National Standard for Information Sciences—Permanence of Paper for Printed Library Materials, ANSI Z39.48-1992

Table of Contents

Volume Two *Mapping Bodies in Networked Space*

Foreword

Roberto Cipriani

The acting body is driven by the mind. The mind manages and coordinates all moves and aims of the body. However, the mind is part of the body itself and is the thinking engine as well as the centre of all decisions.

Even in case of a confused or debilitated state of mind all decisions about how, why and what to do, are taken by the operational mechanism inside the skull. No matter, how irrational or unpredictable it may seem for others. Technology is also a product of the acting body, which recourses to previous experiences, accumulates knowledge in order to apply it for finding new solutions.

Such solutions are usually connected with adjustments after failed attempts, but also with some positive results. Both kinds of solutions merge into the final result, which is the new finding. Such new finding is the basis for further development: from choppers used by primitive man to cut, dig or catch preys to the modern choppers which are helicopters cutting air with sharp and rapid propellers (also called choppers for this reason).

The relation between body and mind is not limited to an interior dimension, but it is able to go beyond, to escape and reach extra-bodily dimensions, which nevertheless remain human elaborations, feelings, emotions, pain and joy. Such perceptions would not be possible without a starting point and reference to the body. For each individual the body is situated in space and time, and it comes to an end with death. Other bodies can keep on having experiences of time and space, even if within limits of endurance which are variable but not infinite. However, the mind, unlike technology, has the possibility to overcome space and time, therefore it is capable of surfing the net in both dimensions much better than any powerful search engine, present or future. Moreover, the mind is powerful enough to make projections into the future and capable of transcending its contingency and imagine the potential future for other bodies (and minds), even when the person imagining all that will not be alive anymore. In other words, the acting body or thinking mind has a boundless potential because it can transcend itself, overcome technology by inventing the future.

The body-and-mind ensemble is therefore able to realize utopias; its bodily icons become a sort of future object-subject which is able to lay the foundations for the future, thus fixing referring meanings that can build future knowledge. The extraordinary potential of the acting body is therefore able to anticipate the future preparing its shape in the present. The strength of acting-thinking can carry out complicated elaborations much better than the most powerful electronic computing machine in use. The fact is that such possibilities cannot be fully implemented. Sometimes we remain astonished before scientific results obtained without particularly advanced technologies. Furthermore, certain bodies seriously damaged by various pathologies can develop an extraordinary content of reflexivity and reasoning suggestions that can produce highly complicated connections and are capable of solving basic problems of living and surviving. In the scientific field, as well as for arts, there are many cases of socially disturbed subjects who prove to be cornerstones or turning points for the development of thought and culture, both artistic as well as material and technological

A rough division of typology concerning attitudes and behaviors would induce to distinguish between those who merely repeat what is existing and already known and pioneers of knowledge who are experimenting to find new solutions and assume an unorthodox behavior if compared to the presumed norms. As a matter of fact, a follower is never just repeating knowledge as well as a pioneer is never only experimenting new possibilities. More often subjects are both pioneers and followers at the same time, because subjects have imitative inputs thus can experiment everyday new adaptive and potentially innovating paths.

This is why it makes sense to say that an acting body is interactive by definition, that is, by its unique characteristic. It is the field in which the interactions take place in space and time, where it generates memories, affections, emotions, canalizes knowledge, suggests solutions for living and fosters unpredictable developments, which may be of the stone-age and of the ancient cultures and technologies or of the present age of short messages from mobile phones that act as prosthesis or extension of human minds.

At last, also the use of red ochre, an earthly variety of hematite, represented for the ancients a sort of chromatic appendix of their body. Not even death could stop such use. In fact, many buried bodies still hold traces of such color after thousands years. This way the body-mind realizes utopias of a prolonged life after death: such a particular and extreme construction of an eternal life is actually a decision of the mind, the decision of maintaining clearly the vital energy of the blood with the red ochre color. So vitality can remain clearly visible after death.

Something similar can be said for the recent widespread of tattoos, which recurs to a tradition started in the Upper Paleolithic Period, continuing in the Neolithic and the Copper Age, between 2,500 and 2,000 BC, especially in the Mediterranean area. However, tattoos disintegrate with the decay of the flesh after death; nevertheless some chromatic traces can be found on the bones, where the color stays longer. The relational value of tattoos has to be underlined, because with a simple form, permanent and usually not expensive assumes the role of interactive tool communicative also on an aesthetic level. It is a sort of diffused "Facebook" ante litteram, with a strong importance for identity belonging, just like a distinctive, almost an individual logo, or an exclusive "trademark" for the subject, who wants to distinguish himself from others, even if he tends to get in contact also through such form clearly perceivable.

Today, also glasses, personal passwords (as many as the different kind of services), pacemakers, dentures, earphones, bypasses, and much more as nanotechnologies represent new frontiers of the acting body. Something is certainly changed compared to the past, but to what extent? Which are the characteristics? And this change towards which goal is heading?

This work presents in both parts many possible answers. Reading it through is deeply intriguing: starting from the titles of contributions which are promising strong reflexive inputs up to the entire work that can be considered a valid new mental stimulator.

Introduction

Bianca Maria Pirani, Ivan Varga

Bodies in technology

The two volumes of the Selected Papers presented at the Research Committee 54 "The Body in the Social Sciences" Program of the International Sociological Association's First World Forum of Sociology held in Barcelona on September 4-8, 2008—including the papers by some invited scholars who did not attend that meeting—analyze the complex interactions of mind, body and microelectronic technologies. They bring together the most promising and theoretically fruitful research developments by internationally renowned scholars, whose work is at the cutting edge of research. Experts from both fields—research into the mind/body relationship and studies in science and technology—provide an up-to-date overview, from an original integrative perspective.

In particular, the volumes aim at contributing to the understanding of mind under the present conditions. For conceptualizing "mind" we usually mean the aspects of intellect and consciousness manifested as combinations of thought, perception, memory, emotion, will and imagination.

According to this definition, mind is the stream of consciousness and includes all of the brain's conscious processes.

Today, we need to discover that although the world may be shrinking in terms of physical access, its horizon continue to expand across the distributed assemblages in context creating mobile boundaries between sensory experience and technologies. While corporeal existence of human beings is linked to space and time, electronic technologies overcome the limitations caused by the spatial and temporary existence and enable the users to expand the horizon of space and time, and thus influence the relationship between mind, body and technology. One could argue whether technologies could substitute for emotions or will. Some theorists (Levy: 2007) give a positive answer to this problem, but even those who deny this possibility acknowledge that microelectronic technologies could influence emotions and will, and often do. None the less, the complexity of the problem requires a broader approach. Mind, as such, constitutes a link between the body and thought (often called "soul", even in non-religious texts). Namely, the mind cannot exist without that specific part of the body, called brain. The philosopher Hannah Arendt (1971/78: 30-37), expounds on the relationship between "body and soul, soul and mind". One can add that with the invention of script (which is also a combination of body and mind) it can "freeze" those thoughts beyond the moment they have been uttered, even if the interpretation of the thoughts depends on the culture of a given society and historical period (suffice it to consider that the holy texts of great religions are subject to interpretations that could contradict one another).

Therefore, the questions arise: Does human memory change as 'mapped' by these mobile, shifting boundaries? How should we best analyze the 'acting bodies' in the context of the enormous complexity of social, scientific and cognitive problems arising from the shifting boundaries between bodies, brain and technologies, between humans and machines and between nature and culture?

The human brain—unique in the organic world—has the infinite capacity to adapt itself to the changing conditions of life. Over the last few years popular books about the brain have become a literary genre. As advances in information technology and communication supply us with information at an ever accelerating rate, the limitations of our brains become all the more obvious. The modern work situation, with its pace and simultaneous demands, often gives us the feeling of having attention difficulties. The torrent of information increases not only the volume of data we are expected to take in, but also the volume we need to shut out. The issue of simultaneous performance is particularly interesting right now, as technological progress seems to encourage or even require it. Thanks to the wireless revolution, we can take technology pretty much anywhere we want to. We chat on the phone while walking, driving or watching television. We can have little displays in our cars showing maps that are continually updated and direct us while we drive. While in meetings we can text people or read e-mails in our BlackBerry. We can sit on the sofa with our laptop while watching television, wirelessly connected to the Internet. New findings in psychology and brain research suggest that the difficulties we find with simultaneous performance and distractions converge onto one central limitation: *the ability to retain information.*

The emergence of the complex, compound contemporary technologies that involve *virtuality, simulation* and *computer modeling (including tomography)* has special implications for embodiment and perception. Technologies, inventions of the human brain, obviously have a great impact on the functioning of brain, on the working of mind and on memory. The issues invention of the human brain, and especially the microelectronic technologies, obviously exert a great impact on the functioning of brain, on working of mind and on memory. The issue are complex and often confused, in part because there is overlap and loose usage concerning all the VSC technologies. In popular form, one can enter virtual reality in a virtual reality arcade. Unfortunately, the description and claims made for this arcade experience are usually cast in the now antiquated frame of early modern epistemology.

According to Turkle (2005: 287), computers become "thinking tools" offering new models of what it means to know and to understand:

> computers, with their reactivity and interactivity, stand in a novel and evocative relationship between the living and the inanimate. They make it increasingly tempting to project our feelings onto objects and to treat things as though they were people—an impulse I called the "Eliza effect" after the early AI program that was designed to seem like a solicitous psychotherapist.

Our connection with things, consequently, is becoming an "object relations" perspective "that takes *objects*" as its *subject*. Relational artifacts ask their users to see them not as tools but as "companions", as "subjects" in their own right. The simple robots are marketed as toys and the more advanced robots remain largely confined to research settings or used for complicated tasks, such as flight simulation or flying unmanned aircraft. In the presence of relational artifacts, people feels attachment and loss: it is at work the metaphor that Turkley calls "Eliza effect".

The present two volumes—volume one, *The Body as Social Icon*, and volume two, *Mapping Bodies in Networked Space*—intend to give answers to these problems bearing in mind the observation of Leroi-Gourhan (1983:85) that "the very problem is the *link* that connects the technologies with the brain of mankind".

Volume one is divided in two parts: Part one, "Icon as Utopia", is the theoretical grounding of the problem. Part two, "The Body as Social Icon" illustrates the social constructionism concerning the body by empirical studies of "social icons" and explores the evolution of human perception and subsequent modification of thought and behavior in response to our increasing reliance on provided data in order to experience the world. The volume extensively reviews the ways by which we can measure the brain and the body to understand the person and social behavior. Each chapter defines a relevant construct, traces its historical development, discusses recent findings, entertains controversies, draws connections with other relevant constructs, and points to new research directions.

The structure of volume two is also contains two parts: "The Body and its Prosthetics: Laboratories"; and "Technology, Physiology and Memory". This volume is a methodological experiment aimed at investigating into the evolution of practices, social formations and experience that underpin the current "knowledge-based society". The first part explores the social potential of location-aware devices, inspired by the use of tracking technology and wireless media, human relationships, movement and identity, seeking to extend and reappropriate the functions of locative technologies by exploring how they can be socially constructive or facilitate new dynamics to occur within everyday life. The second part deals with the embodiment in relation to the cultural evolution of biological memories with a special reference to the networked space.

Based on contemporary theory and research of media and networks, the volume provides an overview of the complex way in which embodiment is enacted across the "recombinant spaces" within which the fast changing physical landscapes and spaces of contemporary social life are being shaped and re-shaped. In short, the essays provide a comprehensive, authoritative examination of the full range of the current "changing bodies"(Shilling, 2008: 1-7) associated with the new computer cultures. The reader will find a representative sample of outstanding research in this emerging and exciting field. By focusing on "acting bodies", both volumes address these central themes along the temporal axis of information processing. They include basic operations like perception, emotion, embodiment, social comparison, memory, language and culture.

Embodiment is a complex phenomenon that envelops both the locative, perceiving active body I am and body permeated with the cultural significances that are also experienced. Human bodies are an intriguing pivot for theory (Shepherdson,1999) and it is difficult to imagine any geography that would matter without them. They straddle the dichotomy between nature and culture, their space is both influenced by social relations and influencing what forms these social relations may take. Incorporating human bodies into understanding of social relations allows a broader, more defensible, if continually shifting, material base from which theory might develop. Virtual reference points, narratives, fictional and non-fictional, permeate every aspect of our lives by "moving the interface off the screen into the real world" (Dourish, 2001: 101).

Synthetically, the work describes research emerging at the interface of two of the newest areas in social sciences: social neuroscience and bodies regulation by micro-electronic technologies. Recent years have seen an explosion of research into the physiological and neural bases of social behavior. This state-of-the-art science book is unique in approaching the topic from the perspective of developmental social neuroscience and sheds light on how complex social abilities emerge from basic brain circuits, whether there are elements of social behavior that are "hard wired" in the brain, and the impact of microelectronic technologies.

The invention of technology

Man is inseparable from his tools. This is why some social scientists call man a tool-making animal. The evolutionary record affirms this. *Homo habilis* used crude stone implements as far back as 2.5 million years ago, more than two million years before *Homo Sapiens* first appeared on the African savannah. Knowledge of this long overlap of human evolution with tool-making produced significant changes in evolutionary thinking. Earlier views, long held in many disciplines, maintained that evolution first effected biology and only later culture. In the current view, it worked on them together. Tool-making and *Homo sapiens* evolved jointly, tools —their making and use—played a part in the development of the large brain that separates present-day humans from their proto-human predecessors.

Man is a species bundled together with technology. The order of tools defines the relationship between the body and things: that is to say, the interfacing mechanisms which enable its orientation in territorial space, its survival and the construction of social life. The deep meaning of technology lies in this pragmatic relationship that allows the body to construct its permanence in the world. Technical relationships are embodied in social relationships and constitute the relational matrix of social intelligence. Werner Sombart (1911: 311-347) considered technology as a great historical modelling force. Technology is a driving force in social change, since its brings about creation and shapes symbolic activities. The development of technical skills, which the body learns by manipulating objects as well as using and perfecting utensils, explains the peculiarity of human cognition compared to animal intelligence as a creative process connecting pre-existing elements in new, and useful, combinations. A "toolbox" fit to invent, after every new interaction, the "able practice" that Marcel Mauss (1936: 394) had already perceived was the "primary tool" of cultural reason. Of all human activities, technique is the only one which does not go back to the starting point. Techniques are learnt and they cumulatively advance with occasional leaps forward. Their aim is to lessen human physical—and lately mental—energy output in understanding and using the natural world for human purposes.

The development of technical skills, which the body learns by manipulating objects and using utensils, explains the peculiarity of human cognition compared to animal intelligence as a creative process connecting pre-existing elements in new, and useful, combinations. A "toolbox" fit to invent, after every new interaction, the "able practice" that Marcel Mauss (1936: 394) had already perceived was the "primary tool" of cultural reason. The practice of a tool brings about the capacity to repeat a performance with a variation depending upon the changing contexts. This capacity is an issue which implies a sort of correspondence and of stabilization between environmental reactions and conditions. Therefore the technological practice makes the body the *interacting field* connecting the individual with society trough the neurophysiological processes generated and regulated by experience. All types of experience and learning modify the neural maps. Neuroscientists have demonstrated that just three months of activity or passivity can affect the cerebral structure. Indeed, the experiences stimulate the formation of synapsis, activate neural circuits, consolidate them and create new ones. The human brain is an open and dynamic living system shaped by the changes provoked by the constant variation of the environment, able to receive specific stimulation due to its functional connection to other systems, in particular to other brains. Relational neurobiology (Siegel, 2001) reinterprets individual differences with reference to the interaction between hereditary traits, environmental influences and accidental events, and it shows, with increasing evidence, that the brain functions which are at the foundation of human experiences—memory, affection and emotions—are not functionally separated from cognitive ones. The relational need drives human beings to create connections with the external world in order to keep an adequate neurophysiological balance. The physiological, sensorial and cognitive mechanisms which are put into action in relational contexts mediated by present digital technologies are the same which can be found in the traditional relationships between human beings and technology. According to the Swedish neuroscientist Torkel Klingberg (2008: 10-11)

> the ability of the contemporary brain is identical to that of the anatomically modern *Homo sapiens*.

The changing elements are represented by the quantity and complexity of the information constituting the "technological environment". Technology is an *activity* rather than a manufact. The definition of technology by D. L. Sills (1968: 58) implies in particular an *activity*.

Technologies are the body of capacities, of knowledge, and of the practice to create, to use and to do useful things.

According to Sills (1968: 58), technology is a modality of creation derived from the *knowing how* (Ryle, 1949), that is to say from the ability to perform specific procedures based on the identification of specific *stimuli.* As a procedural instrument, the basic function of each different technology is to position the body of human beings in their environment, creating those forms of exchange which identify the social space. The adaptive procedures which granted the body its self-preservation and the transmisson of its genetic heritage correspond to the means of intervention on matter. In human beings the nervous system, and in particular the brain, evolved in order to specialize in problem solving, from the ways to prevent drowning to the research of food or of a partner. Each different technological invention is basically an *active manifestation of a new relational product* whose appearance is determined by the unrepeatable nature of its materials, of its events and of the human beings which constitute it. It basically springs from our imperious need to expand, extend, develop the sensorial abilities of our body until this activation enhances and improves it. The main primary justification for technological inventions is represented by this tendency, since the body establishes new relationships with the *habitat* in the unceasing attempt to gain identity in a more complete way. The direction of technological innovations is not independent of the type of the society in which they happen and of the goals for their use. Max Weber added to Marx's concept of "means of production" the concept of "means of destruction". For example, nuclear technology can be used for both production (say, electricity) and destruction. Technology can improve and destroy the *habitat.*

Therefore, the term "technological invention" defines *the adaptive creative process through which human beings intervene on the surrounding materials according to their own needs.* First of all, it is the chemico-physical use of matter which the hand performs in direct motility during the "complex operations of prehension-rotation-translation which characterize manipulation" (Leroi-Gourhan, 1977, vol.2 : 284). Being the first ones to appear, these operations are the privilege of an archaic and very little specialized human hand in relation

> to the wonderful apparatuses to grab and to run such as the lion's or the horse's paw (Leroi-Gourhan, 1977, vol. 2: 285).

In the course of the evolution of the culture of the *Homo sapiens*, the hand has modified and has improved the operational process through the addition of elements which do not reciprocally nullify and regulate the technical action through the activation of a sequencing process which is both biological and mental. This process aims at determining, in the chassis as well as in computers, the consistency of the technical action as a *relational organization of social intelligence.*

The *transition to the tool* is functionally justified by the field of relationships which the hand embodies and transforms. The order of tools testifies to the casual evolution of the "repertoire of recalls" which connects the social life of the first

Homo who inhabited the savannah until the emergence of the symbolic behavior which constitutes one of the main features of *Homo Sapiens*.

The manufacturing of lithic tools is the most evident expression of the acquisition of a technical intelligence: an aim to attain, the planning, the final performance and then the tool. Although it is true that the use of natural tools such as pebbles, sticks and leaves can be also found in other animal species, from otters to anthropomorphic apes, its systematic use and the resulting technological development is peculiar to the human species. The demands of the first human beings can be traced back to food and defence. Modern anthropology assumes that both men and women collected fruits and food was essentially dependent on what they could find in nature, such as animal carcases, fruit and tubers, so that it is reasonably possible to hypothesize that one of the first problems was the stripping of flesh from animals through the use of sharply edged tools deriving from the chipping of silicon rocks (*chips*) and the break of long bones in order to extract the marrow through their percussion with partially "carved" stones (*choppers*).

The *intentional* use of tools is one of the cognitive faculties of hominids which has been already proved. Hominids get excited and use their emotions to enhance the use of the stone utensils they need in order to obtain the sharply edged splinters and the more robust stone tools which are necessary to cut tendons and animal carcases. Through their emotional intelligence, they have an active role in the techno-social negotiating process which integrates their sensorial faculties to the places which have to be explored and covered. *Homo habilis*, who lived in Africa 2 to 1.5 million years ago, was the first hominid who exploited splintered pebbles.

The *pebble culture* is the oldest existing technology and it is based upon the manufacturing of pebbles. Hominids broke pebbles so as to obtain tools which could dig, cut, cut off and make holes. They soon improved their manufacturing techniques so as to remove splinters on both sides of pebble blocks in order to have sharper edges (bifacial pebble tools). The *pebble culture* presents some primary typologies which summarize its technological evolution.

Today we certainly know that our direct ancestors, defined by paleoanthropologists as "anatomically modern" humans, came into being in Africa 200,000 to 165,000 years ago and genetic evidence has demonstrated that all populations living nowadays are connected to those African ancestors. According to the estimate of the team of geneticists headed by Wolfang Enard from the Max Plank Institute of Evolutionary Anthropology in Leipzig, the moment the human form of the gene pool was fixed in the population (through a process of natural selection following a chance mutation) has to be traced back to the last 200,000 years in human history. The "anatomically modern" *Homo sapiens* began with an irradiation to the rest of Africa around 100,000 years ago and was followed by an expansion from Asia probably through a Southern itinerary towards Oceania and a Northern itinerary towards Europe and America. In the course of these itineraries the linguistic abilities of *Homo sapiens* evolved in those referential expressions which are at the basis of symbolic behavior. In Europe the indicators of a linguistically mediated behavior suddenly appear forty thousand years ago, when *Homo sapiens* begins to replace the Neanderthal forms and manufactures highly-sophisticated lithic tools and throwing weapons. He establishes even long-distances commercial exchange networks. He begins decorating his own body with painting and carves small female, male, and animal figures. He engraves animal bones to create musical instruments and he paints images of extraordinary beauty on the cave walls of Chauvet, Lascaux and Altamira. This is an unprecedented cultural explosion in hominids' history which takes place during their emigration to Europe on a territorial area going from Asia to Cantabria.

There has not been a commonly shared view on the connection between the African origin of modern anatomy and the origin of modern behavior and there is not any agreement on the meaning of "modern behavior". Synthetically, there are two opposite theories:

1. Modern behavior makes its appearance in Africa 50,000 years ago and this event is strictly connected to the behavioral modernity noticed in European Superior Palaeolithic sites;
2. Modern behavior developed in Africa in an older historical period and it is connected to the anatomic modernity through a gradual process which probably originated 200,000 years ago.

Despite the fact that genetic and fossil evidence concerning anatomy has been accepted, scholars cannot still agree on the coexistence of bodily and cognitive behavior.

The recent African archaeological finds pre-date by thousand years the presence of modern behaviors. In Congo, for instance, jagged bone harpoons dating back to 80,000 years, and whose level of sophistication has nothing to envy to similar harpoon discovered in Europe dating back to 25,000 years, have been discovered. In Tanzania beads made of ostrich-egg shells estimated 70,000 years old have emerged. Finally, the proof in favour of the "gradualist" thesis has discovered on the South African coast 200 miles far from Cape Town between 1999 and 2000 at the Blombos cave, the Medium Palaeolithic site which has already revealed to its discoverer—Professor Christopher Hensilwood, from the Centre for Development Studies of the University of Bergen, Norway—the sophisticated bone tools mostly destined to fishing which, on their own, testify to the presence of modern kinds of behavior. From an undisturbed deposit of ashes and sands—which the luminescent technology successively help dating 70,000 b. C.—have emerged two fragments of red ochre each nearly five inches long, finely chiselled with a grid of intersecting lines, next to the remains of a fireside and to tens of pierced molluscs.

What does the new South African discovery suggests us?

According to Hensilwood, the presence of carved material surfaces denotes cognitive skills and operational capacities ascribing them to modern behavior. He assumes a mosaic-like "panafrican" developmental model capable to record ages of innovation and regression in specific contexts. In the whole Africa there is no possibility of a single trajectory towards behavioral modernity, as well as a single area in which it is possible to find an anatomically modern human behavior.

Red ochre, a ferrous mineral, has been frequently found in Palaeolithic sites dated less then 100,000 years and was probably used to protect the skin and to tan animal skins. The boards at issue were first scraped and then ground in order to create flat surfaces, where it was possible to carve the refined grid which leads us to infer the presence of an associative habit. One fact certainly emerges: we are in the presence of a tangible product endowed with an interactive function which has still to be determined, or that it will be impossible to determine because of its strong connection with the spatial and temporal context in which it took place. Carved upon the tangibly present grids in red ochre, this function is a *procedural* one and it is *limited* to its performing context. It is an operational and at once cognitive function: through the dynamic action of one's hand it related the human body to the Blombos findings in the reticular engraving of the red ochre fragments. In this sense, we could be in the presence of an expression of social intelligence. In agreement with the definition given by John Kihlstrom and Nancy Cantor (2000: 359-79), the term "social intelligence" defines "an effective action in social relations": i.e., an "acting body".

The consideration of the human body as an "acting body" will be able to conceptualize technologies as both "technical objects" and "cultural artefacts" and their associative function within the joint activity systems constituting the practices of social interaction from which individuals and groups are continually moulded.

This question emerges from the archaic fragments found in the Blombos cave and remains unsolved, after 70,000 years, in the places of common space shaping the interactive bricks of the current "cosmopolis".

In our times science and technology have created many of the problems besetting us at the turn of the century; yet, paradoxically, we cannot address them without their assistance. Systematic reflection about object-producing devices relates technology not only to the effective production of functionally determined correlations (i.e. technology as teleological operator) but also to the subjective action. This "de-objectivizing" of technology is rooted in philosophy (e.g. Heidegger and Ortega y Gasset) and well established in "social construction of technology" approaches.

On these premises, technology presupposes technique (i.e., any systematic action addressed to the organization and adaptive transformation of environment) as significant human action. Technology denotes thus an operational coherence or, in other words, the organization of social order. Technology enables the social system to cope with environmenttal complexity and increases its internal complexity. It also increases and diversifies the complex nature of the interdependences between society and individual actors. Simultaneously, the increasing societal complexity demands new forms of technology to be designed and implemented. In one way or another, technology is in the backstage of any relevant discussion on the contemporary nature of society both in the side of problems (risk, control and global surveillance, bio-ethics and bio-engineering, media manipulation, ecology) and in the side of solutions (instant communications, ubiquity, accessibility, safety, knowledge availability).

We can therefore conclude that the body—technology—society relationship is historically changing. This makes even more important to conduct further studies into these relationships.

The body is the medium of knowledge

For millennia architects have been concerned with the skin bounded body and its immediate sensory environment: they have provided shelter, warmth, and safety; cast light on the surfaces surrounding it; created conditions for conversation and music; orchestrated the touch of hard and soft and rough and smooth materials; and breezes and scents. In current social networking the boundaries of the body and the limits of the nervous system have become "less definite, metaphysicians" at the point that

> we will be tempted to reformulate the mind/body problem as the mind/network problem (Mitchell, 1996: 31).

Before we turn to theoretical issues, we must not forget that *the brain is embodied and the body is embedded*. First, the embodiment. All brain activities depend on signals to the brain from the body and from the brain to the body. The brain maps and connections are altered not only by what *you sense* but also by what *you move*. Synchronism between the elementary self-consciousness which regulates somatic processes and the interaction taking place is responsible for the unrepeatable activation that produces the state of active consciousness. Second, the embeddedness. Your body is *embedded* and *situated* in a particular environment that influence it and is influenced by it. This set of interactions defines your "econiche" as the changeable combination of bodily confines and cultural practices. The development of the whole person-organism is carried out by these boundaries, at the same time body and mind, in a continuous *relational field*. The flows at issue make up the core of somato-sensory experience.

Sensation and perception provided the starting points for modern research into our mental processes. In the early nineteenth century the French philosopher Auguste Comte argued that the study of behavior should become a branch of biological sciences and that the laws governing the mind should be derived from objective observation. Comte's new philosophy, which he called *positivism*, was influenced by the British empiricists John Locke, George Berkeley and David Hume, who maintained that all knowledge is obtained through sensory experience, that is what we see, hear, feel, taste and smell. At birth, according to Locke, the human mind is a *tabula rasa*, a blank slate upon which experience leaves its mark. It was this view that led to the emergence of psychology as a separate discipline apart from philosophy, which had long monopolized the study of human mind. Thus, in its early days, psychology came to focus on the experimental study of mental processes by emphasizing sensation as the key of the mind. How does a stimulus lead to subjective experience? By what sequence of physiological events? For the fathers of experimental psychology—Ernst Weber, Gustav Fechner, Hermann Helmholtz and Wilhelm Wundt—those were the central questions. Their findings gave rise to the fields of psychophysics and sensory physiology. Some of the most exciting advances in our understanding of perception have come from merging these two approaches, for example, in recent human experiments that use positron emission tomography (PET) and functional magnetic resonance imaging (fMRI) to scan brain function. Early findings in psychophysics and sensory physiology, however, exposed one weakness in the empiricist argument: A new-born's mind is not blank, nor is our perceptual world formed simply from passive encounters with the physical properties of objects and stimuli. In fact, our perceptions differ qualitatively from the physical properties of stimuli because the nervous system extracts only certain pieces of information from each stimulus, while ignoring others, and then interprets this information in the context of the brain's intrinsic structure and previous experience. Thus we *receive* electromagnetic waves of different frequencies, but we *perceive* them as the colors red, blue, and green. We receive pressure waves from objects vibrating at different frequencies, but we hear sounds, words, and music. We encounter chemical compounds floating in the air or water, but we experience them as smells and tastes. In short, our perceptions are not direct records of the world around us. Rather, they are constructed internally according to constrains imposed by the architecture of the nervous system and its functional abilities.

Our sensory systems are the way by which we perceive the external world, remain alert, form a body image, and regulate our movements. Sensations occur when external stimuli interact with receptors. Sensory information is conveyed to the brain as trains of action potentials travelling along individual sensory neurons and by populations of such neurons acting together. All sensory systems respond to four elementary features of stimuli: *modality, location, intensity and duration* (Kandel, Swartz, Jessel, 2000: 428).

The complex quality of sounds, visual images, shapes, textures, tastes and odors require the activation of large ensembles of receptors acting in parallel, each one signaling a particular stimulus attribute. For us to savor the richness and diversity of perception, the central nervous system must integrate the activity of an entire sensory sensory complex.

Perception is not a pre-given structure, neither according to its gestalt nor according to its sense. It is constantly in the state of developing and changing. A reason for this is that the manner in which sensible objects of perception are given is inadequate. Every phenomenon implies horizons that are empty and undetermined, and that strive toward concrete fulfillment and arrangement. In this way sensory consciousness is constantly held in motion by tensions that provoke regroupings on the objective side and new attitudes on the subjective side. The special correlation of perspectival aspects and horizon found in perception of material things is taken as a preferred model in the phenomenological theory. A thing is directly intuited by us only from one side. But this side remains in a reciprocal, functioning *interrelation* within a network of referential implications with the side of the thing that is turned away from our gaze and is not directly intuitive. According to Husserl, part of the essence of spatial objective experience is the fact that the body is not only presupposed objectively, as it is for a realistic philosophy, but it is also phenomenally co-present (Husserl, 1952: 56, 81, 144; 1966: 298). Merleau-Ponty (1945: 117) adopted this thesis in his critique of gestalt psychology:

> One's own lived body is the constantly co-present third moment in the structure of figure-background.

In Sartre' view (1962: 388), it is the lived body, although it can be identified in the analysis as the underlying center of the mundane nexus of referential implications, which is always "the already crossed", the "silently happened in the execution of perception and action".

Sartre appropriately compares that consciousness which we have of our own lived-body to the consciousness of the sign, which, in a similar way, is passed over for its meaning and neglected in favor of its sense. In a similar way, Sartre distances himself from Husserl. The spatial orientation of perception proves to be utterly relative. Each dominant formation of perception draws toward itself the "zero-point of orientation" (Holenstein, 1999: 57). One's own lived-body neither occupies the orienting dominance in the world of perception constantly nor does it so even originally.

The location and spatial dimension of a stimulus are conveyed *topographically*, through each activated receptor's position in the sensory epithelium, called its *receptive field.*

The cognition of space constitutes an enormous conceptual challenge to the "homeless mind" (Heynen, 1999: 17) supporting the development of the modern thought. Space is indeed a *field in motion* above all if referred to the *dynamic function of bodily space in cognitive acting.* Our infinitely varied motor behaviors are governed by the integrated actions of the brain several motor systems. The primary purpose of the elaborate information processing and storage that takes place in the brain is to enable us to interact with the environment. We are continuously bombarded by molecules released into our environment. Through the sense of smell and taste these molecules provide us important information that we use constantly in our lives. They inform us about the availability of foods and the potential pleasure or danger to be derived from them. The multitude of different flavors that one can experience derives from a combination of gustatory, olfactory and somato-sensory components.

Bodily space is a determining factor in the framing of social relations and is also reciprocally interconnected in the making of space by those very social interactions that occur in space. George's Simmel's inclusion as a significant and early sociological analysis of space and indeed much of Simmel's work contains within it a keen awareness of the interrelationships between forms of sociability and the spatiality that is expressed within and through those "webs of interaction" that, for him, constitute a society.

Henri Lefebvre's analysis on the rhythms, times and places of everyday life, of the routines and alienation inherent in the socially produced spaces of leisure and recreation, domestic life as well as those of work and the urban life has also been influential for developing research on gendered space and on the spatialising of sexual and ethnic identities.

The spatial dimension has been taken for granted too many times and has not yet been recognized as a *crucial area* in social sciences and, in particular, in sociology. Until a few decades ago, the word "space" had a mainly geometric meaning, generally accompanied by epithets such as "Euclidean", "isotropic" or "infinite", with the aim to describe the "empty area" identifying the concept of space. The key to understand this concept lies in the Cartesian thought. Being an "object" juxtaposed with a "subject" and a *res extensa* juxtaposed with a *res cogitants*, the Cartesian concept of "space" was the container of all meanings and all bodies. The research field known as "epistemology" has inherited and adopted, from the revised and corrected philosophy of space of mathematicians, the notion of space as a "mental object" or a "mental place".

According to Lefebvre (2007: 34),

> from the point of view of knowing (*connaissance*), social space works along with its concept as a tool for the analysis of society.

To accept this much is at once to eliminate the simplistic model of a one-to-one or "punctual" correspondence between social actions and social locations, between spatial functions and spatial forms. Precisely because of its crudeness, however, this "structural" schema continues to haunt our consciousness and knowledge (*savoir*).

Science and technology have had a profound effect on the way humans perceive space and time.

One of the most important functions of cyberspace has been to force a re-evaluation of our conceptions of space: since the Enlightenment the reification of physical, geometric space via Descartes, Newton and Leibniz has been an important contribution to our sense of the real world as operating in three dimensions.

Classical conceptions of space have been problematized by a number of cyberspace theorists (Benedikt, 1991; Heim, 1991; Oswald,1997; Wertheim, 1999) drawing on post-structuralist or postmodern theorists, such as Foucault, who have sought to explore space as not merely the geometricisation of symbolic mathematics, but also as territorialisations of power by dominant social and economic groups. Put simply, for many writers and thinkers space is not only that void through which we move but also the *place* in which we live and experience. Benedikt (1991: 4), drawing on Karl Popper, distinguishes

> three types of space—the physical, the subjective and the structural—and sees cyberspace as the third type of world, the structural spaces that provide "patterns" for our experience.

Locative and other technologies, of course, have their own evolutionary time-scale, one that typically moves faster than the time-scale of biological evolution. A decade ago, technologies that could provide information about the location of a motor vehicle, or a computer, or a person, were in their infancy. A wide range of tools are now in use and in prospect, which threaten to strip away another layer of the limited protections that individuals enjoy. An understanding of the landscape of location and tracking technologies, and of the issues that they give rise to, depends on establishing a specialist language that enables a meaningful and reasonably unambiguous discussion to take place. Those technologies are now well-established, and lack any form of regulatory framework although there is a growing pressure to establish it. IP-address location remains still quite inaccurate. Cellular triangulation and signal-differential techniques, and self-reporting of GPS measurements, are also error-prone, but their accuracy and precision appear to be improving. RFID and NFC devices identify and locate chips with reasonable reliability, and, because of their short range, with considerable accuracy. The 'space' within which an entity's location is tracked is generally physical or geographical. All of the above examples relate to location within physical space. Other kinds of 'space' exist and location within such spaces may be defined in other terms. For example, a location may be virtual, as in the case of a person's successive interactions with a given organisation. A particularly important example is 'network space'. An IP-address records the location in network space of a software process entity (which necessarily is running in a computer entity). Location can be ascertained with varying degrees of accuracy and reliability. The location of installed devices, such as fixed ATMs and EFT/POS terminals, may be quite exact and reliable. Devices such as cellular phones, and portable and handheld computers, are designed to be mobile, and additional information is needed in order to draw inferences about their location at the time of a particular event. Some kinds of locative definition may be limited to a line or cone (e.g. those relying on directional mechanisms), or an area bounded by three or more lines (e.g. those relying on triangulation). Measures of location may be available with varying degrees of timeliness. By this is meant the lag that occurs between the event, and the availability to a person undertaking surveillance of the transaction data reflecting that event. By tracking is meant the plotting of the trail, or sequence of locations, within a space that is followed by an entity over a period of time. Owing to timeliness limitations, data may only be available for retrospective analysis of a path that was followed at some time in the past. A 'real-time' trace, on the other hand, enables the organisation undertaking the surveillance to know where the entity is at any particular point in time, with a degree of precision that may be as vague as a country, or as precise as a suburb, a building, or a set of coordinates accurate to within a few metres. Moreover, a person in possession of a real-time trace is in many circumstances able to infer the subject's immediate future path with some degree of confidence.

They are mostly oriented towards entities, and their effective operation depends on the collection of entifiers that distinguish a particular entity and enable transaction data to be reliably associated with the appropriate entity and perhaps with other transactions. Some technologies are relevant to spaces other than physical (especially net space), and some ones to identities rather than entities. Mobile, location-sensitive devices, semantic mapping and biometrics are all locative and wearable tools that are being deployed in non-authoritarian ways on the street and in the body. Even the critics would agree that locative media art at its best enhances locative literacy and augments the ambiance of the physical places. It carries the ability to intervene on local level whether by creating an on-site experience about the place, transporting us to another site or choosing to abstract us from a given location.

Across a broad range of contexts, the interface between data environments and location has emerged as a central concern, reversing the trend towards digital content being viewed and placeless. In a move away from the virtual reality research, mixed or augmented reality constitutes its phenomenological inverse, where the ultimate goal is to create a system such that the user maintains his his/her presence in that world.

It is extremely simplistic, and contributes to a passive approach, that our choice as consumers at most consists of accepting or adapting to new technologies as they come on line, by. The fact that many people using the Internet—the virtual reality—take it for real, thus there is the possibility to disembody the real world in which we are embedded.

Once again we have the camera as the explicit model for knowledge. What Descartes adds is another analogue: the "mind" or mental substance": the modern subject who is now inside the camera.

Lee Bailey (1989: 66-67) points out

> The *camera obscura* began as an experimental model for the eye and became a ruling metaphor for the mind. By offering a way of picturing the Cartesian inside *Cogito* with a sensory channel admitting pictures from the outside *extension*, the image of skull's darkroom shifted from a suggestive experimental analogy to a concealed methodological paradigm.

The *Cogito*, or the "*homunculus* in the camera", utilizes an artifact (a machine that was taken as a paradigm for the human activity of acquiring knowledge (a cyborg activity), to produce a picture and thus self-understanding of that activity. By taking the subject out of the camera and finding him/her in the world, one has deconstructed the concepts of early modernity and established an embodied, situated knowledge.

Phenomenology takes the *Cogito* out of the camera and finds it, *embodied*, in the world. Merleau- Ponty (1962: 82) makes the point more succinctly:

> I am conscious of my body via the world... I am conscious of the world trough the medium of my body.

As technological systems become more sophisticated, physical systems become devalued and distrusted. The natural and biological have become pathological. This book suggests the body and the senses as the *medium* through which the world is accessible and available and proposes a view of the body as a productive material agent that generates practices in its ongoing materialization. It advocates not simply for a sociology of the body, but maintains that sociology should itself be fundamentally embodied, theorizing not so much about bodies but from *bodies as interacting entities* in space and time.

Embodied action is the result of the interaction of the biological substratum and the totality of relations and emotional experiences that shape our social habits. What really has to be considered in a knowledge based society is the pragmatic constitution of what Husserl calls the "life-word" (*Lebenswelt*) rather than the projections of the brain upon a world supposed already to be organized.

In this sense, the full understanding of the embodied action is one of the most formidable challenges for social sciences in the twenty-first century. We emphasize action, and not only bodily movement that has too long been considered a reaction to stimulation from a fixed environment known in advance by the observer. By the term "action" phenomenologist mean the mental schemas that govern motor intentions, which in the final analysis regulate motor commands, observable in the form of bodily movements. These mental schemes are experienced by the actor as basic abilities, that is as capacity for intervention which become available once one has learned new physical movements, new maneuvers in the practical sphere. Here one encounters Merleau Ponty's notion of "I can". What the phenomenologists object to it is the primacy attributed to the environment, that the experimentalists consider as a world wholly made up of things from which messages emanate and to which replies are given. It is necessary to step back from the "laboratory situation", where experiment assumes an already constituted reality, and take into consideration the role played by the agent himself in building his environment, as Jean Luc Petit (1997: 1-37) has shown.

The human agent does not content itself with being informed about his environment in order to try to modify it afterward; from the beginning he shapes it or better—to use Husserl's strong formulation in his last unpublished writings—he constitutes it as the world that surrounds by projecting onto the aims of his action and his demands for interaction.

Acting bodies: a bridge between technology and working memory

Intelligent infrastructure is a new horizon for computer technology. Computers have proven themselves to be powerful tools for calculation and communication. The next step, experts say, is for computers to become intelligent instruments of control, linking them to data-generating sensors throughout the planet's infrastructure. In his book *The Technium* Kevin Kelly (2009) forecasts the emergence of a global brain—the idea that the planet's interconnected computers might someday act in a coordinate fashion and perhaps exhibit intelligence. We are entering in a new phase of computing, in which computers will be interacting with the physical world as never before. With the ability to send attribute data from personal information devices worn on the body to computers embedded in the environment, one-to-one services could be implemented that are tailored to user's individual needs.

The common thread throughout the research is the remarkable cortical plasticity of the brain, which often adapts to BCIs. With recent advances in technology and knowledge, pioneering researches are now conceivably attempt to produce BCIs that augment human functions rather than simply restore them. The study of how our interactions shape our emotional habits and sculpt the brain has indeed enormous importance for how we do in life. Donald Hebb in his well known study *The Organization of Behaviour* (1949) demonstrates how "experience shapes the brain".

All types of experience and learning modify the brain. You never, step into the same river twice. As the Greek philosopher, Heracleitos said, one cannot step twice in the same river. This means that the constantly changing circumstances and experiences influence the ways our brain is reacting to the impulses from the outside world, and is capable to develop new synapses.

The neurological studies on the plasticity of somatosensory areas (Kaas, Merzenich and Killackey, 1983: 325-56; Kaas, 1991: 137-167) and the Sur's results on visual behavior (Sharma, Angelucci and Sur, 2000: 841-47) show how important sensory stimuli are for determining the way the brain is organized, which in turn underpins the importance of environment.

The primary purpose of the elaborate information processing and storage that takes place in the brain is to enable us to interact with the environment. Research into trained-based brain plasticity seems to confirm that the brain is able to adapt itself to its environment and the greater demands it entails. The ongoing modification of synapses throughout life means that all behaviors of an individual are produced by genetic and developmental mechanisms acting on the brain. Therefore, everything brain produces, from the most private thoughts to the most public acts, should be understood as a biological process that is, however, shaped by cultural factors. Environmental factors and learning bring out specific capabilities by altering either the effectiveness or the anatomical connections of existing pathways (Siegel 1999).

Advances in neurosciences, combined with some new forms of approaching the body-mind-society interrelations, have changed our conceptions of the natural and social body, thus giving rise to the possibility of introducing, by "embodied action", a new paradigm. We are beginning to know more about the links between the development and exploration of the body and the use and growth of intelligence and between social experience and creative thought.

We have a better idea of the species-specific capacities of human organism, and know how much humans depend on interaction with fellows to develop them. Since human bodies are the instruments that both discover and make decisions about self, others and the worlds of nature and cultural tradition, the embodied action is concerned with the interface between the body and society, the ways in which the physical organism constrains and inspires patterns of social interaction and invention of culture.

Take the power of emotions to disrupt thinking itself. Neuroscientists use the term "working memory" for the capacity of attention that holds in mind the facts essential for completing a given task or problem. Working memory is a temporary storage system that underpins our capacity for coherent thought. Some 40 years ago, Baddeley and Hitch (1974) proposed a way of thinking about working memory that has proved to be both valuable and influential in its application to practical problems (also see Baddeley, 2007, 2009). The term working memory refers to a brain system that provides temporary storage and manipulation of the information necessary for such complex cognitive tasks as language comprehension, learning, and reasoning. This definition has evolved from the concept of a unitary short-term memory system. Working memory is involved in the selection, initiation and termination of information-processing functions, such as encoding, storing and retrieving data. Working memory has been found to require the simultaneous storage and processing of information. It can be divided into the following three subcomponents: (i) the central executive, which is assumed to be an attention-controlling system, is important in skills such as chess playing and is particularly susceptible to the effects of Alzheimer's disease; and two slave systems, namely (ii) the visuo-spatial sketch pad, which manipulates visual images and (iii) the phonological loop, which stores and rehearses speech-based information and is necessary for the acquisition of vocabulary of both native and other languages. Memory for plan and actions relies primarily on working memory, in particular the operation of the central executive component.

By memory we usually mean our capability to remember certain events of our past or to retrain and retrieve data and knowledge. Classical theories of memory tended to portray it as a thing-in-itself, something *deposited* in the brain and mind—an impression, a trace, a replica of the original experience, like a photograph. Fortunately, we now know enough about memory to relate laboratory studies to the world beyond. In other words, our scientific knowledge of memory and how it works can help us to explain those aspects of memory that most people find of greatest interest. The neurobiological study of memory has yielded three generalizations: memory has stages, long-term memory in multiple regions throughout the nervous system, and explicit and implicit memories involve different neuronal circuits. Different types of memory processes involve different regions and combinations of regions in the brain. Explicit memory underlies the learning of facts and experiences. Declarative or explicit memory contains single recollections or informations that may be reported and described; it may also be called a "knowing that". Declarative memory relies on a medial temporal lobe system, whereas habit learning relies on the striatum (Cohen & Eichenbaum, 1993). Implicit memory processes include forms of perceptual and motor memory—knowledge that is stimulus-bound is expressed in the performance of tasks without conscious effort, and is not easily expressed verbally. Implicit memory flows automatically in the doing of things, while explicit memory must be retrieved deliberately. It is now appreciated that memory is not a unitary mental faculty but the operation of many separate systems. This raises the possibility that different cellular mechanisms may be involved. One broad classification scheme distinguishes declarative (explicit) memory from non-declarative (implicit) memory (Squire 1987). Declarative memory includes the facts, episodes, faces and routes of everyday life and is accessible to conscious recollection. Declarative memory is affected in amnesia, and it depends on the integrity of the damaged structures and connections. Non-declarative memory includes skillful behavior, conditioned responding and priming. Priming refers to facilitation in identifying perceptual objects, including words, that is produced by their recent presentation Non-declarative memory affords the capacity to acquire knowledge without conscious efforts. In the case of skill learning, knowledge of the skill is embedded in procedures, and the skill is expressed in performance by engaging the procedures. In the case of conditioning, dispositions are gradually acquired to respond to stimuli. In the case of priming, perception and processing of stimuli are facilitated (Tulving *et al.* 1988).

Studies on the biology of memory show that the so-called consolidation process (the transformation of unstable, short-term memories into a long-lasting code) is *not permanen*t as it was hypothesized. If we consider the meaning of memory in an evolutionary view we should look at the organisms for their potential to change as the result of events that occur during the lifetime of the individual. The experiences of an organism can modify the nervous system, and the organism later behaves differently because of these experiences. This abi lity to change gives organisms the capacity for learning and memory .

The capacity for memory is a special case of the ability of neurons to exhibit plasticity. Plasticity is a general term describing the fact that neurons can grow or shrink, or otherwise change their input and output characteristics, as a function of their local histories.

Reflections about the physiological basis of memory led most earlier biologists to consider some type of growth or change in the existing structure of neurons. Thus, the idea that existing nerve cells can grow seemed a reasonable way of accounting for the persistence of memory. This basic idea was made more specific by Hebb (1949), and it has been restated several times since, making explicit the idea that the synapse (the tiny gap between neurons) is the critical site of plastic change. Experimental evidence has become avai lable only in the past few decades. Not surprisingly, some of the most direct evidence comes from animals having relatively simple neuron systems. For example, the *Aplysia,* a sea slug examined extensively by Erik Kandel (1976), has about 18,000 neurons distributed among nine cell groups called ganglia. This animal exhibits several simple forms of learning and memory, including habituation (the weakening of a response through repeated stimulation), sensitization (the facilitation of a response by a strong or novel stimulus) and classical conditioning (wherein a previously neutral stimulus comes to elicit a response as a result of repeated temporal association between the neutral stimulus and a second stimulus that ordinarily elicits the response). The short-lasting varieties of habituation and sensitization have been shown to depend on changes at synapses in the nerve circuits. Specifically, there is either a decrease (in the case of habituation) or an increase (in the case of sensitization) in the amount of chemical transmitter released at these synapses. The important conclusion is that the neurons of vertebrates show considerable capacity for morphological growth and change in response to experience: these principles were proved to fit for humans as well. One kind of neuronal plasticity involved in memory is long-term potentiation (LTP). LTP is a long-lasting increase in the strength of a synaptic response following stimulation of an input pathway. It occurs prominently in the hippocampus, a structure that is important for memory. LTP also involves morphological change in the structure of neurons (Lynch and Baudry, 1984).

A large body of experiments points to the conclusion that memory is stored by changes in synaptic efficacy, and that these changes are brought about by common cellular events that may be found in many or all of the neurons that change. What makes memories specific depends largely on which neuronal connections are altered, and on the architecture of the specific networks in which alterations occur. Questions about how individual neurons change in response to experience are fundamentally important. In recent years, a number of neurophysiological techniques, mostly brain imaging, such as functional magnetic resonance (fMR) and positron emission tomography (PET), brought new knowledge on the relationships between memory and the brain.

The main result is that there is no specific memory "center" where the whole memory is stored.

The most likely site of memory storage is the set of particular cortical processing systems that are ordinarily engaged during the perception, processing, and analysis of the material being learned. Memory can be considered to be distributed in the sense that no single memory center exists and in the sense that many parts of the nervous system participate in the construction of a single event. Despite this holistic view of memory, specific brain structures and connections that participate in memory functions have been identified through the study of the human amnesic syndrome and animal models of the syndrome. Amnesia is a severe impairment in the ability to acquire new knowledge or to recall recently acquired knowledge, which occurs when there is bilateral damage to the medial temporal lobe or the midline of the diencephalon. The point is that there are brain structures, which are not themselves repositories of permanent memory , that are important for memory functions. The use of magnetic resonance imaging has made it possible to detect abnormalities in the hippocampal formation of some amnesic patients. Further studies indicate that the full medial temporal memory system consists of the hippocampus and the adjacent, anatomically related structures. While the stability of long term memories has been assumed for a long time, an increasing body of knowledge suggests that long-term memory is *not as stable and impermeable* as it was hypothesized in the classic models of memory. In particular, it appears that recalling a memory renders it fluid and unstable—able to be changed before being re-fixed into the circuitry of the brain—and that change could include altering its meaning or even deleting it completely.

Until recently it was thought that the memory stays in the backwaters of the mind, aging but essentially remaining the same. Yet, recent data refuted this idea. To study the process of consolidation, researchers interfere with the steps involved in fixing a memory in order to test their influence on long-term recall. Joseph LeDoux and Karim Nader (2000) trained rats to associate a darkened box with an electric shock to their paws. The rats learn the box is to be avoided and freeze the next time they are put into it (fear conditioning). If, a few days after this conditioning, the animals were given a drug to prevent protein synthesis before being reminded of the box, there was no difference in their ability to remember it as a bad place. The memory seemed fixed and safely stored in line with the classical view of memory consolidation. But, if the rat had a brief reminder of the box just before the drug was administered, the rat lost its fear conditioning. The memory had somehow been erased. LeDoux and Nader labeled this phenomenon as *reconsolidation*.

Traditional consolidation theory suggests that memories are fixed locally by protein changes within a few hours of the event and then filed to long-term storage in the cortex after a period of days or weeks. After conditioning rats in the same way, they left them for 45 days, by which time the memory should have been fixed and immune to interference. As before, the rats given no reminder of the box before being injected showed their conditioned response to the box. But the rats that were given a reminder of the box before being given the drug did develop amnesia. The consolidated memory—which conventional theory said was permanent and stable—had been removed by the action of recalling it. It appears that memory moves from the hippocampus to the cortex during consolidation, but is returned to the hippocampus for reconsolidation by the act of recall. This dynamism would be unnecessary if the brain just wanted a photograph album, but it fits perfectly if memories exist to make sense of the present, and distorting some memories while generalizing about, or even deleting, others serves to improve the mind's recognition and understanding of the world in order to in order to successfully act within it. If a memory becomes plastic every time it is recalled then it can be re-filed in a usefully updated way. The mind can make choices about whether to merge old and new or to reinforce their differences. In brief, then, reconsolidation theory suggests that when memories are recalled they become vulnerable to change. The amount of information that can be stored in the long-term memory is virtually boundless. *The capacity of working memory is instead limited to the location of space-time.*

The mapping process is aimed at determining this location relatively to spatial contingency. The mapping process was first identified in the late 1940s by University of California-Berkeley professor Edward Tolman, and, it began with laboratory rats. In his experiments, Tolman challenged each rat with a maze that offered food at the end. He noticed that each time the rats passed through the myriad small paths and blind alleys, they made fewer mistakes. Eventually, they were all able to move swiftly to the goal with no false starts. This told Tolman that the rats had internalized the makeup of the maze in their brains, which Tolman called "the central office". Similarly, human infants come to realize through experience that crying will bring food and/or attention. A child learns not to touch a hot stove. A person who has been blinded can still find his way around his house.

In 1971 John O'Keefe and John Dostrovsky made the remarkable discovery that hippocampus contains a cognitive map of the spatial environment in which an animal moves. The location of an animal in a particular space is encoded in the firing pattern of individual pyramidal cells, the very cells that undergo LTP when their afferent pathways are stimulated electrically. The animal is thought to form a "place-field" *map* of the space that it occupies. When the animal enters a new environment, new "place fields" are formed within minutes and are stable for weeks to months.

The mapping process refers to the attributes and relative locations of people and objects in the environment, and is an essential component in the adaptive process of spatial decision-making, such as finding a safe and quick route to and from work, locating potential sites for a new house or business, and deciding where to travel on a vacation trip. Thus, cognitive mapping is a form of memory, but it is also more than that. Retaining the sequence of streets in the directions to your house is memory; seeing these streets in your "mind's eye" as you speak is cognitive mapping. One working definition of cognitive mapping comes from Downs & Stea (2005: 34) in their textbook *Cognitive Mapping and Spatial Behavior:*

> a process composed of a series of psychological transformations by which an individual acquires, codes, stores, recalls and decodes information about the relative locations and attributes of phenomena in their everyday spatial environment.

The mapping process should be hypothesized as the spatial location of working memory: it encodes information about location, orientation, distance, and direction. In everyday life it has numerous functions. It enables us to remember places and to remember how to find our way around. We can recognize places as familiar and recall routes from one location to another. We can devise novel routes or shortcuts for reaching a goal. Mapping is also used for locating objects and for remembering how to find them. It organizes the interaction between bodies and the immediate environment through the ability of social actors to search locations, to find shortcuts and novel paths, to estimate distances between remembered places, and to draw sketch maps of the explored environment as functional factors in the evolutionary development of the brain.

The map, in all possible ways, constitutes a unique contribution toward error detection and its use not only in the forms of geographical maps but in the form of the *chemical and electromagnetic flows* connecting the body to its environment. Maps preceded both language and counting. Mapping practices are contingent intersections of technologies and the cultures and processes of social life. Historians and anthropologists have now examined a wide set of cultural contexts in which mapping has been practiced and have documented a historical range extending into prehistory and across the vast majority of human cultures. From the stick charts of the Marshall Islands to the Aboriginal map-paintings of Australia to the modern Western topographic devices, the map is central to human activities. According to Matthew Edney (1993, 1996, 1997), maps are *social constructions* as well as each mode of mapping is intimately tied to social, cultural and technological relations, which are contingent on particular times and places. For example, after the Renaissance the three primary modes were chorography, charting and topography, reflecting mapping activity at various scales. By the early eighteenth century, however, these modes had merged into a single mode of mathematical cosmography (i.e., the geometrical and astronomical processes of mapping). Synthetically, maps are paradigmatic examples of spatial knowledge that is produced in the knowledge space we inhabit. Not only do we create spaces by linking people, practices and places, thus enabling knowledge to be produced, we also assemble the diverse elements of knowledge by spatial means.

Fifty years ago, Marshall Mc Luhan (1964: 157) made a claim which simultaneously was a challenge:

> Maps are a prime vehicle for repositioning, reframing, rethinking science because theories are maps, maps are science instantiated, without maps science would not have been possible.

In order to meet McLuhan's challenge and rethink the self-evident rationality of science and the map/science relationship, we need to ask what is so fundamental in the map.

The primary purpose of the elaborate information storage that takes place in the brain is to enable us to interact with the environment. To perceive how the world impinges on our bodies, the brain is organized to map the tactile sensory system of the skin. The receptive fields of cortical neurons become progressively more complex with each stage of information processing, thus extracting more cohesive features of a stimulus at each stage. The primary purpose of the elaborate information storage that takes place in the brain is indeed to enable us to interact with the environment. To perceive how the world impinges on our bodies, the brain is organized to map the tactile sensory system of the skin. The receptive fields of cortical neurons become progressively more complex with each stage of information processing, thus extracting more cohesive features of a stimulus at each stage. Mapping is an expression of an "invisible college" which constitutes almost a new discipline cutting across the old disciplines of geography and sociology, with a considerable dash of other social sciences, and even a tantalizing flavor of history, as we move *from space into space-time.*

The map concept is perhaps the real key to the renewal of the social sciences. This is an aspect of the theory of scientific epistemology which has not been sufficiently explored.

In the composition of relations or capacities between different bodies, you do not know before what a body or a mind can do in a given encounter, a given arrangement, a given combination. Borrowing terms from the Middle Ages or from geography, we can define a body by *longitude* and *latitude.* A body can be anything; it can be an animal, a body of sounds, a mind or an idea; it can be a linguistic corpus, a social body, a collectivity. We call *longitude* of a body the set of relations of speed and slowness, of motion and rest, between particles that compose it from this point of view, that is between "unformed elements": what Spinoza calls "the simplest bodies". We call *latitude* the set of affects that occupy a body at each moment, that is the intensive states of a galvanizing force generated by the acting body in interaction with the context (territorial, social, geographic). In this way we construct the *map of a body.* The longitudes and latitudes together constitute the *locative memory* which is always variable and is constantly being altered, composed and recomposed, by individuals and collectivities. This memory constitutes the plane of immanence or consistency, which is always variable and is constantly composed and recomposed by individuals and collectivities.

The contribution the body makes to the brain is not limited to supporting vital operations, but includes regulating the space and time which organizes the contents of a normal mind. There is no bodily existence without or outside the space-time-frame. The mapping process provides the *localized acting of the space-time frame* relatively to the working of system specific patterns of wanting, feeling, doing and interacting. Embodied action consists of this mechanism that we define *acting body*, to mean the *active field* generated by the synchronization of biological memories with the execution of the contingent action.

Both of this book consider the acting body as an *arc of action* that concerns with embodied practices in contemporary culture.

By incisive theoretical contributions and empirical works, the authors of volume one *The Body as Social Icon* seek to shed light on the notion of "socially constructed body". By focusing on "social icons of the body", they suggest a revisioning or repositioning of the disembodied sense of "vision" that has dominated sociology, and much of Western social thought. The disembodied sense of vision in sociology may be seen in the relationship between the researcher (observer) and the researched (observed), which traditionally have been viewed in term of distance. We are our body, in the sense in which phenomenology understands our motile, perceptual and emotive being in the world. But we are also bodies in social and cultural sense, and we experience that too. The focus of volume one is on the individualized contingency of the embodiment analyzed by the authors in different cases studies. In developing different foci instead of broad categories of subjectivity the focus has moved firmly towards the complexity of individual body.

Reaching the body: stability or change? is the capital question that connects volume one with volume two, *Mapping Bodies in Networked Space*.

The second volume is concerned with the ways humans are "tied to" the technological frame. Consequently, its key task is to capture the ways that new technologies enter everyday life and how the process of such integration is at once practical and symbolic. The focus of attention becomes how orderly social conduct emerges from the detail of each setting in which it is undertaken, and orderliness is achieved in the face of the endless contingencies to which it is subject. Rather than humans and non-humans, we are beginning to think about flows, movements, arrangements, relations. It is trough such dynamics that the human (and non-human) emerges from the densely interwoven system of social networks "within which the knee bone is connected to the I-bahn" (Mitchell, 1996: 173).

The first part of this volume, "The Body and its Prosthetics: Laboratories", is a socio-technical assemblage aimed at exploring the increased surveillance of bodies through the development of new technologies. This is a complex analysis of how both gun and human are transformed. The second part, "Technology, Physiology and Memory", looks at the body as the result of the interactions among the biological substratum and the whole of relations and experiences with the new technological environment.

Synthetically, both volumes draw together these themes and methodologies as a means for developing a methodological frame able to determine the basic functions that bodies play in the extensive mapping of the human memory.

The modern cultural tendency to separate body from mind, and to elevate the mental over the corporeal, has trivialized the extent to which the body is the obvious point of departure for any process of knowing, especially participant observation. This cultural tendency, as well as a weakness for mistaking words for things and for viewing material objects as neutral and mute, contribute to the neglect of the material world. This disconnection can become a disconnection between the body and the sense of self, emotions, thought, sensations and memory.

And yet, it is through our bodies that we understand and act within this world. Our ability to experience is directly related to our embodiment in a given context. These experiences involve the mapping process. The mapping process is related to the biological basis of consciousness and the mental processes by which we perceive, act, learn and remember.

The actions of the brain underlie not only relatively simple motor behaviors such as walking or eating, but all complex cognitive actions that we believe are quintessentialy human, such as thinking, speaking and creating works of art or philosophical and scientific ideas.

Humans are vastly superior to other animals in their ability to exploit their physical environment. The remarkable range of human behavior—indeed the complexity of the environment humans have been able to create for themselves—depends on a sophisticated array of sensory receptors connected to a highly flexible neural machine, brain, that is able to discriminate an enormous variety of events in the environment.

Humans are remarkable for the complexity and pace of their evolutionary history. No other mammal has spread over such a large geographic and ecological range, and evolved so many similar species and radically new forms of behavior, within just two million years. The origins of this variability are fundamentally genetic—in the human genome—and natural selection produces new species entirely out of the genetic variation within existing species. But the accumulation of early hominid technology and culture gave our biological variability an accelerating push. The brain with which we are born today are almost identical to those with which modern *Homo sapiens* were born forty thousand years ago. At the "information portal", the same brain's capacity is loosing its ability to retain information. In 2004, a number of well known neuroscientists including Nobel laureate Eric Kandel, wrote a review about the ethical dilemma computer technologies raise. The article begins:

> Humanity's ability to alter its own brain function might well shape history as powerfully as the development of metallurgy in the Iron Age.

The review was entitled "Neurocognitive Enhancement: What can we do and What should we do? " (2004: 421-425).

By introducing the concept of *acting body,* this book intends to contribute to answer to the question concerning the future development of *Homo sapiens.*

We consider the acting body as a *bridge* between technology and working memory. Viewing the body in memory and action makes for a radical change in our approach to the social sciences.

Bibliography

Arendt H. (1978), *The Life of the Mind*, 2 vols., London, Secker & Warburg.

Baddeley A.D. and Hitch G.J. (1974), "Working Memory", in Bower G.A., ed., *Recent Advances in Learning and Motivation*, volume 8, New York, Academic Press.

Baddeley A.D., and Wilson B. A. (2002), "Prose Recall and Amnesia: Implications for the Structure of Working Memory", *Neuropsychologia*, 40, 1737–1743

Baddeley A.D. (2003), "Working Memory: Looking Back and Looking Forward", *Nature Reviews Neuroscience*, 4.

Baddeley A.D, Eysenck M.W. and Anderson M.C. (2009), *Memory*, London, Psychology Press, Routledge.

Baddeley A. (2007), *Working Memory, Thought, and Action*, Oxford, Oxford University Press.

Bailey L.W. (1989), "Skull's Darkroom: The *Camera Obscura* and Subjectivity", in Turbin P.T., *Phylosophy and Technology*, Boston, Kluwer Academic Publisher.

Benedikt M. (1991), *Cyberspace. First Steps*, Cambridge MA, MIT Press.

Cohen G., Conway M. (2008), *Memory in the Real World* (third edition).

Dourish P. (2001), *Where the Action Is: the Foundation of Embodied Interaction*, Cambridge MA, MIT Press.

Downs and Stea (eds.) (2005), *Image and Environment. Cognitive Mapping and Spatial Behavior*, Transaction Publishers Rutgers, The State University of New Jersey.

Farah M.J., Illes J., Cook-Deegan R. *et al.* (2004), "Neurocognitive Enhancement: What Can We Do and What Should We Do", *Nature Reviews, Neuroscience 5.*

Hebb D.O. (1949), *The Organization of Behavior: A NeuropsychologicalTheory*, New York, John Wiley & Sons.

Heim M (1991), "The Erotic Ontology of Cyberspace" in Benedikt (ed.), *Cyberspace: First Steps*, Cambridge MA, MIT Press.

Hensilwood (2002), *The Art behind Modern Behavior, The Chstian Science Monitor*, http://www.csmonitor.com/2002/0321/p13s01-stss.html

Heynen H. (1999), *Architecture and Modernity*, Cambridge MA, MIT Press.

Holenstein E. (1999), "The Zero-Point of Orientation", in Welton, D. (ed.), *The Body*, Oxford, Blackwell.

Hoogland R. (2002), "Fact and Phantasy: The Body of Desire in the Age of Posthumanism", *Journal of Gender Studies*, 11 (3).

Kaas J.H., Merzenich, M,M and Killackey H.P. (1983), "The Reorganization of Somatosensory Cortex Following Peripheral Nerve Damage in Adult and Developing Mammals", in *Annual Review of Neuroscience*, 6.

Kandel E. R. (1976), *Cellular Basis of Behavior: An Introduction to Behavioral Neurobiology*, San Francisco, Freeman Press.

Kandel E. R., Swartz J.H. and Jessel T.S. (2000), *Principles of Neural Science*. New York, Mc Graw Hill.

Kandel E.R. (2005), *Psychiatry,Psychoanalysis and the New Biology of Mind*, Arlington, APA Publishing.

Kelly K. (2009), *The Technium*, www.kk.org/thetechnium/.../1000_true_fans.php

Khilstrom J. and Cantor N., "Social Intelligence", in Sternberg R. (2000), *Handbook of Intelligence.*
Klingberg T. (2009), *The Overflowing Brain*, Oxford, Oxford University Press.
Lefebvre H. (2007), *The Production of Space*, Oxford, Blackwell.
Leroi-Gourhan A. (1977), *Le Geste et la Parole. Technique et Langage*, Paris, Albin Michel.
Leroi-Gourhan A. (1983), *Le Fil du Temps. Ethnologie et Préhistoire*, Paris, Fayard.
Lynch G. and Baudry M. (1984), "The Biochemistry of Memory: a New and Specific Hypothesis", *Science*, 224, 1057–63.
Mauss M. (1936), "Les Techniques du Corps", *Journal de Psychologie*, 3-4: 385–409.
Mc Luhan M. (1964), *Understanding Media: The Extensions of Man*, Cambridge MA, MIT Press.
Merleau-Ponty M. (1962), *The Phenomenology of Perception*, New York, Humanity Press.
Mitchell W.J. (1995), *City of Bits. Space, Place and Infobahn*, Cambridge MA, MIT Press.
LeDoux J. E., Nader K. and Fear G. E. (2000), "Memories Require Protein Synthesis in the Amygdala for Reconsolidation after Retrieval", *Nature* 406.
O'Keefe J.-Dostrowsky (1971), "The Ippocampus as a Spatial Map: Preliminary Evidence from Unity Activity in the Freely-Moving Rat", in *Brain Res.*, 34: 1, 171–175.
Oswald M. (1997), "Virtual Urban Futures", in Holmes, D. (ed.), *Virtual Politics*, London, Sage.
Petit J.L. (1997), http://jlpetit.chez.com/
Ryle G. (1949), The Concept of Mind, Chicago, The University of Chicago Press.
Sartre J.P. (1960), *Critique de la Raison Dialectique*, Paris, Gallimard.
Shepherdson C. (1999), "The Epoch of the Body: Need and Demand in Kojève and Lacan", in Weiss G. & Haber H.F (1999), *Perspectives on Embodiment: The Intersections of Nature and Culture*, New York, Routdledge.
Shilling C. (2008), *Changing Bodies. Habit, Crisis and Creativity*, London, Sage.
Siegel D.J. (1999), *The Developing Mind*, New York, Guilford Press.
Sharma C. (2000), *Wireless Internet Enterprise Applications—A Wiley Tech Brief.* John Wiley & Sons.
Sills D.L. (1968), *International Enciclopedia of the Social Sciences*, New York, Mc Millan.
Sombart W. (1911), *Technik und Kultur*, in "Archiv für Sozialwissenschaft und Sozialpolitik", 33: 305–347.
Squire L. R. (1987), "The Organization and Neural Substrates of Human Memory", *International Journal of Neurology*, 21, 218–221.
Tulving E., Schacter D.L., McLachlan J. and Moscovitch M. (1988), Priming of semantic autobiographical knowledge: A case study of retrograde amnesia. *Brain and Cognition*, 8, 3–20.
Turkle S. (2005), *The Second Self. Computers and the Human Spirit*, Cambridge MA, MIT Press.
Wertheim M. (1999), *The Pearly Gates of Cyberspace*, London, Virago.

Volume One

The Body as Social Icon

First Part

Icon as Utopia

Chapter 1
Utopian Bodies

Mark Featherstone

Abstract
In this paper I consider several iconic bodies that may be seen to capture the core ideological values of what I consider to be the four principle utopias in western social and political thought. In order to ground the paper, I return to ancient Athens in order to consider the Socratic-Platonic philosopher body that inhabited the polis, struggled to manage the embodied conditions of reason and passion, and inhabited the original utopian city, Plato's *Republic*. Following this work, I focus on two modern utopian bodies, which evolved in dialogue with the tyrannical Platonic body, in order to consider the key elements of Soviet Communism and Nazi Germany. In this, the second section of the paper, I explore the significance of the Communist worker body and the Nazi soldier body through reference to the works of Marx and Jünger and centrally show these iconic utopian bodies reveal everything in the elements of human embodiment that they exclude and deny. Finally, I turn off my study of the ancient body of the philosopher and the modern bodies of the worker and solider in order to consider the contemporary utopian body of the voracious consumer, ironically through reference to Deleuze and Guattari's revolutionary schizophrenic body, which was meant to explode the repressive body of the modern state represented by Freudian psychoanalysis.

What is the utopian body?

What is the relationship between the human body, the image, which I take to refer to the visual sign, and the icon, or image become object of veneration? In order to understand the link between the human body and the image we should refer to the works of anthropologists such as Mary Douglas (2003) who explains the body as the original metaphor for the production of complex culture, a kind of culture machine, or psychoanalysts such as Jacques Lacan (2006) who proposes a similar relation between body and image, through the notion that proper psycho-social development entails the alienation of the self-identical body in the body image which then becomes the screen for reflexive thought and the engine of human cultural development. But what about when we lift the meaning of the concept of the image from basic representation, a sign that signifies something else, to consider the relationship between the body and icon, which I take to refer to an object of veneration or adoration, that signifies some lost state of unity or future perfection? In this paper I propose to answer this question through a consideration of the relationship between the body and particular iconic representations of corporeal form. More specifically I intend to explore this relationship through a historical study of what I propose to call the iconic, or utopian, body in western social and political thought. This is, of course, an ambitious project that could lead us from the study of Byzantine theological art to contemporary celebrity culture and debates surrounding transitions from image, through icon, to idol, where the iconic object is no longer a sign of some other body, but rather the lost perfect body itself. Given the enormous historical scope involved in the study of the iconic, utopian, body, I examine key examples from Athens to America in order to explore what it means to idealise particular representations of the body. But before we begin let us consider the idea of the iconic, utopian, body in more detail.

Since I want to explore the social and political significance of the iconic, utopian, body I will refer to the idea of the utopian body, rather than the iconic body, because I want to emphasise the secular focus of my paper, avoid the theological connotations implicit in the notion of the iconic body, and foreground the view that the utopian body, which includes ideal theological bodies, is the product of social, symbolic, systems projecting ideological theories of an ideal human state onto the screen of the body. Given that body writers, such as Turner (2008) and O'Neill (1985), agree with Douglas that the human body serves as an originary metaphor, and that human culture is largely anthropomorphic in nature, it may be the case that the study of the ideal, or utopian, body in history is the privileged position to understand western conceptions of the ideal state of human society and that what the study of utopian social and political ideology loses when it does not address the centrality of the body are the basic corporeal themes that the anthropomorphic thesis would suggest are key to all social and political thought, never mind its utopian variant. Of course, the anthropomorphic association of body and society and body and politics, which as I will show clearly conditioned both ancient and modern social and political thought in the form of the cosmological body and the body politic, no longer really functions in western post-modern, post-industrial, society. Both O'Neill and Turner bemoan the passing of our recognition of the body as metaphor because of the ways in which this alienates those who inhabit high tech western societies from their own bodies and the bodies of others. Since, for both O'Neill and Turner, the body is the basic form of humanity, which conditions every cultural achievement ever made, it is problematic when we can no longer see our bodies in the world because this means that we can no longer understand the common basis of our humanity which is premised on the fact that despite all our cultural differences we inhabit the same basic human body. This idea of the common body of humanity is, of course, a utopian theory itself, which recalls Merleau-Ponty's (1969) idea of 'the flesh of the world', and conditions the theory of human rights, but it is not a view which is embedded in our everyday self-understandings of the world we inhabit. The common human body may still form the basis of our understandings of our world, but the metaphorical body is largely unconscious in the contemporary west, simply because we occupy high tech, post-modern, post-industrial societies where cold technology has become the central metaphor for thinking about existence.

The mechanised, post-human, body is, of course, another, or we might even say, *the*, utopian fantasy of the body (Rabinbach, 1992). Although this utopian body is one neither O'Neill or Turner want to see escape the pages of science fiction and become central to our understandings of the human world, I would imagine they would both recognise that it is largely too late and that the idea of the mechanised body largely dominates western views of corporeality, society, politics, economics, and perhaps even reality itself. If this is the case, and the idea of the mechanised body is largely dominant in contemporary western culture, then I would suggest that the current paper *must* take the form of a psychoanalytic excavation of the history of the body in western culture for one significant reason. What is this reason? I think the reason I must employ the principles of psychoanalytic methodology in this paper is that the failure of the anthropomorphic body metaphor, which no longer functions in western culture, and the related idealisation of the mechanical body, which simply works, means that the notion of the body is totally absent or unconscious in our thinking about the world which is premised, in modernity at least, on solving problems and addressing malfunctions. Simply because the ideal mechanical body is a body that never malfunctions or breaks down it becomes unconscious in a culture set on progress through the dialectical overcoming of technical problems. What is this body, the body that never breaks down, and consequently becomes a kind of no body in a culture centred on solving problems, if it is not the perfect body, the utopian body, par excellence? It is under these conditions that the body, which may remain the privileged metaphor for thinking about the world, even in its industrial / post-industrial mechanical or cybernetic form, is rendered completely unconscious and appears to vanish. As O'Neill (1985) and Turner (2008) both show, this situation is enormously problematic because it leads the west to repress the truth of the body, pathologise the different body that is incapable of living up to the post-human standards of the mechanised body, and simply ignore the body in pain on the basis that it represents the horrific real of the vulnerable body. Perhaps this is why, despite O'Neill's view that the body is the privileged metaphor for understanding abundance and poverty, we occupy a world where abundance and poverty are etched onto hyper-visible mass mediated bodies, but remain silent about the enormous inequalities we see embodied before us. It may be the case that the rise of the mechanical, cybernetic, body over the course of the long history of the west, which stretches back to Athens and seems to have ended in post-modern America, and the simultaneous repression of the real body to the shadowlands of the unconscious, means that westerners are perhaps psychologically unwilling for deep cultural reasons to truly recognise the body in pain or empathise with the suffering, starving, tortured body of those who cannot measure up to the mechanical, phallic, utopian body of western culture.

This is why the present paper must function as a psychoanalytic excavation of the history of the body in western thought, and more especially the utopian body, from the archives of western culture which has sought to repress, not only this history, but also the very recognition of the mortal human body itself, behind the surreal fantasy of a utopian body that never malfunctions, breaks down, or ultimately expires. After 2500 years in the making it is not surprising, but still rather shocking, that this surreal, fantastic, body has become normal and an everyday reality in western culture. Given this history it should be clear why this excavation work is important. Despite the fantasy of the utopian body, which represses the body in pain for the sake of the perfectly mechanised absent body, the suffering body is the truth of embodiment and, to refer to Lacan, the real that western culture must come to terms with, if it is to not only traverse the fantasy of the utopian body, which means that we must suffer in silence, but also recognise that it is the very fantasy of the utopian, mechanised, body that produces suffering bodies in pain in the first place, precisely because it is completely insensitive to the realities of what Turner (2008) calls the vulnerable body of humanity. The purpose of this paper is, therefore, to excavate the history of the utopian body in western culture in order that we may come to terms with this fantastic construction and remind ourselves of the vulnerability of the real body. My use of the term utopian body throughout the paper reflects not only the imaginary perfection of the iconic bodies under consideration, but also makes use of the etymological significance of Thomas More's Greek term utopia, which simultaneously means good place and no place, to suggest that the utopian bodies under examination are similarly good bodies, where good refers to the ancient notion of the good as ideal or perfection, and no bodies, as in absent bodies that are both perfectly functional and therefore not present and totally imagined and consequently rooted in the ethereal realm of ideas.

Apart from employing the etymology of More's term to reflect on the imaginary nature of particular social and political iconic bodies, the term utopia is also particularly useful for thinking about the nature of the idealised bodies under consideration because in much the same way that More imagined his island paradise as completely enclosed in terms of time and space, the imaginary ideal bodies of western culture are always totally self-identical and seem to exist outside time and space. Following this equation of More's paradise island and the imagined perfect body of western thought it would be possible to make the claim that More's early modern utopia was in fact a metaphor for the idealised body of the Greek utopians, but I would rather not take this route through the examination of the utopian body because I am concerned to examine concrete representations of the ideal body in particular social and political contexts. Given the timelessness of the utopian body, which is self-identical in time and space, and the formal similarity of the utopian bodies I propose to examine, which I take from Ancient Greece, early Christianity, Revolutionary France, Soviet Russia, and Nazi Germany, I think we might speak of the utopian body as a kind of ideal type that has recurred through western cultural history and culminated in the emergence of the contemporary globalised capitalist absent body or no body. In the section that follows I propose to briefly survey the utopian bodies of Ancient Greece, early Christianity, Revolutionary France, Soviet Russia, and Nazi Germany, emphasising the ways in which these perfectly sculpted, disciplined, or mechanised ideal bodies, presupposed the existence of some other, excremental, suffering body in pain. I will conclude the paper with an analysis of what I consider to be the culmination of the history of the utopian body in the emergence of the globalised, western, cybernetic no body, which is no less dialectical than the other examples of utopian bodies under examination, despite the massive weight of psychic repression screening out the multitude of suffering bodies which characterise contemporary globalised society. The objective of the production of this history, which we might call following

Mirzoeff (1995) a 'bodyscape' of western culture, is to illuminate the emergence of a dialectic of, on the one hand, imagined bodies characterised by hard muscle, cold machines, and ethereal presence and, on the other hand, suffering, starved, and tortured bodies in pain, so that we can better appreciate the problematic nature of the ideal body in western social and political thought.

Pumping iron, or, the utopian body in history

The notion of a dialectic of, on the one hand, utopian, built, bodies defined by hard muscles, cold technology, incorruptible virtue, or ethereal presence and, on the other hand, excremental, ruined, bodies characterised by starvation, suffering, disease, or pain presupposes the existence of corporeal progress or bodily evolution through a Platonic-Hegelian process of conflict, synthesis, and eventually overcoming. In the utopian bodies under consideration in this paper, the notion of dialectical progress is reflected in the idea of embodied combat with, and overcoming of, the excremental other body, which is seen to crumble or collapse before the monumental challenge of the utopian body. As I will show it is apparent from the various cultural representations of the utopian figures I want to examine that combat, or struggle, is necessary for the evolution of the utopian body to its ideal cosmic, divine, universal, pure, or free state, and that the excremental body in pain is seen as an archaic remainder that must be cast off if the new man, Homo Novus, is ever to become reality. Thus the utopian bodies I want to explore are all engaged in combative struggles with excremental others, struggles that are represented as being of enormous social and political importance for the future of the city, society, or even humanity itself. For example, Plato's (2003) philosopher body, waiting for the moment of death which will transform his miserable earthly body into first a lifeless phallic statue and second an ethereal form identical to the cosmological body of nature, was caught up in a struggle for the soul of the city with the passionate body of the demos, which was driven by mortal desire for debased earthly objects, rather than immortal reason and cosmological justice. Consider Paul's early Christian body. In this instance the determination of the believer to maintain a pure body, suitable for fusion with the divine body of Christ when his earthly sinful flesh expires, was endlessly tested by evil invasive forces set on preventing his transformation to this ideal corporeal state (Martin, 1999). The same story maps onto the modern cases of the French Revolutionary, Soviet, and Nazi utopian bodies. Whereas the incorruptible virtuous French Revolutionary citizen body was engaged in perpetual combat with the monstrous aristocratic body represented by the mythological Hydra (De Baecque, 1997), the ideal metallised Soviet body, most especially under Stalin, was forged in the fires of rapid industrialisation to fight the lustful capitalist body and the conspiratorial saboteur body, which were both diametrically opposed to the ideal metallised communist body in that they were similarly largely feminised, portrayed as soft, and driven by desire, passion, and lust (Hellebust, 2003). The same claim may be made about the ideal Nazi body. In much the same way that the perfect Soviet body was metallised to oppose the conspirator bent on preventing the completion of the revolution, the utopian body of Nazi Germany took the form of a steel-hard soldier able to contribute to the party's cosmic struggle to eliminate the excremental other, the degenerate Jew, the rotten Slav, and every other body that did not conform to the perfect Aryan type imagined by Arno Brecker, sculptor of the neo-classical Nazi body (Theweleit, 1989).

Does the same dialectical logic hold for the contemporary globalised capitalist no body, who roams through virtual cyberspace without care or concern, and closely resembles Deleuze and Guattari's (1972) fantasy of the schizophrenic surfing across the smooth space of the body without organs? Although democratic ideology would tell us otherwise, pointing out that there is no dialectical relationship between the progress of the body building voracious consumer of the west and the starved, suffering, body in pain of 'the rest' who largely inhabit the global south, Hardt and Negri's (2000) analyses of global capitalism and neo-colonialism suggest that the utopian body of contemporary global capitalism is completely bound up with the excremental body in the dead zones on the world map. If this is the case then perhaps we must conclude, in line with Marx (1988), that capitalism is an enormous machine for the production of excessively wealthy and absolutely poor bodies and that the position of the wealthy, healthy, body is totally reliant on the miserable state of the absolutely poor, suffering, body in pain. Alternatively, we could resist, or at least modify, this claim by explaining that the objective of capitalism is the production of profit, and the wealthy body, rather than poverty, and the starved body, which is simply a by-product of economic inequality. In many respects this is correct. Capitalism is completely driven by the pursuit of surplus value, which supports the evolution of the new utopian body of the voracious consumer, and is largely indifferent to the condition of miserable worker bodies, which are simply understood in terms of labour-power and their capacity to produce surplus value.

But if we recognise that capitalism is guilty of no more than a kind of malign neglect where the wealthy have no concern for excremental others, which are understood as either objects ripe for exploitation or hazardous human waste to be held in secure zones (Bauman, 2003), then we must also be aware that in many ways the construction of the utopian body was similarly a priority in each of the other utopian movements under consideration, and that production of the excremental twin was equally understood as little more than a problematic by-product of the drive to corporeal perfection in each of these cases. Despite the clear similarities between the key utopian bodies of western culture, there are however important difference relating to the ways in which they reacted to, and treated, their excremental twins. Although the original social, political, and theological utopian bodies of Ancient Greece and early Christianity could not rely on the enormous economic mechanism of global capitalism for sorting utopian from excremental bodies, perhaps because such an economic machine was not available, they made use of similar stratifying systems, such as caste and ideology, to ensure that some bodies were considered fit to progress towards the ideal state through contemplation and exercise, whereas other bodies were condemned to either toil on behalf of the utopian bodies who had no time for muscle work or simply suffer in silence and wait for the apocalyptic moment when the chosen people would unite in the body of Christ and everybody else would be cast into Hell. In other words, neither the Ancient Greek nor the early Christian utopian body was reliant on the complete elimination or obliteration of the excremental other body to secure their realisation. The security of caste and the exploitation of slavery was enough for the Greeks, while Paul's cultish utopian body was neurotic in its fear of invasion and construction of prohibitions concerning sex, food, and contact with others, but never sought to wipe out the other body (Martin, 1999). This is not, of course, to suggest that violence was not prevalent in Ancient Greece or early Christianity, but rather that the ideal bodies of both cultures were not genocidal in their attitude to the other body.

We know that the same cannot be said of the modern revolutionary bodies of the French Revolutionary, the Soviet Communist, and the Nazi that were more or less totally reliant on terror and violence to subdue the invasive, conspiratorial, body of the other and construct their own self-identical corporeal forms. While the French Revolutionary body was able to rely on the rationality of the guillotine to eliminate the excremental aristocratic other (De Baecque, 1997), the metallised Soviet body was protected by the Gulags, which were either the making or more usually the breaking of Stalin's utopian / excremental body (Getty and Naumov, 1999). Similarly, the steel hard neo-classical Nazi body evolved through the extermination of the degenerate body that was bent on its destruction in the death camps, such as Auschwitz, which were meant to eliminate the threat of the biological contagion and contamination of the perfect Aryan form (Agamben, 1998). Given that Auschwitz was called 'Anus Mundi' by senior Nazi figures illustrates that in the Nazi case in particular the association of the degenerate body with excrement was explicit and not simply understood in metaphorical terms (Esposito, 2008). The key point to take here however is that in each of these modern examples the emergence of the utopian body was enabled and hastened by the elimination of the excremental body which, in the Nazi case, produced the classic example of embodied liminality, the famous Muslim or Muselmann, who was the polar opposite of the Nazi muscleman imagined by Brecker, but was similarly caught somewhere between life and death, on the line separating the Aryan living God from the Jewish sub-human (Agamben, 2000). The Ancient Greek and early Christian examples seem tame by comparison, even though it is possible to suggest that the tortured, tormented, body has always been a key symbol of Christianity (Moore, 1996), simply because neither culture was reliant on the complete destruction of the other to maintain the fantasy of a utopian body. But how can we explain this difference between the ancient and modern utopian bodies on the issue of the other body?

The reason for the difference between the Ancient Greek and early Christian attitude to the other and the modern French Revolutionary, Soviet Communist, and Nazi view of the same figure is not completely clear, but I believe it may be related to two factors: the revolutionary status of the modern utopians, which led them to meet the other who was thought of in terms of a counter-revolutionary threat with genocidal violence, and the earth-bound status of their utopian bodies, which left them with no room to project their idealistic bodies into some other-world, leaving them free to suffer earthly bodies in pain in the knowledge that they would overcome their miserable corporeal conditions in the next life. Regarding the first point, the difference between the modern and Ancient Greek utopian bodies is clear. While the former was a revolutionary figure, who needed to secure his position vis-à-vis the ideal body of the previous regime, the utopian body of the latter was the product of the ruling elite, which was more or less secure in its position, and in fact employed the image of the perfectly formed body to try to naturalise its position. In other words, the ideal body of the modern revolutionary was born from a kind of existential anxiety, which explains why his form was reliant on the elimination of the other, whereas the perfect body of the Ancient Greeks was a conservative symbolic of hierarchy and order, which was simply meant to reassert the cosmic order of things (Martin, 1999). But what about the second key point, relating to the earth-bound status of the modern revolutionary ideal bodies which separated them from both the Ancient Greek and early Christian perfect bodies that were both ethereal and other-worldly in their idealised forms?

I think that this issue pertains particularly to the differential use of corporeal violence in Paul's early Christianity and modern French Revolutionary, Soviet Communist, and Nazi ideology. Whereas Paul's revolutionary thought explained that the weak would overturn their oppressors in the next world, thus making it possible for them to endure their painful situation in this life, the moderns could not fall back on the luxury of the idea of some other worldly utopia. On the contrary, the onus was on the modern revolutionaries to produce their utopian body and consequently their utopian society immediately through the destruction of those archaic remainders of their traditionally miserable situation, the bodies of the excremental other (Sagan, 2001; Featherstone, 2007). Given that the early Christian felt no need to obliterate the body of the other, who was already damned by virtue of his lack of belief, the task of life became about ensuring his own ascension to the mystical body of Christ when his mortal body expired. What this means is that the early Christian, and the same principle applies to the Ancient Greek utopian body which was similarly characterised by the need to prepare for the other world, turned violence inwards in the form of self-denial, renunciation, asceticism, and strict obedience to the word of God for the sake of making sure that his body was fit to make the transformation to utopian body come judgement day (Martin, 1999). The Greek practice of 'melete thanatou', which refers to the notion of the preparation for death, functioned in the much the same way as early Christian obedience in that it was designed to ensure that the ethereal soul was prepared to leave the fleshy body behind and become one with the cosmological body of nature when the earthly body finally expired. This belief in the need to prepare oneself for the other world is exactly what defined the Socratic attitude to death which was characterised by the view that life itself was simply preparation for the ascent to cosmological perfection (Plato, 2003).

Although it would be difficult to make an argument for the value of pain and torment in Ancient Greek thought, there is a sense in which those who support the notion of a link between the philosophy of Socrates and the cynic Diogenes, who famously refused to seek worldly comfort or obey the normal rules of Greek society, in the name of refuting the Nietzschean idea of the Socratic monster of reason and affirming the truth of the embodied Socrates of the people, confirm that the master had little interest in this world and saw life as preparation for death and the fusion of the soul with the cosmos (Hadot, 2002). The importance of renunciation, self-discipline, self-control and the refusal of the fleshy body is clearly present in Paul's early Christianity and the crucifixion and consequent resurrection are classic symbols of the need to endure the body in pain in order to overcome the apocalyptic moment of death and ascend to the mystical body of Christ (Moore, 1996). As we have seen the notion of the value of apocalyptic, transformative, corporeal violence is similarly present in the modern examples of the French Revolutionary, Soviet Communist, and Nazi ideal bodies, but in each of these cases it is not enough for the new man to assert his incorruptibility. In each instance the new utopian body in the making is reliant on the destruction of the old degenerate body to ensure its realisation. This much is evident from the history of revolutionary terror in the cases of both France and the Soviet Union and the Holocaust in the example of Nazi Germany. But what is the root cause of the revolutionary body's reliance on the elimination of other body and how can we differentiate the combative nature of this body from the combative nature of the earlier Greek / Christian body?

Apart from restating that the Ancients saw corporeal transformation occurring in some other world, and could feel certain about this process simply because the idea of the relation between the body of man and the body of the cosmos was rooted in ancient thought (Martin, 1999), and that the revolutionary position of the new man, his status as becoming, rather than being, and perhaps his unconscious sense of inferiority, led him view violence against the body of his historical enemy as essential to his defence in ways that the ruling elite in Ancient Greece never would have, simply because they were more or less secure in their position, we must also make two other related points. The first point to make is that the modern revolutionary's need to eliminate the other body, rather than simply defend itself against it, may be related to his own very modern sense of unconscious uncertainty about processes of modern progress and his insecurity about the idea of the quasi-theological mission of history, which may be similarly related to both his unconscious view of his own inferiority vis-à-vis his historical oppressor, and his knowledge of modernity as the age of both constant disruption and Promethean man who makes his own world. This latter point is essential because it suggests that what separated the modern revolutionary from the ancient in his view of the construction of the utopian body was his sense of the need to actively make himself in history, rather than simply conform to some pre-existing notion of cosmological balance. Moreover, I would suggest that herein resides the answer to the second question posed above, relating to the difference between the combative nature of the ancient and modern utopian bodies under consideration. In my view the combative nature of the Ancient Greek and early Christian utopian bodies was largely defensive in nature and related to their need to either maintain the status quo or maintain their integrity vis-à-vis the potentially dangerous other safe in the knowledge that if they could maintain these states they would successfully merge with the divine cosmological body when their time came (Martin, 1999). The key ideas here are belief in hierarchy, the view that hierarchy will endure unless it is upset by revolutionary forces, and the notion of balance which should characterise the microcosm of the utopian body, the community of citizens or believers united in the ideal city, and the macrocosm of the cosmological order, represented by oneness in Plato's philosophy and God in Paul's early Christianity.

By contrast to these utopian bodies, which I think we might call conservative despite the revolutionary nature of Paul's political theology, simply because their was no attempt to overturn the hierarchy of worldly bodies in his thought, the combative nature of the modern revolutionary body was offensive, or perhaps we should say pre-emptive, in its approach to defensive action. What this means is that modern revolutionary utopian body was combative in an offensive sense, although in each case this was presented as defensive action, because modern revolutionary consciousness was characterised by the view that although history was in some respects planned, it was totally reliant on the men of the future carrying out its work and preventing the men of the past from stalling its progress towards utopian perfection (Featherstone, 2007). However, that both forms of utopian body, what we may call the ancient conservative and modern revolutionary utopian bodies, appear aesthetically very similar, illustrates their exclusionary nature. There is, therefore, no real sense that the transition from the ancient conservative utopian figure to his modern revolutionary utopian brother entails a shift to real inclusivity or recognition of the diversity of bodies in the world. On the contrary it appears that the modern revolutionary body simply built on the platform of its ancient predecessor in the construction of a perfect white, male, body that could not tolerate otherness (Agamben, 1998). What is more, the issue of the identity of the utopian body is not simply an iconographic issue, confined to the study of aesthetics, because what the iconic, utopian, body offered, and continues to offer, through the provision of an idealistic object of veneration, is an ideological text which functions, in much the same way as other utopian models for the social, political, and economic transformation of city and community, by engaging with reality and producing effects on real bodies. In this way the utopian body has always offered, and will continue to offer, an ideational measure to understand real world bodies and categorise them on the basis of sex, gender, race, ethnicity, or class.

From Ancient Greece, through early Christianity, to the modern French Revolutionaries, Soviet Communists, and Nazi Germans, the utopian body has always been understood as a predominantly white male body concerned about pollution by others who were primarily female, racially or ethnically suspect, or part of an exploitative class (Bataille, 1985). Of course, one may respond to this point with the claim that the utopian body is simply an aesthetic, or iconic, symbol employed to stratify real bodies on the basis of a particular world view and that all world views stratify bodies in some way or other. Although there may be some truth in this claim, what is problematic about the utopian bodies under consideration, and over-codes concerns about the identity categories they employ to measure bodies, is the way that they respond to the relative measurements made about real worldly bodies. In other words, it is not that the utopian bodies under examination categorise bodies on this, that, or the other basis that really matters, but the fact that when they categorise bodies they do so through a totally inflexible complex of normality and pathology that renders them completely unable to recognise otherness, or other bodies, that are condemned as animalistic, sinful, lustful, corrupt, rotten, or in need to extermination. Under the sway of this complex of normality and pathology the impossible utopian body, which it is never possible to attain, becomes normal, while the fleshy body in the world becomes pathological and in need of either modification or extermination (Canguilhem, 1991). Herein we encounter the final step in the construction of the utopian body complex which occurs when some real bodies are considered, and marked, suitable for entry into the category of potential utopian bodies and others are cast out as worthless refuse and condemned to lives of either labour, pain, and suffering, or immediate extermination.

I believe that this logic holds for the Ancient Greek and early Christian utopian bodies and their French Revolutionary, Soviet Communist, and Nazi German successors. In each case an ideological complex of normality and pathology was constructed and then real bodies were sorted on basis of their relation to either category. The only qualification I would add to this claim is that it may only be in the case of the Nazi German utopian body and its excremental twin, the degenerate Jew, that the full bio-political inference of Canguilhem's original normality / pathology structure takes full effect (Esposito, 2008). But having recognised this point, I believe that each of the other examples may be explained through Canguilhem's binary structure understood on a metaphorical level. Beyond Canguilhem, the classic author of corporeal normalisation is, of course, Michel Foucault. Although the concept of the utopian body is implied in nearly every one of Foucault's (1977, 2003b, 2008a) works on corporeal normalisation, by virtue of the fact that for every pathologically insane, unhealthy, or unfree body, there must exist the utopian image of a perfectly sane, perfectly healthy, perfectly free body, he never wrote a substantial work about the construction of the utopian measure, the utopian body, preferring instead to concentrate on the evolution of the pathological twin. What we have from Foucault on the notion of the utopian body is limited to a short radio presentation from the 1960s called 'Utopian Body' (Foucault in Jones, 2006). In this essay he explains that the real body is the opposite of the utopian construct, because it is always present and never in some other place. In other words, one can never escape one's own body. Following the logic of the utopian bodies we have considered above, Foucault tells us that utopia is a place where one possesses a no body, an incorporeal body, that never malfunctions, breaks down, or expires. But, he continues, the problem with this model of immaterial embodiment is that the only place it is found is the land of the dead.

Given the ancient utopians' commitment to the preparation for death, and the modern utopians' tendency towards terrorism and mass murder, it is difficult to disagree with Foucault's conclusion that the only place one is likely to find the utopian body in reality is the graveyard, the land of tombs, mummies, and corpses. Following this conclusion, he supports the view that the utopian body is reliant on violence and death to enable transformation to its ideal state of rigor mortis when he explains that the reason the graveyard is the privileged space of the utopian body is because it enables the translation of the fleshy mortal body into the immortal, incorporeal, body of the utopian colossus, who we can identify with the cosmological one of Platonic thought, Paul's body of Christ, the French Revolutionary Herculean citizen body, the Soviet metallic man, and the Nazi Aryan God. But for Foucault the problem of the utopian body, which emerges from the graveyard like one of George Romero's (1968) flesh eating zombies, is precisely that it is totally self-enclosed and totally uncommunicative. Like Romero's zombie, the living dead utopian body is incapable of human relations, primarily because it is no longer human. The only relation it knows is one of endless murder and cannibalism. In other words, it lives by killing, eating, and transforming live bodies into zombie bodies. It is, therefore, the utopian dimensions of the utopian body that problematise its existence in Foucault's eyes. He explains that it is this condition, the condition of utopian closure, that results in the inability to recognise the other, and mortifies the utopian body, depriving it of what is simultaneously the essential weakness and fundamental power of the real body, the utopian power to relate to the other.

The global no body and the vulnerable body in pain

It is difficult to read Foucault's account of the living dead utopian body without recalling Adorno and Horkheimer's (1997) image of the voracious consumer who has no mind of their own, but has been transformed into a completely reactive body by the machinations of the late capitalist culture industry. Akin to Foucault (1977), who wrote about the destruction of the thinking body and the rise of the docile body in modernity, Adorno and Horkheimer sought to expose the industrialisation, or mechanisation, of humanity in the totally administered society they associated with modern capitalism. The difference between Foucault and Adorno and Horkheimer however, and the reason the latter thinkers may be particularly relevant to the study of the utopian body at this point, is that they enable us to think through the relationship between the status of the body in Nazi Germany and the short American century that followed in the wake of the liberation of Auschwitz, precisely because they lived in and theorised both societies (Rickels, 1991). Although Foucault's various studies of the production of the perfectly disciplined body is, of course, relevant to the study of the shift from the image of the perfect totalitarian body to the utopian capitalist body of the late 20th century, primarily because the Structuralists emerged in order to study the unfreedoms of liberal post-World War II society, I think Adorno and Horkheimer capture this shift more particularly because it mirrored their own move from Germany to America in the 1930s and was as a result reflected in their thinking about the connections between European totalitarianism and American liberalism. The reason the liberation of Auschwitz on 27th January 1945 is particularly important in respect of the status of the Nazi utopian body is because it was this event that most notably resulted in the exposure of the horrendous under side of Hitler's experiment to produce a master race. Following the Red Army's discovery of the camp, complete with the skeletal bodies of the Muslims who could no longer think for themselves but had been transformed instead into reactive robotic men, the fantasy bodies of Arno Brecker's neo-classical super-hero, represented by sculptures such as 'Readiness', and Leni Riefenstahl's blonde Gods, reflected in her 'Triumph of the Will', became terrible images of the other side of the utopian body in the making, the excremental other body which necessarily shadows perfect corporeal form. History shows that in response to the exposure of the horrendous nature of the dialectic of the utopian body at Auschwitz the European model of utopian thought fell into disrepair and the fantasies of the steel hard, virtuous, body were condemned to the category of inhuman totalitarianism in favour of a new brand of American market humanitarianism which purported to recognise the humanity of the body by recognising and catering for its earthly desires (Featherstone, 2007).

However, as Adorno and Horkheimer (1997) would show in their work on the culture industry, the problem with new utopian body of desires, the figure Foucault (2008b) calls Homo Economicus, is that he was a utopian body in the worst sense of the term. That is to say that he was reduced to the status of body by a social, political, and cultural system that produced consumers through the creation of false needs and reduced every aspect of life to economic calculation. In this context, free thought was replaced by reaction to mass media stimulation, the potential of the construction of independent sign systems to enable free thought was limited by the pervasive nature of the culture industry, and democratic man was transformed into an economic body by his integration into the metabolism of the body of capitalism. In this respect, then, Adorno and Horkheimer's (1997) equation of Nazi Germany and Capitalist America is not only reliant on the similarity between propaganda and public relations, and the view of the fantastic nature of American democracy, but also, and this is perhaps the central point, the connection between the reduction of both the Nazi totalitarian body and the American consumer body to the status of what Agamben (1998) calls bare life that is characterised by the simple reproduction of economic metabolism. It is well known that Hannah Arendt (1973) saw the terrible reduction of humanity to the status of machines or robots, which would suffer none of the imperfections of human life, as the goal of the European totalitarianisms, and understood the camps as laboratories for the production of this robotic new man. Although Brecker's steel hard men reflected the utopian face of this Nazi God, it is, therefore, the Muselmann who represents the truth of the Nazi body because it is only in the case of the Muslim, who has lost his mind, that the human body is reduced to the basic metabolism of bare life (Agamben, 1998; 2000).

In light of this view, the terrible conclusion of Adorno and Horkheimer's (1994, 1997) various studies of American consumer capitalism is that the classic American capitalist utopian body, which is characterised by self-realisation and the freedom of desire, is simply a fantasy and that the truth of the American consumer is a stupid body following imprinted desires and attributing these to some drive towards self-realisation because that is exactly how the dominant ideological position frames them. What is even worse is that the comparison between the Nazis' Jewish Muselmann and the American consumer is not simply confined to the ways in which both figures are reduced to stupid bodies. Instead, we must recognise that in much the same way that the Muselmann's reduction to the status of stupid body entailed a return to a state of nature where all that matters is metabolism, the ideal body of consumer capitalism is characterised by his microcosmic connection to the macrocosmic body of capitalism which is defined by the ceaseless metabolism of production and consumption. Finally, the problem here is not only that the utopian body of consumer capitalism is a stupid body, which cannot think and has been reduced to the status of a simple metabolic system, even though that would be bad enough. Rather, the issue is that the modern version of the ancient anthropomorphic notion of the micro-macrocosm, which implicates man's body in the body of the whole, in this instance relating to the notion of a capitalistic universe where reality is defined by metabolism, takes no account of the organic nature or the vulnerability of the human body because it is based on the sci-fi fantasy of a mechanical man.

Herein resides the true horror of the consumer capitalist utopian body. It is not only that man is reduced to the status of body or that the body is reduced to the function of economic system but that that system is no longer modelled on the organic body of man that experiences fatigue, suffers injuries and pain, and eventually expires, but rather the prosthetic, technological, body of humanity and the fantasy of the body that runs perfectly and lives forever. It is this idea, the idea of the technological, mechanical, cybernetic body Rabinbach (1992) explains through the notion of the human motor, that passes for anthropomorphism in consumer capitalism and enables us to continue to think through the ancient notion of the micro-macrocosm. But what kind of macrocosm, what kind of universe, is it that Rabinbach's human motor inhabits? The answer is that the new steel hard man, who is the ideal type of the utopian body, inhabits a cold mechanical universe that has no place for the frailty or vulnerability of humanity. This vision of consumer capitalist civilization as a cold, post-human, void is, I think, a good approximation of the dystopic other side of Deleuze and Guattari's (1972, 1984) utopian texts on capitalism and schizophrenia that make so much of the anti-oedipal freedoms of the schizophrenic who surfs across the smooth space of the endless desiring machine, the body without organs. The essential problem with Deleuze and Guattari's work is, of course, that they want us to believe that the fragile, shattered, body of the schizophrenic is in some way dangerous to consumer capitalism and that her desire transgresses the boundaries of this new paranoid utopia that cannot tolerate others.

The naïve utopianism of Deleuze and Guattari's generation, the generation of 1968, resides in this belief. I think this is more or less clear today. Given the benefit of hindsight we can see that the schizophrenic-consumer capitalism complex is not defined by a relation of opposition and transgressive, but rather a process of endless assimilation whereby the broken body of the schizophrenic is confronted with the utopian image of perfection available through the consumption of commodities and promised that this will repair both her fractured body and the deep alienation she feels from both herself and other bodies. The ideological twist here is that the fractured, lacking, body is a real body, a utopian body, that can relate to others, and that this body is only made to appear lacking by a mechanical civilization that is totally inhospitable to real bodies. It is this situation, conditioned by the violence of modernity to real bodies, which alienates the schizophrenic from herself and drives her through the endless chains of commodities produced by consumer capitalism in search of a perfect body suitable for living in mechanical civilization and relating to others who are playing the same miserable game of mechanisation or death. This dystopia is Deleuze and Guattari's (1984) utopia, characterised by the schizophrenic's endless flight across the capitalist body without organs, which has today become a globalised utopia by virtue of the endlessness of the task of repairing the broken, alienated, body through commodities that will never solve the constructed issue because the essential problem resides in the mutilation of the body by mechanised civilization, not the flawed nature of the real body that cannot match the technical perfection of the clockwork universe.

But the success of the consumer capitalist civilization resides in the construction of this problematic body. It is this flawed body, the body in need of prosthesis, driven into the endless commodity future by the construction of the idea of fundamental human lack, which has produced the total, globalised, body of the consumer capitalist utopia-dystopia. It is also this body that has finally realised the Platonic (2008) fantasy of globality, the perfect micro-macrocosm that links the seat of human reason, the round head, to the spherical oneness of God. But the irony of this realisation, which has finally completed the long history of utopian thought and the utopian body, is that Plato's dream of the sphere of reason, the sphere of head and the sphere of cosmos, has resulted in the production of a total body devoid of reason that is at once totally unified and torn asunder, simultaneously completely stuffed and starved to death (O'Neill, 1985). The contemporary utopian of brains may complain that surely Plato's fantasy has been completed by the realisation of today's information super-highway and the elevation of the mind in the knowledge economy, but I would resist this claim noting that the significance of the invention of cyberspace resides less in its elevation of the life of the mind and more in its impacts upon the body that is liberated from frailty, vulnerability, and the embodied reality of humanity and projected into the total body of the post-modern, globalised, capitalist cosmos in the form of code. The primal scene of the knowledge economy and the spiritual home of the information super-highway is, of course, Silicon Valley, California. But California is not only home of the industry that liberates the utopian body of code from the fleshy body that must feel pain. It is also the birth place of the cinematic built body, represented by the body of Arnold Schwarzenegger, who once starred as an indestructible action hero in movies such as *The Terminator* (1985) where he waged war on other fleshy bodies, and is now Governor of the Golden State, which boasts one of the largest economies in the world, and is the home of Adorno and Horkheimer's (1997) nightmarish consumer society. But why is this important?

Despite my apparent detour, these facts are not insignificant for our study of the utopian body in history because in many respects Schwarzenegger represents the bridge from the Ancient Greek utopian body-builder ideal through the Nazi neo-classical body of Brecker, because Arnie was famously said to have told an interviewer he admired his famous countryman Hitler, to the American super-hero body, which is at once represented by the mechanical killing machine of *The Terminator* and the traditional monosyllabic muscle hero of *Predator* (1988). But if Schwarzenegger represents the archetypal utopian body, this body was only made possible by the fact that California is simultaneously the home of techno imaginary in the form of Hollywood, body modification in the shape of the cosmetic surgery industry, and the porno body of the San Fernando Valley, which parodies the iconic body of the mainstream movie industry and the perfected, modified, sculpted body of cosmetic surgery in a cold mechanical parody of Deleuze and Guattari's desiring body, where in body is simultaneously hard fucking metal and totally commodified meat. But none of this technology, none of this constructive genius has been able to alleviate the pain of estrangement felt by Californians who live on the hard edge of the high tech society, and consequently experience the greatest pressure to become utopian bodies, a point which is evidenced by the claustrophobic relationship between California and therapy culture (Rickels, 1991).

The result of this failure to resolve the problem of the alienation of the body in California, the home of western body modification, is that nowhere else in the west has been able to solve the issue of estrangement, and that the entire culture has become locked into viewing the body as a project best enacted and solved through consumer capitalism (Turner, 2008). Although a more positive solution to the obvious failure of the prosthetic model of the utopian body model would have been to embrace the vulnerable body and accept the real of the body, the contemporary globalised capitalist body has become Deleuze and Guattari's tragic desiring machine with the related effect that it has been transformed into a kind of war machine allergic to the other body, which has been consigned to the sweat shop of global capitalism and excluded from the programme of utopian body building (Bauman, 2003). This is a truly horrendous situation, whereby the waste product of global capitalism, those who must struggle to survive on a daily basis, have become the other bodies that the western utopian built body at best ignores and at worst seeks to destroy or terminate in the name of its own continuing inability to conform to an impossible imperative to become a tech-no body. This is how I think we must interpret the corporeal aspect of the contemporary neo-colonial wars in Iraq and Afghanistan. That is to say that I think that beyond the standard interpretations of political economy, which clearly feed back into theories of commodity consumption and self making, the western capitalist utopian body of the colonizer confronts the other body of the colonized as a resource to be exploited and a largely insignificant stumbling block to be overcome en route to a higher form of corporeal existence represented by the dreamworld of consumer capitalism (Memmi, 2003).

That the key issue in Iraq remains energy is important because what this signifies is the utopian body's desire to maintain its status as perfectly functioning human motor. The case of Afghanistan, the new front line of the war on terror, is similarly illustrative because the issue here is security. What does security signify if not the utopian body's desire to continue to live a life of fantastic luxury on the basis of the exploitation of the world's other body which is considered insignificant and unimportant? This view, which paints the utopian body of the west as the body of Nietzsche's (2003) last man who wants to live a lazy life of luxury for ever more, is confirmed by the neo-colonizers' obsession with their own militarised cyborg body count and more or less complete ignorance to the civilian fleshy body count of the other. In this context, it is clear that human concern for real bodies has left the scene and that all the western colonizers are interested in is supporting their own transformation into utopian bodies by securing essential raw materials to ensure the proper functioning of consumer capitalism and making sure that the other body is in no position to resist their position as consumed objective body in the capitalist relations of consumption. In response to the imposition of this exploitative and humiliating situation by the western neo-colonial war machine, the other has no means of resistance but their own body. As Fanon (2001) suggested in the 1950s in relation to the French-Algerian colonial war, when the other is reduced to the status of bare life, transformed into a body in pain, and forced to live in a world defined by death and suffering, one should not be surprised when they respond by meeting their exploiter, the colonizer, with savage violence.

Since Deleuze (2003) made the same point in his work on Francis Bacon, noting that suffering meat is the core of revolutionary politics, we may conclude that what is taking place in the contemporary neo-colonial wars of Iraq, Afghanistan, and I think the same may be said of Palestine-Israel, is a struggle between the other body, which has been reduced to employing the tactic Feldman (1991) called body as weapon to try to reassert its basic humanity, and the war machine of the western utopian body, which trades off what Mbembe (2003) refers to as necropower in the name of eliminating the real body that wounds, suffers, and feels pain. In other words, I think it is possible to understand the current, and final, phase of the history of the utopian body through the notion of contemporary neo-colonial war as corporeal war between the other fleshy body reduced to the last resort of employing the suicidal tactic Feldman refers to through the concept of the body as weapon in his analysis of the IRA hunger strikers of the early 1980s, but the same idea can clearly be used to explain the tactic of the suicide bomber, and the high tech western utopian body driven to exploit and terrorise the other in order to sustain its own paradigm of body as project and ensure both resource and corporeal security in the face of the suicidal other who has nothing to lose but their tortured body in pain (Scarry, 1988). While Cavarero's (2008) explanation of the phenomenon of the indiscriminate destruction of the defenceless civilian body in pain by the high tech war machine through the idea of horrorism foregrounds the necropower of the utopian body, Feldman's notion of the body as weapon makes use of the Foucauldian (2003a) idea of counter-power to show that for every expression of power, no matter how awesome, there is an adequate form of counter-power or resistance. Unfortunately, however, whether we consider the necropower of the utopian body or the suicidal counter-power of the other body employing the body as weapon strategy to try to oppose the awesome military machine of the colonizer, the result is the same, the destruction of the real vulnerable body in pain.

That recognition of the terminal nature of the relationship between the emergence on the scene of the utopian body of the western war machine and indiscriminate death is everywhere in what we might call following Mbembe (2003) the death worlds of contemporary neo-colonial war, from American soldiers' choice of the ghoulish Drowning Pool record 'Bodies' (2002), with its screamed mantra of 'let the bodies hit the floor', as their soundtrack to war (Wilson, 2008), to the logics of grim survival and glorious martyrdom that characterise the lives of the local population, should lead us to conclude that we need to find a different approach to dealing with the lived reality of the human body today that does not translate the deeply held human fear of death into a violent drive to found an immortal utopian body. Surely the appearance of the death cult of the other body, which leads it to try escape the intolerable pain and humiliation of life through suicidal attacks on the western utopian body that may transform it into a utopian body of peace in the kingdom of God, should lead us to recognise that we must rethink our entire view of the human body in the world. In short, I think that we need to recognise the pain of the other body that survives rather than lives, and understand its formal connection to the utopian body of the western consumer, who has been similarly reduced to the status of a pure body that simply metabolises nature in order to exist, in order that we can start to see why we must escape the history of the utopian body and recognise the utopianism of the real body which, as Foucault (Foucault in Jones, 2006) explains, revolves around the ability to relate to other bodies in the world. In order to achieve this objective it may be that what we need today is a new political theology of the suffering body in pain, which recognises that vulnerability is the essence of humanity, and that this can never be hammered out of our species by fantastic thanatopolitical schemes set on the production of mechanical, cybernetic, utopian no bodies (Turner, 2008).

Filmography

Romero, George, 1968, *Night of the Living Dead*. Walter Reade Organisation.

Cameron, James, 1985, *The Terminator*. J. Arthur Rank.

McTiernan, John, 1988, *Predator*. 20th Century Fox.

Bibliography

Adorno T. W. and Horkheimer M. (1997), *Dialectic of Enlightenment*. London: Verso.

Agamben G. (1998), *Homo Sacer: Sovereign Power and Bare Life*. Palo Alto, CA: Stanford University Press.

Agamben G. (2000), *Remnants of Auschwitz: The Witness and the Archive*. New York: Zone Books.

Arendt H. (1973), *The Origins of Totalitarianism*. New York: Harvest Books.

Bataille G. (1985), 'The Psychological Structure of Fascism' in his, *Visions of Excess: Selected Writings: 1927-1939*. Minneapolis: University of Minnesota Press. Pg. 137–161.

Bauman Z. (2003), *Wasted Lives: Modernity and its Outcasts*. Cambridge: Polity.

Canguilhem G. (1991), *The Normal and the Pathological*. New York: Zone Books.

Cavarero A. (2008), *Horrorism: Naming Contemporary Violence*. New York: Columbia University Press.

De Baecque A. (1997), *The Body Politic: Corporeal Metaphor in Revolutionary France: 1770-1800*. Palo Alto, CA: Stanford University Press.

Deleuze G. (2003), *Francis Bacon: The Logic of Sensation*. London: Athlone.

Deleuze G. and Guattari F. (1972), *Anti-Oedipus: Capitalism and Schizophrenia*. Volume I. London: Athlone.

Deleuze G. and Guattari F. (1984), *A Thousand Plateaus: Capitalism and Schizophrenia*. Volume II. Minneapolis: University of Minnesota Press.

Douglas M. (2003), *Natural Symbols: Explorations in Cosmology*. London: Routledge.

Esposito R. (2008), *Bios: Biopolitics and Philosophy*. Minneapolis: University of Minnesota Press.

Fanon F. (2001), *The Wretched of the Earth*. London: Penguin.

Featherstone M. (2007), *Tocqueville's Virus: Utopia and Dystopia in Western Social and Political Thought*. London: Routledge.

Feldman A. (1991), *Formations of Violence: The Narrative of the Body and Political Terror in Northern Ireland*. Chicago: University of Chicago Press.

Foucault M. (1977), *Discipline and Punish: The Birth of the Prison*. London: Penguin.

Foucault M. (2003a), *Abnormal: Lectures at the College de France, 1974-1975*. New York: Picador.

Foucault M. (2003b), *The Birth of the Clinic: An Archaeology of Medical Perception*. London: Routledge.

Foucault M. 'Utopian Body' in Jones, C. A. (Ed.) (2006), *Sensorium: Embodied Experience, Technology, and Contemporary Art*. Cambridge, Mass: MIT Press. Pg. 229–235.

Foucault M. (2008a), *History of Madness*. London: Routledge.

Foucault M. (2008b), *The Birth of Biopolitics: Lectures at the College de France, 1978-1979*. Basingstoke: Palgrave.

Getty J. A. and Naumov O. V. (1999), *The Road to Terror: Stalin and the Self-Destruction of the Bolsheviks: 1932-1939*. London: Yale University Press.

Hadot P. (2002), *What is Ancient Philosophy?* Cambridge, Mass: Harvard University Press.

Hardt M. and Negri A. (2000), *Empire*. Cambridge, Mass: Harvard University Press.

Hellebust R. (2003), *Flesh to Metal: Soviet Literature and the Alchemy of Revolution*. Ithaca: Cornell University Press.

Jones C. A. (Ed.) (2006), *Sensorium: Embodied Experience, Technology, and Contemporary Art*. Cambridge, Mass: MIT Press.

Lacan J. (2006), 'The Mirror Stage as Formative of the I Function as Revealed in Psychoanalytic Experience' in his, *Ecrits: The First Complete English Translation*. London: Routledge. Pg. 75–82.
Martin D. B. (1999), *The Corinthian Body*. London: Yale University Press.
Marx K. 'Estranged Labour' in Marx K. and Engels F. (1988), *The Economic and Philosophic Manuscripts of 1844 and the Communist Manifesto*. New York: Prometheus Books. Pg. 69–85.
Mbembe A. 'Necropolitics' in, *Public Culture*. 15 (1), 2003. Pg. 11–40.
Memmi A. (2003), *The Colonizer and the Colonized*. London: Earthscan.
Merleau-Ponty M. (1969), *The Visible and the Invisible*. Evanston: Northwestern University Press.
Mirzoeff N. (1995), *Bodyscape: Art, Modernity, and the Ideal Figure*. London: Routledge.
Moore S. (1996), *God's Gym: Divine Male Bodies of the Bible*. London: Routledge.
Nietzsche F. (2003), *Thus Spoke Zarathustra*. London: Penguin.
O'Neill J. (1985), *Five Bodies: The Human Shape of Modern Society*. Ithaca: Cornell University Press.
Plato (2003), *The Last Days of Socrates*. London: Penguin.
Plato (2008), *Timaeus and Critias*. Oxford: Oxford University Press.
Rabinbach A. (1992), *The Human Motor: Energy, Fatigue, and the Origins of Modernity*. Berkeley: University of California Press.
Rickels L. (1991), *The Case of California*. Baltimore, MD: The Johns Hopkins University Press.
Sagan E. (2001), *Citizens and Cannibals: The French Revolution, the Struggle for Modernity, and the Origins of Ideological Terror*. Lanham, MD: Rowman and Littlefield.
Scarry E. (1988), *The Body in Pain: The Making and the Unmaking of the World*. Oxford: Oxford University Press.
Theweleit K. (1989), *Male Fantasies*. Volume II: *Psychoanalysing the White Terror*. Minneapolis: University of Minnesota Press.
Turner B. S. (2008), *The Body and Society*. London: Sage.
Wilson S. (2008), *Great Satan's Rage: American Negativity and Rap/Metal in the Age of Supercapitalism*. Manchester: Manchester University Press.

Chapter 2
Obama as charismatic icon

Jeffrey A. Halley, James P. Ordner

Abstract

This article examines the election of Barack Obama through the lens of Weber's concept of charismatic authority. We also explore the relationship between charisma and icon as a form of ideological representation projected upon Obama from different segments of the U.S. population. We examine what Obama represents for blacks, liberals, conservatives and other social groups as indicated in selected news articles and develop the notion of Obama as charismatic icon. Although very new on the scene and not well known, Obama is seen to have certain extraordinary properties (charisma). And his image embodies, more than a representation, certain iconic qualities. In fact, this image, although iconic, also represents different qualities for different audiences: a polysemic icon.

Introduction

The election of Barack Obama in November 2008 represents a landmark in American politics. For many black U.S. citizens the time for a black president was long overdue, while for others the thought of having a black president seemed impossible in their lifetimes. The political sphere in the U.S. has traditionally excluded blacks from the upper echelons of power, with some exceptions. This racial exclusion is in part due to institutionally designed constraints, but more importantly there has been a lack of political will in the general public to rally behind a black nominee for president. So what makes the election of Barack Obama different from Jesse Jackson's run for the presidency in 1984 and 1988? The cultural climate of the times certainly plays a role in election cycles, but there seems to be a qualitative difference in the reception of Obama in the U.S. populace. Below, we examine the nature of charismatic authority as understood by Max Weber and its relation to iconic imagery, and we use several newspaper articles as a cultural sounding board for discussing the perception of Obama in terms of race, political orientation, generational differences and cultural reception. We find the case of Obama to be politically compelling in that he symbolizes a variety of different iconic images that correspond to different subject locations.

Max Weber and charismatic authority

In *Economy and Society* (1968), Max Weber introduced three ideal types of political authority for describing different forms of leadership: charismatic, traditional, and legal-rational. Weber describes these ideal types as crystallizations of certain properties found in leaders at various historical moments. There is a developmental order correlating social complexity and the appearance of the three types of leadership, although they "cannot be placed into a simple evolutionary line: they in fact appear together in the most diverse combinations" (Weber 1968:1133). Weber identifies charismatic authority as the earliest and simplest form of leadership and it essentially relies on the deference bestowed upon a leader by his/her followers. Power is imputed to the charismatic leader because of what is perceived as exceptional qualities or skills. Traditional authority is established through customs and social conventions found within well-established social systems, such as a monarchial system of governance. Legal-rational authority represents the most advanced or modern form of leadership, best represented by liberal-democratic political systems. While each of these ideal types of authority have distinct characteristics, and historical contexts may produce a variety of leadership forms, most exceptional leaders exemplify one or more of Weber's ideal leadership styles.

In order to understand the appearance of charismatic leadership we must examine the social conditions the give rise to a desire for this particular kind of authority. At any historical moment leaders may possess charisma, but having exceptional qualities or skills alone does not guarantee leadership. Historical, cultural and social conditions must be right for the emergence of the charismatic leader. Weber indicates that legal-rational and traditional forms of authority rely on "a permanent structure and, with its system of rational rules, [are] oriented toward the satisfaction of calculable needs with ordinary, everyday means" (1968:1111).The stability and mundane everydayness of legal-rational and traditional authority represent the routinization of leadership where authority is based on predictability and the steady developmental progression of established custom and legal/economic institutions.

What makes charismatic authority distinct from legal-rational or traditional authority is how charismatic leaders emerge as an extraordinary phenomenon in opposition to the regularity and stability of legal-rational and traditional authority. Weber writes how "the 'natural' leaders in moments of distress - whether psychic, physical, economic, ethical, religious, or political - were neither appointed office-holders nor professionals in the present-day sense, but rather the bearers of specific gifts of body and mind were considered supernatural (in the sense that not everybody could have access to them)" (1968:1112). When the stability of traditional or legal-rational forms of authority fail due to social unease, such as economic crisis or war, people may seek charismatic leadership outside of the normal, business as usual political order. Weber associated charismatic leadership with a lack of irrationality. The deference projected upon a charismatic leader is based on emotion and personal belief rather than any rational, traditional or legal basis. Here we find the potential danger in charismatic leadership due to a fostering of emotional investment and potentially irrational/unrealistic belief in a leader's extraordinary powers or skills. Weber also indicates the ephemeral nature of charismatic leadership, noting how "charismatic authority is naturally unstable... [and] his divine mission must prove itself by *bringing well-being* to his faithful followers; if they do not fare well, he obviously is not the god-sent master." (1968:1114). There is a window of opportunity available to charismatic leaders when social conditions are tested by some crisis, but this window can easily close if they lose their charismatic appeal.

Although Weber was critical of charismatic leadership, he was equally if not more critical of legal-rational and traditional authority due to their routinization and rationalization of society through bureaucratic processes. Weber wrote that effective political leaders should possess the qualities of "passion, a sense of responsibility and judgment" (Weber 1978:212). Traditional and rational-legal leaders may display objectivity and proportion, yet often lack passion and vision, while a charismatic leader may display great passion without objectivity or proportion. The tension existing between these characteristics found in Weber's ideal types of leadership is hard to overcome or synthesize within one leader, yet for Weber the ideal leader possesses a mixture of charisma and a deep ethic of social responsibility. More precisely, genuine political leadership is a "matter of boring down strong and slowly through hard boards with passion and judgment together" (1978:225).

Weber underscores that charisma is not exclusively a property of the person, but rather "the mere fact of recognizing the personal mission of a charismatic master establishes his power" (Weber 1968:1115). The relation between charismatic leader and follower is dialectical in nature and both necessarily depend on each other's mutual recognition in order to function. It is recognition from followers or disciples that activates charisma, and while the charismatic leader may display some extraordinary quality which attracts disciples and followers, according to Weber "the effectiveness of charisma rests on the faith of the ruled..." (1968:1125). It is as if charisma is a 'black box' phenomenon, and the person's existence is not of great importance. The arbitrary nature of charisma points to the significance of historical and social contexts and the mindset of the populace as necessary preconditions for an emotional/ideological investment in a charismatic leader. When the routine world is seriously threatened or a society's political institutions are perceived as corrupt or illegitimate, collective desire for something radically different is created within the populace that can seemingly be fulfilled through a charismatic figure who offers a way out of bad times. In this way, the aura surrounding a charismatic leader can be seen as a projection of an individual or groups' desires for a more adequate form of (political) representation. Without the subjective projections of deference, belief, allegiance, hope or faith imputed upon a charismatic leader, charismatic authority will fade and wane as it "flows back into the channels of workaday routines..." (1968:1121).

Charisma and icon

There are certain similarities between the way individuals and groups project deference upon charismatic leaders and the social formation of iconic imagery. An icon is defined in Charles Sanders Peirce's notion of semiotic as a "...non-arbitrary intentional sign- a designation which is to a significant degree representational of, or has some isomorphism with, or bears an intrinsic resemblance to the object it designates" (Landsberg:93). While charisma fuses power and persona through a follower's projected belief or faith, the iconic image collapses a signified object into its signifier as a self-identical image: an identification between sacred notion and sacred object. The effect, as we will see, is ideological. The iconic image attempts to escape the encapsulating, routine rules of semiotics, thereby excluding its content from the arbitrary nature of symbolic systems *as a universal signifier*. In this manner, making an image into a thing-in-itself represents a pure form of ideology. For example, the image of Christ in the icon is not a representation of Christ, it is the thing-in-itself: it is actually Christ. In this sense, the likeness stands in for the object. In today's terminology, an icon may appear in the sense of a well-known symbol, name, or face that embodies certain more general qualities, such as the image of Marilyn Monroe standing in for sexuality itself, or Ché Guevara's iconic image standing in for revolution.

In the history of representation, we can distinguish two extreme and polar ideal types: the Jewish notion of the ban on the image, and the icon. The icon, as noted above, suggests a union between an object and a concept, while the prohibition on the image derives from the Second Commandment, "Thou shalt not make unto thee any graven image." In terms of semiology, the ban on representation is a refusal to provide signifiers to represent a signified. The signified, classically God, but for Marx, and Horkheimer and Adorno, utopia, is by its nature ineffable, so that it cannot be portrayed or identified through naming or symbolic rationalization. Representations of God in the concrete form of idols or images are not adequate, as the Golden Calf story of the Old Testament attests.

Against the Judaic tradition prohibiting representation, the icon represents a closed-off, 'identical' image. The identity of signified and signifier found in iconic imagery is ideological because the fusion between signifier and signified is subjectively forced upon the object and is thereby made unforgettable. The subject dresses the sacred object as it dresses itself, as an idealized image. The subjective nature of iconic identification that suggests the sacredness of the icon does not come from the object itself but is bestowed, as with charismatic authority, upon the object in the form of an ideological projection emanating from the subject. Charisma and icon both depend on the subject's acquiescence to or acceptance of ideological identification with a designated object; and as Weber indicated in the case of charismatic leadership, without subjective recognition the sacred object, the icon could never rise above the mundane nature of everyday life.

Creating a social space for charismatic authority

Before we examine Obama as charismatic icon, it is helpful to describe the social and cultural conditions that paved the way for his election. As Weber notes, charismatic leaders tend to emerge in times of crisis or social instability. The historical development of U.S. elections can be considered a two-way process, with people attempting to move away from something they find distasteful (which creates a sense of *lack*) while at the same time moving towards something they find more appealing (a perceived *remedy*) (cf. Bloch 1986; Lacan 1998)[?]. The current (2008-2009) disenchantment of many traditionally conservative people with the Republican Party's political/ideological position and the subsequent movement toward centrist/liberal political ideology is a timely example of this *lack/remedy* process of shifting sentiments found in the U.S. political process.

Continuing with this notion of political process as *lack/remedy*, the question becomes what deficiencies can we identify in the U.S. political system that people were trying to overcome or leave behind through the election Barack Obama? For what question is Obama the answer? Looking at the past eight years (2001-2008) in terms of general confidence/approval U.S. citizens placed in the government system we find several indications that large segments of the population were feeling disillusioned with *status quo* operations in Washington. Over the course of the Bush administration, polls indicate a steady decline in approval ratings of Bush's handling of domestic and foreign affairs (Pew Research Center Poll 2008). The war in Iraq, the response to Hurricane Katrina, the Walter Reed incident and the financial crisis of 2008 all contributed to creating a sense of unease in the general population and a decrease in national morale. Within the general public's perception, blame for these crises and failures targeted the Bush Administration.

The social climate leading up to the 2008 presidential election indicates unease and anxiety concerning the future direction of U.S. leadership, and the Bush Administration's national and international policy failures created a vacuum of political legitimacy. Due to this perceived lack of leadership and vision, the social and political environment in 2008 was ideal for the emergence of a charismatic leader to remedy the lack/desire within the population. There was a desire for a leader significantly different from Bush's authoritarian leadership style, a person offering a new vision, someone for people to look up to and believe in.

Obama as polysemic icon

General public response

To examine some the different perceptions of Obama as charismatic icon, we selected a sample of news articles relating to the reception of Obama within the U.S. population. We begin by discussing Obama's general appeal, and then look at Obama as icon through the lenses of race, politics, generational difference and popular culture. While there are many news articles related to the issue of Obama's reception in the U.S., we have selected those that provide examples of Obama's charismatic appeal.

Barack Obama's rapid ascent within national politics occurred under everyone's radar. A product of Chicago politics, rather than Washington, he had little time to establish a reputation as Illinois senator from 2005-2008. Without an extensive political track record, how did Obama garner the support needed for securing the presidency? The first stirrings of Obama's charismatic appeal on a national level can be traced to his 2004 Democratic National Convention speech.

This speech, given in support of John Kerry's bid for president, set Obama apart as an inspirational orator. The quote below taken from the speech illustrates Obama's hopeful rhetoric and his vision for America: "Hope in the face of difficulty, hope in the face of uncertainty, the audacity of hope: In the end, that is God's greatest gift to us, the bedrock of this nation, a belief in things not seen, a belief that there are better days ahead... I believe that we have a righteous wind at our backs, and that as we stand on the crossroads of history, we can make the right choices and meet the challenges that face us" (Transcript: Illinois Senate Candidate Barack Obama, 2004).

The use of hopeful language as an inspirational/rhetorical tactic is not new. Neither is the utilization of charisma as a means for garnering political support. Charisma and politics seem to go hand-in-hand, but a person either has charisma or they don't depending on how they are perceived. As noted, people must *project* deference onto an individual in order for charisma to operate effectively. Barack Obama seems to have charisma attributed to him, or he reflects certain qualities that register with a wide spectrum of people. An International Herald Tribune article mentions how, "Of course, there's more to Obama's appeal than his elevation of hope to a mantra. There's the simple fact that he's a remarkably attractive figure, extremely articulate and smart, and he represents a moderate sort of economic liberalism that many Americans yearn for after these years of tax-breaks-for-the-rich conservatism" (Bernstein 2008). It is the generally positive reception of Obama within the U.S. population that created a social/cultural space for the formation of Obama as charismatic icon. The fact that there was a spectrum of people from different demographic categories attracted to Obama's presentation who were willing to believe in his vision for America highlights our notion of the polysemic icon.

Polysemic icon and ideology

The image of Obama as political centrist and economic liberal has a populist appeal, one that operates across a broad range of people from different racial, political and economic demographics, consistent with the idea of Obama as polysemic icon, which is an image or symbol capable of having multiple meanings, each of which could stand on its own. There is as of yet no dominant or preferred meaning or reading of the Obama phenomena (cf. Hall 1980:134 for a discussion with a different emphasis).

In order for an image to be a polysemic icon, something in it must be reflected within the population as well as in the individual subject. As polysemic icon, the charismatic leader is perceived as different things by different people, symbolizing contradictory ideals or characteristics from the variety of people who identify with him. The ubiquitous iconic image of Jesus is a classic example that is in contrast with the notion of polysemic icon. The ideas of love, salvation and redemption encapsulated in Jesus' iconic image represents a unifying type of iconic representation that is transferable across arbitrary divisions such as racial identity, political party, or national identity, while Obama's iconic status is divisive and problematic. In the case of Obama's broad appeal, the same process of iconomic identification holds true, but the Obama image has multiple, contradictory meanings depending on subject location. Joseph Nye from the Huffington Post mentions how "...another part of his charisma is in the eyes of his followers. We know it when we see it, but in the case of charisma, we are also looking in a mirror. As the economy worsens and old racial divisions re-emerge, the national debate changes. This will be the true test of Obama's charisma. Charisma tells us something about a candidate, but even more about ourselves, the mood of the country, and the types of change we desire" (Nye 2008). The isomorphic aspect of identification between follower and charismatic leader is the source of the ideological power of iconomic imagery. Followers see something of themselves, possibly an ego ideal, in the icon, therefore collapsing the normal distance between representation and signified object. In this way the polysemic icon not only fuses signified and signifier within a single image, it eliminates the gap between subjective identity and the object-other. This is its universalistic aspect and it gives the attractiveness of the icon its foundation in faith. This is, however, unstable because perfect union between subject (individual) and object (other) cannot be sustained without a mediating force itself subject to question. Like the Second Commandment's prohibition against representing God (totality), the ideological effect of the polysemic icon bears a risk of irrational belief in the impossible; the utopian desire for reconciliation of subject and object. It is no coincidence that irrational belief in the possibility of this reconciliation is one way of describing hope, an important catchword in Obama's rhetorical toolbox.

Obama and race

There are a variety of ways to examine Barack Obama's reception as charismatic icon in racial terms. Obama's bi-racial parentage and uncharacteristic upbringing certainly affects how people perceive Obama within the context of race. While many U.S. presidential candidates' biographies are rather traditional and straightforward (and white), Obama's history is complex and in many ways open to interpretation. A New York Times article mentions how Obama's "...biography gave him a credibility no one else could have: he was able to identify with the poor and the Arab-American, with immigrants, with African-Americans. In other contexts, he has identified with white farmers – people like his mother's family...Most people can find something to identify with in Barack Obama, and he can find something to identify with in them. We have never had a politician quite like this" (Malcomson 2004). The wide variety of individual perspectives that mesh with Obama's life-story is unique within American politics and indicates the presence of the necessary social recognition required for elevating an individual to popular iconic status. John F. Kennedy and Martin Luther King acquired iconic status in much the same way as Obama, yet neither had the broad and contradictory range of Obama's charismatic appeal. Obama's reception within white communities was unprecedented in American politics, especially among white youth. Obama's racially mixed heritage makes any racial designation questionable, and this racially open quality seems to appeal to white people.

Obama's reception within the black community seems to vary depending on generational differences. Some see Obama as a new hope for black representation, while others question his racial authenticity (c.f. Coates 2008), and for some blacks Obama may not be 'black enough.' Columnist Matt Bai suggests the notion of racial authenticity within black communities was a significant issue during the 2008 campaign: "This point about whether Obama was "black enough," a senseless distinction to most white voters, came up often in my discussions. It referred to the perception among some black leaders that not only had Obama not shared their generational experience, but also that he hadn't shared the African-American experience, period. Obama's father was a Kenyan academic; his family came to America on scholarship, not in chains" (Bai 2007). The racially ambiguous nature of Obama's history also allowed him to distance himself from the Civil Rights Era community and political leaders of the 1950's and 60's, an identification that could have hurt him politically within white communities. Despite the question concerning Obama's racial authenticity, the general response to Obama within black communities was overwhelmingly positive and supportive. Again, the open nature of Obama's racial identity reinforces the polysemic effect of his reception as charismatic icon.

Liberal/conservative response

During the campaign, diverse ideals and fears were projected onto upon Obama, creating the sense of an opposition between the sacred and the profane. On the liberal side of the political spectrum, Obama was the anti-Bush candidate and possible savior of the Democratic Party. After eight years of conservative leadership in Washington, many democrats looked to Obama as a candidate who was progressive/centrist in political ideology, rather than the unilateral/authoritarian approach taken by the Bush White House. During the election, left-leaning segments of the U.S. population were enamored with Obama, and it is within the liberal population that Obama most stands out as a charismatic icon. No other segment invested as much hope for change in Washington by electing Barack Obama than politically left-leaning Americans. The danger here, as noted above, is the ideological nature of such iconic status: totalistic hope is a feature of the icon, and not consistent with rationally grounded expectations.

The conservative reception of Obama was interesting. The idea of polysemic icon does not necessarily imply a positive reception or a positive iconic status. As a negative form of iconic imagery, the fears of many conservatives concerning Obama's potential election were palpable: Obama is a socialist, he is going to only help blacks and punish whites, he is a Muslim, he is a Muslim sympathizer, he is going to destroy American values and usher in a new era of radical secularism. During the election, these fears were spread via a network of conservative media sources, as noted in a 2008 Associated Press article in reference to Obama's religion: "The false rumor that the Democratic presidential candidate and Illinois senator is Muslim has been fanned on the Internet and conservative talk radio... Conservatives, less educated voters and white evangelical Protestants are likelier to believe Obama is Muslim, as are people from the South, the Midwest and rural areas, the poll showed" (Pew Research Center 2008). In fact, during the election one out of ten voters thought Obama was a Muslim. The issue of race was especially problematic within conservative circles. The popular conservative talk-show host Rush Limbaugh was the first to air the song "Barack the Magic Negro", a satirical song many suggested promoted a racist message. From the conservative point of view, the song implied a cynical recognition that whites might assuage their racial guilt, seemingly 'magically,' by voting for Obama (Ehrenstein 2007). In this manner, Obama's iconic power seems to stir fear and cynicism within conservative circles, and inaccurate or false projections concerning his religion or race represent a negative icon formation.

Generational response

The generational response to Obama can be divided into three categories: The WWII generation, baby-boomers and the emerging youth culture. The history of racial tensions within the U.S. has had a lingering effect in terms of generation. Older Americans (60-80, especially white males) were more likely to view the election of Obama with apprehension due to his lack of experience (McGillis 2008). The generational gap between Obama and the WWII generation, not to mention the racial problematics associated with the 1950's and 60's, is perhaps too great for creating enthusiasm in the oldest generation for a young, charismatic black candidate. Obama is not a charismatic figure for this group. But for baby boomers, representing a large segment of the U.S. population, the reception of Obama was more mixed. The idea of what it means to be 'old' has changed since the 1950's, and this shift in perception did influence how baby boomers identified with Obama. It seems Obama's youthful character appealed to many baby boomers, a generation known for their valorization of youth and remaining youthful (Nagourney 2008). For many baby boomers, Obama was perceived as both charismatic and iconic.

While race and age may have played a more central role in the WWII generation and baby boomers receptions' of Obama, this concern does not translate to the younger generation, for whom the election of Obama seems to be 'not that big a deal' in racial and cultural terms. A Time Magazine article notes how, "According to surveys of voters entering the caucuses, young voters preferred Obama over the next-closest competitor by more than 4 to 1" (Drehle 2008). If there is any segment of the population in the U.S. today that is multiculturally minded, it is the youth of the country. Through various forms of media exposure, especially music and film, the knowledge and interaction between Anglo, Black and Latino cultures have significantly effected how youth in the U.S. think about race. Obama's iconic status coalesced out of the massive enthusiasm and support generated for his candidacy in high school hallways and university campuses across the country (Nagourney 2008). The grassroots mobilization of Obama's campaign energized an army of U.S. youth, and the iconic image created by artist Shepard Fairey helped created a "cool" image of Obama for youth culture to latch on to.

Visual iconography: Shepard Fairey's icon of Obama

The artist Shepard Fairey has fashioned an iconic image of Obama, which is perhaps Barack Obama's most well-known and recognized image. Charles Sanders Peirce has noted that iconic images have a correspondence (isomorphism) between signifier and signified (cf. Peirce 1991). The pictorial sign is related by similarity, resemblance or 'naturalness' (cf. Gombrich 1972). But an icon is really not based entirely on resemblance or identity, since it is almost always stylized, idealized, or indebted to convention, as in the saint's halo in a painting. Even the most motivated icons only achieve resemblance, which maintains a gap, a nonidentical space, between signifier and signified, much like Adorno's notion of the nonidentity between object and concept (cf. Adorno1966).

Shepard Fairey has been accused of using a photo taken by a photographer. In fact, one needs to say that he utilized that photo, but altered in a number of ways. First, he cropped it, making it more of a head and shoulder image than in the original photo, which displays Obama with more of his body visible. Second, he saturated the image with three colors: red, light yellow, and blue. The yellow is close enough to white to denote the American colors of red, white and blue. But the image is also polysemic in that it can point to the many skin colors of the world. Different versions have a different word below the image: progress, hope, or change. So in the final analysis, we have an image derived from an original photo, but radically idealized, abstracted, colored, and given a text.

There has been a controversy about whether the artist has the right to appropriate the image. This can be examined from a legal point of view. But, for our purposes, we want to underscore that Shepard Fairey wanted to create an icon. The debate over who owns the image may be legally significant, but it leaves out the most important dimension: icons by their very nature are universal and cannot in this sense be property. So in this sense Fairey is right. He appropriates the image in fashioning an icon.

Fairey's Obama image reminds one of Andy Warhol's image of Che Guevara. Warhol's work reproduces nine images, three by three, in vivid and contrasting color. Fairey's Obama is often put up in multiples, like advertising in the French Metro. The effect of seeing many of these images tends to reinforce, stylize, and give a certain iconic power. The image has also been widely reproduced and disseminated by media.

It could be argued that Fairey's Obama icon was designed to function as a progressive image. One notes that the image has been put up in poor neighborhoods to embody hope, and Fairey has been explicitly instrumental in Obama's campaign in that he wanted to use his iconic image as an election tool. According to Al Shaw, "[t]he typography also reiterates populism. As H&FJ (the creators of Gotham) recount, the font actually originated out of New York's "vernacular lettering" and signage– liquor store neon signs, hand painted truck lettering, the sign at the Port Authority Bus Terminal, etc. (not exactly bastions of elitism). It was designed to be quintessentially American–lettering from before the era of "graphic design" (Shaw 2008). But as we note below, even progressive icons contain elements of ideology.

Image at a standstill

In this paper, the ideas of a polysemic icon and charisma are crucial. The first implies multiple interpretations bound to be at odds; the second implies a certain non-rationality and fragility or lack of durability. Together, they suggest an ideological force that has to do with the universality implicit in an identification or confusion of signifier and signified. We discuss the gap that remains, and demonstrate this conflation of signifier and signified, the totalizing aspect of the accounts, and the indifference to rationality that is part of the appeal of charisma.

In this light, we examine the similarities between the social formation of charismatic leadership and the icon, and within this context highlight our notion of the polysemic icon through an examination of the popular reception of Barack Obama during the 2008 election. As noted, the cultural climate in 2008 was favorable for the emergence of a charismatic leader. War, political corruption, incompetence, and the economic recession all contributed to the disenchantment of U.S. citizens with the Bush White House, and created a space (*lack*) and desire for a new vision (*remedy*) for the U.S.'s future direction. This social unease created a window of opportunity for the emergence of a charismatic leader. Weber indicates that in order for charismatic leadership to function people must impute power or sacredness upon the object of reverence. The same necessity for subjective ascription, the giving of power to an object, applies to the formation of iconic images. The mystical/magical action of making an object or person sacred (outside the mundane) through subjective identification is both irrational and ideological in that it attempts to achieve the impossible unification or identity of subjective desire and an object of desire within a designated symbol. Iconic imagery represents an image at a standstill; an attempt to heal the symbolic gap existing between signifier and signified through iconic symbolism.

Historically, sacred or holy icons achieve universal signification/recognition by dissolving *outer* differences and contradictions, rising above the arbitrarily set boundaries of cultural/political/social identifications located between people so as to include all people under a concept or belief (such as Jesus' universal, brotherly love). The polysemic icon, on the other hand, achieves differential recognition and reverence by embodying difference and contradiction *within* its presentation. Due to the unresolved, open nature of the polysemic icon, difference is resolved externally, within the reception of the icon, through individual identification with some contingent aspect located within the icon. Barack Obama does not embody only one identity for all, he has come to represent a sacred icon for some and a profane icon for others; he is white and black, an insider and an outsider, open to interpretation. And due to the unresolved and open nature of Barack Obama's history, he has become many things to many people depending on what they see in him; he has become a polysemic icon.

Bibliography

Bai M. (2007), *Is Obama the End of Black Politics?* New York Times Magazine, August 6, (http://www.nytimes.com/2008/08/10/magazine/10politics-t.html).
Bloch E. (1986), *The Principle of Hope.* Cambridge, MA: The MIT Press.
Coates Ta-Nehisi P. (2007), *Is Obama Black Enough?* Time Magazine, February 1, (http://www.time.com/time/nation/article/0,8599,1584736,00.html).
Ehrenstein D. (2007). *Obama the 'Magic Negro'.* Los Angeles Times, March 19, (http://www.latimes.com/news/opinion/la-oe- ehrenstein19mar 19,0,5335087. story? coll=la-opinion-center).
Hall S. (1980), *Encoding/Decoding.* In: *Culture, Media, Language: Working Papers* in *Cultural Studies.* London: Hutchinson.
Gombrich E. (2000), *Art and Illusion.* Princeton, NJ: Princeton U. Press.
Lacan J. (1998), *The Seminars of Jacques Lacan, Book XX: Encore, On Feminine Sexuality, The Limits of Love and Knowledge* 1972-1973. Trans. Bruce Fink. New York: W. W. Norton & Company.
Landsberg Marge E., *The Icon in Semiotic Theory. Current Anthropology* Vol. 21, Pg. 93.
Malcomson Scott L. (2004), *The Nation; An Appeal Beyond Race.* New York Times, August 1, (http://www.nytimes.com/2004/08/01/weekinreview/the-nation-an-appeal-beyond-race.html?scp =1&sq=Malcomson%20S.,%20%C2%ABThe%20Nation;%20An%20Appeal%20Beyond%20Race%C2%BB,%20New%20York%20Times,%20August%201,%202004.&st=cse).
McGillis A. (2008), *In A Changing Corner of Pennsylvania, A Glimpse Of Democrat's Age Problem.* The Washington Post, August 12, (http://www.washingtonpost.com /wp-dyn/content/article /2008/08/11/AR200808 1102203.html).
Nye J. (2008), *Charisma and Leadership.* The Huffington Post, April 30, (http://www.huffingtonpost.com/joseph-nye/charisma-and-leadership_b_99466.html).
Nagourney A. (2007), *Age Becomes the New Race and Gender.* The New York Times, June15, (http://www.nytimes.com/2008/06/15/weekinreview/15nagourney.html).
—— (2008), *Will Obama Rock the Youth Vote?* The New York Times, October 8, (http://www.nytimes.com/2007/10/08/us/politics/08web-nagourney.html).
Peirce C. S. (1991), *Peirce on Signs: Writings on Semiotic.* Edited by James Hoopes, Chapel Hill, NC: The University of North Carolina Press.
Pew Forum on Religion and Public Life (2008), *One in 10 Voters Mistakenly Thinks Obama is Muslim.* March 27, (http://pewforum.org/news/display.php?NewsID=15273).
Pew Research Center (2008), *Bush and Public Opinion: Reviewing the Bush Years and the Public's Final Verdict.* December 18, (http://pewresearch.org/pubs/1063/bush-and publicopinion).
Shaw A. (2008), *Bitter Has Its Own Poster.* April 12, (http://bagnewsnotes.typepad.com /bagnews/2008/04/bitter-has-its.html).
Transcript: Illinois Senate Candidate Barack Obama (2004), The Washington Post, July 27, (http://www.washingtonpost.com/wp-dyn/articles/A19751-2004Jul27.html).
Weber M. (1968), *Economy and Society.* New York, NY: Bedminster Press.
—— (1978), *Politics as a Vocation.* In: *Weber: Selections in Translation.* Edited by Runciman W. G., New York, NY: Cambridge University Press.

Chapter 3
A social icon: the "squart"

Pierre Bouvier

Abstract

The occupation of some places by persons coming from the artistic and cultural milieu draws attention. The situations that are initiated by them are located notionally at the crossroads of free time and working time. They redefine relations between the spheres of work and leisure. They provide shelter to individuals who are engaged in this type of intervention in a very special time-space. Specific social ties emerge, weaving reciprocity between living and doing. Thus, there is the contribution of varied contexts in such situations. The reasons for this type of action lie in the existence of a sharp disparity between the number of workshops and other places for creative work and the multitude of persons, particularly young artists, seeking a structure for their activities. These difficulties have led some to resort to unwarranted occupation of buildings, multi-storied houses, industrial wasteland and other places currently unoccupied by their owners. These 'squartters' find a shelter in such places, which give them protection as also provide an opening. In these places, they feel protected against instability and find a solution to their unceasing and often futile search for a place to pursue their desired activity. Thus, these places allow artists: visual artists, video directors, musicians, actors, etc. to carry out tasks which is their vocation as also a hobby. As regards the latter, hobby, it is clear that it is a chosen and not an imposed activity and, is an integral part of the development of the individual. As for the former, work/vocation, it also implies a sense of effort, the use of practices and knowledge that is to initiate and acquire a mastery of materials and creative processes. Furthermore, the creative productions that take place in these places are more or less incorporated into the trade. All these aspects reveal the iconic peculiarities of these contexts –utopias-, as much playful as toilsome.

The word squat, used in Anglo-American English, refers, according to the Webster dictionary, to a term whose origin goes back to the old French esquater, esquatir, meaning to flatten, to apply to the ground. These derivations imply primarily the action of crouching, to be in a position close to the ground. Thereafter it began to be used to refer to persons, who occupied a stretch of land and settled themselves without having any legal rights thereto as it happened in the United States or in Australia during colonization. In its contemporary usage a squat means a place and specifically a built up place of any type: a factory, hangar, workshop, building or dwelling that acts as a shelter for more or less a long time. It is illegally occupy by the squatters. First of all, for the applicants, the initial idea grows out of the will to find a way to cope with a situation that is theirs, that is of not being able to find a home. Taking possession of place is to solve immediate problems, including the basic one of ensuring a roof. This means the ability to organize anything remotely livable. The effort involves the person and his family members and / or those who have joined the voluntary and deliberate action to ensure minimum resources as to be able to survive in the short or long run.

The activity of these shelters, after having in the first place secured a roof, centers around the weaving of social ties and first of all of developing interpersonal relationships between the occupants. The getting together of these highly destabilized persons, who have been forced to take the extreme action of occupying premises, requires the working out of an arrangement, which is often difficult to develop given the wanderings and trajectories of each of them. The reasons that have led them to this situation are many and varied. In his study on the Hobos, Nels Anderson has worked out a typology of diverse shelters and trajectories and the intentions of vagrants who squat in public space, gardens and parks and those of unoccupied buildings. Regarding the former, who take shelter in shacks made of canvas, boards and other light materials they seek to protect themselves from bad weather conditions and also to organize in any way their daily life around a fixed and personalized place. These squatters, at least those who do not stay in small furnished hotels but make do by taking over a public space adjacent to railway yards or in the fallow land, are afraid of being noticed by the police, municipal or private, i.e. of the rail companies. When the squatters use buildings, houses, warehouses or other built up space, the perspective is not only confined to the immediate life activities of the migrant or homeless workers. As indicated by this member of the Chicago School, the Hobohème situation harbors a potential of giving expression to symbolic, cultural or even artistic values:

"On the outskirsts of cities, however, the homeless men have established social centers thar they call "jungles", places where the hobos congregate to pass their leisure time outside the urban centers". (Anderson, 1961, 16)

The camps, temporary or permanent, are so many places, especially for the most stable among them, where social communication and ties transform these makeshift shelters into entities of rich cultural interaction.

> "The average man of the road has had a variety of experience and not a little adventure. In the jungles there is always an audience for anyone who wants to talk, whether of his thoughts, his experiences, or his observations."
>
> (Anderson, 1961)

These conditions have been observed and analyzed at the beginning of the twentieth century: "Like community, we can see the labor market as a place, as social relationships, or as an ideological formation. (...) No doubt men and women who worked for wages had a different perspective on the issue. They certainly understood the need to earn wages to ensure the heath of their families and communities. But their experiences with seasonable and irregular work demonstrated that unemployement was an inevitable part of life, and many harbored dreams of escaping the labor market altogether in small business, homesteading, and farming. (...) Ar the same time, communities could provide laborers with an important margin of freedom from the labor market. Family groups, friends, and susbsistence activities such as gardening, hunting, and fishing allowed men and women to remove rhemselves from the labor market temporarily." (Frank Tobias Higbie, 2003)

These situations are not unrelated to the ways that prevailed in Western Europe in the first half of the XX° century and even after such a date. After

World War Two, the shanty towns that surrounded the great capitals were also the makeshift shelters for the migrant populations. One of them surrounded, for some time, the Nanterre University. This implied a better understanding by the students of the every day life of those migrants. At first sight, leisure activities were very rare except in the form of media programs presented via radio and later via television or through meetings between fellow-believers in the surrounding cafes.

The economic and structural crisis of the late twentieth century has recreated, but in other forms, the situation of shortage of available space. This concerns not only constructions of the fifties–sixties which have suffered increasing degradation due to a lack of monitoring but also to possibilities in terms of places offered to artists so that they could practice their art.

As highlighted by Raymonde Moulin, the numbers of visual artists have effectively increased during the second half of the twentieth century (Moulin, 1997). The government, as well as the various bodies involved in the field of culture, has not paid much attention to the needs of this population. The latter, despite its growth, represents only a small number, a few thousand individuals. Its importance is strongly correlated by the previous representation of artists as persons not inclined to social and public commerce. Its relatively unorganized character as a population group, although some among them may have modified, to their benefit, the provisions such as access to social security, do not make them a lobby, a targeted audience of voters (Derivery, Maggiani, 2005). In the 1980s, the policy of the Ministry of Culture acquires a certain dynamism involving a renewed proposal for a democratization of culture. This view, however, was strongly interrelated to the desire to favor a relatively small number of artists, presented as a pool having the potential if not actually the vocation to enter the international art market.

The institution, the museum and galleries make art works. The very young or those who for reasons of choice or because of various circumstances, have been located on the outskirts of these circles of "excellence" derive only a very marginal benefit from these public or private structures. This has been particularly true as regards access to workshops. Many artists do not have their own studios for pursuing their creative activity. They have to make do, often at home, given their very limited financial resources.

Faced with this situation some, tired of waiting for hypothetical building space or holiday workshops (which may sometimes be diverted from their original purpose to the benefit of anybody, not at all linked to artistic creation), opt for other solutions and in particular, occupation of empty premises. To try to find solutions to this lack of workshops in the Paris region but also in many other French or foreign cities, the only solution is to take to squatting without permission or agreement of the owners of buildings and premises (Grésillons, 2005). The choice of these building depends on the fact that these are uninhabited places, often empty for long years.

This may have been an opportunity for the emergence of an adequate practical model at the level of interaction to solicit a population group. This prospect opens up singular collectives of a "living together", in the squart, where work and leisure can be forged together.

Squart Rue de la Bourse (Paris)

One of the oldest squart is located in Ris Orangis, on the outskirts of Corbeil, a town a few dozen kilometers from Paris. This squart, created in 1981, is called the CAES, acronym which means: Centre Autonome d'Expérimentation Sociale. In Paris itself and for several years, many squarts have emerged. Among the best known are those who settled in 1998, rue Thorigny in the heart of the Marais and in front of the Picasso Museum. The occupants chose to call their place: Ssocapi (reversing the spelling order of Picasso). For several years, this wasteland gathers painters, sculptors, designers and other player of the art world. Its location, adjacent to one of the most popular museums, helped it attract an audience. It became one of the icons reflecting the reality and frustration that has been taking place in society. This very fact contributed to its closure. A few years later, without obliterating other places such as, for example la Grande aux Belles, installed in 1995 in the district of Paris X, a dramatic action was implemented by a number of these seasoned squartters. They choose a Hausmannian building located just opposite the Bourse (Paris Stock Exchange) and next to the Agence France Presse, in order to attract attention of the media, public and political authorities about their more than difficult situation. Even the gallery owners of Avenue Matignon came to see what was happening there. Before their expulsion from the squart (in September 1999), the paintings and all deemed important creative art works were removed. The rest was placed in the Paris Stock Exchange Square: a surrealist show of paintings, sculptures made of all materials, varied objects, umbrellas, and wax dummies previously placed in the Igor Balu museum installation/collection. Artists raised a few shacks made of fabric, plywood and other items that could serve as walls or as a roof. On 3 September 1999, the police intervened and evacuated the place. Garbage collectors, in full day light, threw the sculptures, painted objects, frames in the metal cases of the garbage lorries, under the very gaze of the passers-by and the artists, who were non-violent and offered croissants to the police. An arrangement was worked out beforehand: that if artists posed no resistance then there would be no arrests of the squartters.

These artists have continued to insist that these occupations apply only to empty or abandoned buildings; that the activities carried out were to help them perform their creative work and, moreover, formed a part of the neighborhood fabric and undeniably helped in rebuilding social ties (Bouvier, 2005).

These sites combine the artistic production of its authors to the potential of their environment to which they offer various opportunities related to leisure activity. In the first instance, you can discuss and understand, in situ and real time, ways to make a painting or a sculpture. This has been going on until very recently, at 59 rue de Rivoli: Chez Robert-Electrons libre. Secondly, these squarts arouse an interest in artistic activity or at least encourage a certain involvement in the various workshops. Thus the squat of the Rue de l'Echiquier: Le Théâtre de verre, la suite de l'utopie (The Glass theatre, an aftermath of utopia), offers a show in an old disused railway station of the Sernam, ranging from dance, theatre, singing, puppet shows, communal meals, films regarding the creation of mosaic or modeling: "This action-installation is part of the process of occupation, the creation and abandonment of urban wasteland. It is designed as praxis, action-reflection philosophical and artistic, poetic and plastic, ethical and aesthetic, singular and plural." (Pasini, 2005: 15) In September 2006, this activity, more than being effective, despite a large support, was obliged to vacate at the risk of paying a fine of 50 Euros per day.

These sites allow artists: visual artists, actors, singers, video directors, musicians, etc. to exercise what is both a vocation and a hobby in the squarts. As regards the latter, leisure activity is really chosen and not imposed and forms part of development of the individual. As regards the former – work - , it also involves the orientation of effort, the use of practices and knowledge that are to be learnt and brought together in terms of a mastery of materials, physical and conceptual processes. Furthermore these creations carried out in these places, the urban wasteland, are a part of the dynamics of gift exchange or, otherwise, more or less of trade. All of which reveal the uniqueness of these contexts, of these shelters as places of recreational play as well hard work, as also of utopian icons. "But it is necessary to speak about 'work' when one speaks of painting? Wouldn't the term 'leisure' be more suitable? The fact is, for me, work and leisure are inseparable concepts in relation to painting" (Lazar, 2006) In this way they join and participate in the problematic of the relationship between work and leisure and, more fundamentally, to the specific potentialities of any individual.

Bibliography

Anderson N. (1961), *The Hobo, The Sociology of the Homeless Man.* Chicago, The University of Chicago Press.

Boris G. (2005), *Le Tacheles, histoire d'un "squart" berlinois.* Multitudes.

Bouvier P. (2005), *Exister ensemble, les squarts.* In: *Le lien social*, Paris, Gallimard, 314–327.

Bouvier P. (2007), *Arsène.* Paris, Lelivredart, (coll. Artension).

Derivery F. and Maggiani R. avec la collaboration de P. Bouvier et de C. Rédélé (2005). *Les artistes terrassant la CAVAR 1972–1975*, Paris, EC Editions.

Grésillons B. (2005), *Le Tacheles, histoire d'un "squart" berlinois.* Multitudes.

Higbie F. T. (2003), *Indispensable Outcasts, Hobo Workers and Community in the American Midwest, 1880–1930.* Urbana, University of Illinois Press.

Lazar C. (2006), *ART inprogress.* Paris, Pg. 21.

Moulin R. (1997), *L'artiste, l'institution et le marché.* Paris, Flammarion.

Pasina L. (2005), *Age Made in squat.* Vanité n. 10, Pg. 15.

Chapter 4
Philosophy as memory

Gaspare Mura

Abstract

If null is an entity in "potent", the history of man should be the fountain of the identification and of the imagination which are in "act" to flow on, in space-time dimension. On this peculiar process reserved only to the human beings, manifest the necessity and the importance of the memory. Non man can live without memory. Here we can locate the greatness and the misery of man. Greatness, because, from the memory scatters an ontological view of human beings and of the world around. Misery scatters due to the lack of the rhetorical and hermeneutic power. The human existence is not going to limit itself to its rudimental relation with his own physical status and with the nature, instead, it goes beyond all this, and connect via thought processes to hidden dimensions of his own life and of the nature. This is how man understands the truth. But the truth or the identification of hidden dimensions without the memory power is impossible. There are two ways to understand memory: collective and individual. Anybody born to a human family possesses a memory. Therefore the family as a basic collective organ, posses a memory. Here indicates also the uniqueness of any individual and all collective memories we can find in the world. Now, we can also individuate other two universal values: justice and freedom, that is to say, the respect to any human being as he/she is, and respect to the nature. On this article, one can read in part a historical process of transformation of the importance of memory, and in the other, the actual urgent situation of which necessitates mutual "sincere dialogue" among individual and different collectives in the world via memory process, on the back ground of, as we wrote before, respect to the freedom and of justice.

It seems that the philosophy as "memory" is itself a rhetorical figure, due to its origin as a human science. That is to say: if man is for definition a philosopher, he/she ought to think departing from a given social-political and cultural memory. Therefore, it is plausible to deduce that the philosophy as "memory" coincides with the memory in a "pre-atomic" level of thought. Regarding this fact, we can find an interesting story from the Greek mythology which emphasize the importance of memory. This is also an another fact that the hermeneutic tradition stretches all the way back to ancient Greek philosophy. There we can find an "orphic tablet", which is found in a tomb of a maiden, in Magna Greece. According to this story, the soul is recommended not to relate and communicate with the source of Lete, or of forgetfulness. Instead, it recommends the soul to have communication with the second source which "runs from the swamp of Mnemosyne", that is to say, memory. The tablet proceeds saying that: "You will find on the left of the houses of Hades, a fountain, and beside it a white straight cypress: regarding this fountain, it is recommended that, not even to walk closely to it. And you will find an another cold water fountain, that, flows from the swamp of Mnemosyne: there in front, you will find guardians, to whom you have to tell that: "I am the daughter of earth and of starry sky. As well, I am heavenly descendent; this is known to you too. I am parched with thirst and die. Please, give me immediately, cold water that flows from the swamp of Mnemosyne". And they will let you drink from the divine source, and then you will reign with the other heroes.

The religion of mysteries, in fact, believed that the salvation of the soul is not consist in a process of annihilation of personal identity, but, on conserving it for eternity through the memory. Exactly, for this reason, Mnemosyne, mother of the Muse, is also considered the mother of every soul, and, on the other side, since the human civilization is based on historical memory, Mnemosine is also considered as the mother of civilization. On base of this reason, the Italian humanists affirm: "we are dwarfs on the shoulders of giants". Therefore, the awareness that each and every creativity, also in an individual level, may not be authentic, without historical memory. According to Thomas Campanella, the people forget what they are naturally, are being rescued by the memory. But according to him this memory comes in the form of the philosophy, that leads men to realize their identity. The memory is not only an "arts of memory", recovered by humanists like Cicero and Quintilian, which is in its turn, the technique of "place", that is to say as a physical place, connects images. Regarding of these ideas and words, says Pieter of Ravenna , the most famous theologians of the 'arts of memory' of the Renaissance, that the memory must be images that "stimulate the imagination".

Philosophy as memory in her fundamental articulated formulation as we mention at the beginning, can find in classic Greek world. The notion of culture in the sense of formation, is define in Plato as "Paideia" (training). It emphasizes the need for man to "cultivate", to educate through the historical memory; that is to say by the works done in the past. This idea became not only the base of Humanism and of the Renaissance, but also of every human civilization. "The Greeks, Jaeger writes, relative to the great oriental historical cultures, represent a radical 'progress', a new order of everything that concerns human life in the community. This is one and the foremost creation of Greeks. As we highly appreciate the artistic, religious and political front of the people, the history of what we might call culture, in our conscious sense, begins with the Greeks (W. Jaeger, *Paideia*, I, 3). The distinction between "nature" and "culture", which is the basis of philosophy as "memory", scatters out of the extraordinary dialectical thinking of Greeks. This gives man the ability to connect each element to an organic whole, of which history is the record: " The Greeks have an innate sense of what is the 'Nature'. The concept of Nature, that they first coined, is without doubt, springs out of their peculiar spirit. Long before they create this fragment of perception of memory as the unity, they may have seen things with this point of view, that is to say, nothing in the world was isolated in its particularity, but always seen as a whole, by which received its position and its significance. We call this way of seeing as 'organic', because it conceives the individual as a member of a whole. The urgent need of the Greek spirit of conscious understanding of the laws of reality, which its turn, reveals itself in every area of life, in thought, in word and in action as in any kind of artistic creation, is connected with the ability to capture the inherent, original, natural structure of organic matter of being " (W. Jaeger, Paideia, I, 9-11). The philosophy as memory is to the Greeks the same philosophy as the ability to organize every single item in an organic whole, which is impossible without the "memory".

In Plato there is a further consideration in the design of memory. It will remain at the base, albeit, in different ways, for much of the Western philosophical culture. Apart from the myth of *Thot* of *Phaedrus*, where Plato argues that the writing is just a type of " medicine of the memory", because the most important things are already written in the souls by the words of the master. The real writing is realized by the oral word , which in its turn, writes directly in souls, and not on the tablets of wax. Plato conceives of human knowledge as essentially a mode of "recalling" what the souls have learned in their previous existence. Therefore knowledge, for Plato, is *anamnesis*, that is to say, memory. And this conception of knowledge as a "recall" because, it was already impressed on the soul, is the "form" of knowledge, lived in modern philosophy and in the ideas of Descartes. According to them, the knowledge as the memory, is innate and known by the human mind through a process of *reminiscence* and of intuition. These two processes, are two forms of a priori knowledge in Kant, which are in some way a revival and a reformulation of the Platonic theory of knowledge.

As Gadamer says, Plato is the central point to the theory of hermeneutic for one reason: for his dialectic method. It contains two other hidden principles, which are proper to this method: authority and formation. Though we often encounter in his dialogues, as the answers of the interlocutor, only "yes", "no" "maybe" etc. they, relief the other importance facts like the identifying and drawing out new propositions in the form of hypothesis. These in their turn, become the "flat forms" to the other new conclusions and so on. When properly conducted during a dialogue (by a Socratic like teacher) can transform the life of an interlocutor, by rooting out false belief and discovering new truths. This reveals another character of his dialectical method: instead of accept passively what is said, one has to use the dialectical capacity as the dignity of a philosopher, to search for the new knowledge. As Plato says, there are some principles which are not provoke interrogations, and there are some other objects, in a given context with prejudice, that formulate further questions. What influence those objects, according to Plato, are opposite impressions (Cf. Plato, *Republic*, 523 b ff.). This idea also goes with that of Aristotle: the vocational capacity of man to perfect the nature. That is to say, it is obvious that the given dialogue is based on something that exist, and that meaning, the participators to the dialogue, must have the capacity to recognize what is new according to what is given. This moment is called by Gadamer as "fusion of horizons". According to this, regarding a solution to a given problem is possible thanks to this new understanding. Returning to Plato, we can understand why he recommends geometrics for the formation of the young people, that is to say, to put in act and to exercise the dialectical capacity of them (Plato, *Republic*, 539 a-b). Only this way, one can analyze a problem dividing in to parts and putting them again together, and to connect its parts with the context given and according to the background dimension(partly determined). That is to say in this way, a young man achieves the philosophical skills and insights.

As we can read in the dialogue of *Phaedo*, Plato has define sensible objects which are designated to continues change, situating them in an abstract dimension of idea, then to form a perfect world that is not destined to be changed (Cf. Plato, *Phaedo*, 78 d – 79 a; Aristotle, *Metaphysics*, 987 b 4-7). When we are asked to find the similarity among two different lines, according to Plato we use four conditions of the soul: "imaging (*eikasia*), belief (*pistis*), thought (*dianoia*) and understanding or knowledge (*noesis*)" (Cf. Plato, *Republic*, 509 c – 511 d). In this way we can understand that the "Beauty" is just a manifestation of "Good" (Cf. Plato, *Philebus*). Therefore, we can understand how it is important the history and memory to the human life, under the discipline of dialectic as the culmination of a of philosophical education. As we have seen before, according to the process of *anamnesis* of Plato, any fragment of knowledge is a mode of "recalling", what one have seen, in a previous existence of the world of ideas. According to this, to know something means to "recognize" the preexistence of that thing in an ideal, therefore, in a real world. Augustine, contrasts this conception of memory, as of a large container of memories. In this way he gives the boot to the whole literature of autobiography and diarist medieval and modern world. Augustine transforms Platonic anamnesis, because for him the memory is not simply "remembering", but it is the mode of trace lived experiences, particularly experiences of happiness and goodness, and in a special way the experience of God. As Augustine affirm, nobody can discover God in a form of an idea or as a concept, but as "real presence" in one's life. That is to say, God's presence in one's lived memory. On the path of memory, Augustine meets God: "Tardi ti amai, bellezza così antica e così nuova, tardi ti amai" (Augustine, *Confessions* X, 27, 38th). S. Augustine seeks God in his mind, exactly in his lived memory, through the experiences of happiness: "Cercando te, mio Dio, io cerco la felicità della vita" (Augustine, *Confessions* X, 20, 29th). Thus, *Confessions*, becomes for Augustine, the philosophy as memory, which is not only to recall what is past, but also to recognize the path of salvation and happiness. For him, as is indicated in the tenth book of the *Confessions*, the memory is *capax Dei*. Augustine, analogically says that, the memory is "the great hall" of the soul, where deposited our experiences and knowledge. Augustinian concept of memory as path to walk to encounter and to recognize, becomes symbolic of the transition between the ancient culture and Christianity. Precisely because, the Platonic anamnesis which is linked to the Greek conception of time as *kronos*, is transcended and transformed by the new biblical and Christian concept of time as the "event" of salvation. The Augustinian conception of memory as *distentio animi*, based, in fact, on many traces of memories of world beyond, of the past, of the mathematical and conceptual paradigms. The base of all these resides on the presence of living God in the memory as the experience, and not just a memory to recall. For Augustine, the knowledge of God is "a kind of collection", by the thought, elements scattered, disorderly content from memory, and working by the reflection, so that the same memory where first hid here and there neglected, be held, so to speak, within reach, for now look forward to the consideration of family spirit [intentio]" (Augustine, *Confessions* X, 11, 18). Aristotle, the disciple and the antagonist of Plato, distinguishes memory and reminiscence. Memory is to recall the past events, while reminiscence is to search, in the past, what we have lost. The memory is somehow a faculty which acts passively, while the reminiscence involves in active mode in the conscience, vigilant and willing to find what interests us but is forgotten. On the basis of the concept of memory of modern science, of psychology, of neurology and of psychiatry we can find the traces of Aristotelian concept of memory. The great interest of contemporary culture discretion to memory, unlike the Middle Ages where the memory was mainly the Platonic-Augustinian *memoria Dei*, is in fact a scientific and as we said before, directly or indirectly, on basis of the Aristotelian concept of memory. Bacon, at the beginning of modern era, not only speak of an "art of memory", but also of the *idola* in the memory as *volantes phantasiae.* According to him these are the errors and fantasies in the memory, which in their turn, deceive the minds of men, making them stay in the Platonic "cave", to contemplate only the shadows and the sunlight: the idols of the tribe of the forum, of the theater. Even our language, says Bacon, is used with words that "twists their power on intellectual strength". With Bacon, the philosophy begins to consider the memory as a deceive element and not as guardian of the truth. But this vision is developed only within the scientific tradition, from which, soon it took distance to become a new philosophical speculation.

It is not accidental that the idea of Hegel on philosophy, as consciousness of time. Unless this innovative idea, philosophy could be treated only as a memory. Instead, the philosophy, on this process, requires the rational mediation, to go beyond the concatenation of ideas and concepts that have emerged over time. Here we can say that, this is possible only through memory. For this Hegel could write that "all philosophy was necessary", because the whole philosophical truth coincides with its history, therefore, we can say that the memory is the custodian of philosophical truth. " Ogni filosofia è stata necessaria, e tale è ancora; nessuna quindi è scomparsa, anzi tutte sono conservate affermativamente nella filosofia come momenti d'un tutto.[...]. I princìpi si conservano, e la filosofia più recente è il risultato di tutti i princìpi precedenti; in tal senso nessuna filosofia è stata confutata. Ciò ch'è stato confutato, non è il principio di quella data filosofia, ma soltanto la pretesa che esso rappresenti la conclusione ultima, assoluta". (Hegel, *Lectures on the history of philosophy*, I, 48). Hegel thus expresses his conception of philosophy as memory or rather, the ability to illuminate the concept, that is to say, the consciousness and therefore what is deposited in mode of unformed in the memory: "La filosofia non è sonnambulismo; ma piuttosto la più vigile coscienza; e l'opera di quegli eroi consiste appunto nell'aver tratto il razionale in sé dalle profondità dello spirito, dov'esso si trova dapprima soltanto come sostanza, come essenza interiore, e nell'averlo recato alla luce, nell'averlo sollevato alla coscienza, al sapere; consiste, insomma, in un progressivo risveglio". (Hegel, *Lectures on the history of philosophy*, I, 50).

In more recent times the memory as a philosophy has been studied profoundly by various authors and philosophical schools of great importance. We can mention certainly Freud and psychoanalysis which are working on memory, declining, in the positivistic sense of original Augustinian intuition of memory and retrace the new paths of modern science regarding the memory. For this we must recognize Bergson, one of the most important theorists of memory, which underlines in his *Matter and Memory and survey data of immediate consciousness*, his struggle against materialists and positivists of his time, writes that in human consciousness, there is much more elements than of the brain in order to be relief. This "more" could be revealed only by the memory and by the recall. Bergson establishes a fundamental distinction between the time of consciousness, understood as the inner time which is not assimilable to the in a given period of chronological time, and memory, which is the place where we find ourselves. This process of dimensional uplifting, reveals the whole meaning of our existence and opens us to the understanding of our origins. Much of the contemporary literature is inspired, consciously or not, by this bergsonian insight: "We are our past" (Bergson), and the literature becomes the path in search of self. From *La luna e i falò* by Cesare Pavese to the famous *Recherche* of Marcel Proust, we can find this influence, that is to say, the search for the own identity through a path of recalling.

The meditation on the philosophy as memory, is present in particularly in the Hermeneutic philosophical tradition. From the interpretation of modern philosophy as "oblivion of Being" of Martin Heidegger, to the rediscovery of *Oneself as Another*, of Ricoeur, in which knowledge of the self occurs through a process of memory. It is worth, as we understood under all these circumstances regarding the theory of interpretation, to continue a further study on Ricoeur. In a recent volume titled "The memory, history, the oblivion", Ricoeur summarizes a hermeneutic point of view on memory. Departing from a predominant psychological reflection on memory, Ricoeur reads the memory through the "writing", which is in turn a review of time and history through the "narrative". As already explained in the book, *Oneself as Another*, the man needs to revisit their past to discover the dynamics that have directed from the same original consistency to the "ipseità", or to what he has become along the history formed by the personal, existential and free choices. The recognition of this process that leads from the original to the "ipseità" is formed for Ricoeur, exactly by the memory, in which a man discovers *self as another of oneself*, and this type of memory is also rapresents the paradigm of history. The personal memory can infact, be expressed only via story, narration, so that the language of writing becomes the place of true discovery of self. It is this dynamism that describes not only the necessity of narration of artists, writers, poets; but also historians. Infact, for the hermeneutics of Ricoeur, the history does not consist only in the 'res gesta', that is to say, in the description of historical facts according to the model of Herodto, Tucidite, Tacitus. The story is rather a "recall" to the past. It is an interpretation of historical events in the light of this need to revisit events that take part of the past and are not dead in the past of oblivion. They are always with us *hic et nunc,* in the memory, which in turn, reveals all that we were, and what we have become through a dynamic of personal or collective choices that make us what we are.

This view of Ricoeur, the history as memory, accost especially, to the historical hermeneutic of Giambattista Vico. To Vico, *"verum ipsum factum"* in the sense, that the truth was attributable mainly to historical facts, thus coinciding with the philosophy. Ricoeur, on the other side, come near to the conception of the history of another great representative of contemporary hermeneutic, H. G. Gadamer. For Gadamer all our knowledge comes from a particular perspective, formed by the historical tradition and cultural belonging. The memory, for Gadamer, is precisely that process of "application" (Anwendung) through which we relive all our past, because we adopt it and give back the life by applying it to the present.

The "Wirkungsgeschitliches Bewusstsein", or the consciousness of the history in act, forms the heart of the hermeneutical theory of Gadamer. Accordino to this theory of Gadamer, the memory conceives as a process that reveals the influence of our past, not only personal but also cultural and in our present.

Henri Bergson, in his "Matter and memory", had conceived that, the memory, from a scientific point of view, could not be reduced to a psycho-physical processes, because it revealed something "more and beyond" relatively what is analysable from an empirical science. Ricoeur does not deny that even the sciences, particularly psychoanalysis, can contribute to the understanding of history, spcecifically regarding what he calls the "obstructions" of memory: "the memory, he writes, has its impediments which make grounds for past falling and suffering, it has its discharges and its resistance, its worry and its denials, that psychoanalysis and social psychology describe in a small scale of individual level and that in the largest scale of the representation of the history are engaged in power games". However, the concern of Ricoeur focuses primarily on historical narrative, that i sto say on "historiography", which is properly the writing of the story that encloses the historical memory.

A great Italian philosopher of the twentieth century, Benedetto Croce, the same philosophy is identified with the historiography. This is because, Croce, does not have the metaphysical kind of objects, but for him history is formed only by the actions of man. So, the historiography and philosophy, as he writes, in the "*History as thought and as action*", and in "Historiography and Philosophy", both converge in this particular form of historical memory which is the historiography. Ricoeur, more attentive of Croce, realizes the presence of many models of narrative, in the contemporary world, beyond writings from the history: fiction, drama, essays, pamphlets, and even non-writings such as: photographs, paintings, films that make particularly difficult the work of Historians today. Moreover, having carefully studied the importance of the 'ideology' as the instrumant and the tool of the selection that works within the memory, and thus in the oblivien of historical facts, Ricoeur speaks of the "Duty of memory" as a "resposibility to, not to forget". This does not mean to remember the past wounds, sufferings, humiliations, frustrations, but, it means, rather a sort of "elaboration of mourning" that "avoids ruminatine them". On this way he invites,to take this into account through an "active memory, with intelligible and bearable", which scatters from the duty of memory. By this hermeneutical process, it becomes possible to Ricoeur, to heal a memory that wounded by the history.

In this way we can affirm that the whole hermeneutic philosophy is crossed by a deep meditation on the meaning and value of memory. In *Se come un al-*

tro, Ricoeur shows how the man feels the need to "narrate" in a symbolical order of storytelling in order to understand the "sense" of his own existence, which is revealed in the dialectic between the fundamental "mediation" and "*ipseità*" of his free life project: "...the person does not know himself in directly, but only through the signs placed in his memory and his imagination from the great cultures." (Ricoeur). Hermeneutics rediscovers the ontological value of narrative language. Followed by this, starts a revaluation of all the expressive and communicative capacities of language which have been forgotten by the hermeneutics of suspicion and by present, linguistic sciences, including, in very special way, the importance of "metaphor" as defined in *La metafora viva*: a language of revelation, or rather , the figure of "more" that a language can bring. On the other hand this is the key to understand all its religious language, penetrating in the thickness of the ontological symbol. This in its turn would be ignored due to its pure immanent conception of language" (Ricoeur). The value of This fact had already been discovered by Vico as a language of truth. In Vico has its origins in Jewish tradition. The language, for the philosophical hermeneutics, from Heidegger to Ricoeur to Betti, becomes guardian of the memory, that not only goes back to the origins, but also opens the future in a perspective of dialogue, understanding and brotherhood.

Bibliography

Anderson R. Dean (2000), *Glossare of greek rhetorical terms connected to methods of argumentation, figures and tropes from Anaximenes to Quintiliano*. Peters, Leuven.

Aristoteles (1991), *The art of rhetoric*. With an english translation by John Henry Frese, Harvard University press, Cambridge, London.

Aristoteles (1994), *Aristotel's De anima*. Brill, Leiden.

Aristoteles (1976), *Aristotele's metaphysics*. (Mu and Nu) Clarendon press, Oxford.

Bergson H. (1990), *Matter and memory*. (Matière et mémoire 1896), Zone Books.

Bremm J. (1988), *Interpretation of Greek mythology*, Routledge, London 1988.

Chernyakov A. (2002), *The onthology of time, being and time in the philosophers of Aristotele, Husserl and Heidegger*. Kluwer, Dordrecht.

Furley D. edited by (1999), *From Aristotele to Augustine*. Rutledge, London, New York.

Grassi E. (1990), *Vico and humanism: Essays on Vico, Heidegger and rhetoric*. Peter Lang, New York.

Hard R. (2004), *Rutledge handbook of Greek mythology*. Routledge, London, New York.

Hegel G. W. F. (1980), *The philosophy of right: The philosophy of history*. Enciclopedia Britannica, Chicago, W. Benton.

Kennedy G. (1972), *The art of rhetoric in the roman world 300 B.C. – A.D. 300*. Princeton University press, Princeton, New Jersey.

Kotzé A. (2004), *Augustine's Confessione: communicative pur pose and audience*. Brill, leiden, Boston.

Mack M. (2001), *Anthropology as memory*. M. Niemeyer, Tubingen.

Minchin E. (2001), *Homer and the resources of memory*. Oxford University press, Oxford.

O'Donnell J. J. *Augustine's Confessions*. Clarendon press, Oxford.

Overgaard S. (2004), *Husserl and Heidegger on being in the world*. Kluwer, Dordrecht.

Ricoeur P. (1965), *History and Truth*. Trans. Charles A Kelbley, (Evanston, Illinois: Northwestern University Press.

Ricoeur P. (1984–1988), *Time and Narrative*. Volumes 1-3, trans. Kathleen Blamey and David Pellauer (Chicago and London: University of Chicago Press).

Ricoeur P. (1988), *Humans as the Subject Matter of Philosophy*. In: *The Narrative Path, The Later Works of Paul Ricoeur*, eds. T. Peter Kemp and David Rasmussen (Cambridge, Mass: MIT Press).

Ricoeur P. (1991a), *Explanation and Understanding*. In: *From Text to Action*, trans. Kathleen Blamey and John Thompson (Evanston, Ill: Northwestern University Press).

Ricoeur P. (1991b), *From Text to Action*. Trans. Kathleen Blamey and John Thompson (Evanston, Ill: Northwestern University Press).

Ricoeur P. (1992), *Oneself as Another*. Trans. Kathleen Blamey (Chicago: University of Chicago Press).

Ricoeur P. (1995), *Intellectual Autobiography*. In: Lewis Edwin Hahn, ed., *The Philosophy of Paul Ricoeur, The Library of Living Philosophers Volume XXII* (Chicago, Illinois: Open Court).

Shrimpton Gordon S. (1997), *History and memory in ancient Grees*. McGill – Queens University press, Montreal.

Soarez C. (1578), *De arte retorica libri tres, Ex Aristotele, Cicerone, et Quintiliano precipue depropmpti apud Thomam Bozzolam*, Brixiae.

Woodman A. J. (1988), *Rhetoric in classical historiography*. Four studies Areopatice press, London, Sydney, Croom Helm, Portland, Oregan.

Chapter 5
Icons, symbols, cultures

Ivan Varga

This chapter is divided in two parts. The first one intends to give a brief overview of icons, symbols and their changes in different cultures. The second part considers the body as icon and symbol in society under different cultural conditions.

General considerations

The word *icon* (from the Greek εικον —eikon—image) was originally applied to a religious work of art, usually a painting of divinity or saints and angels in the Eastern Orthodox Christianity. In that, and in other icon painting Christian traditions, icons are usually flat panels but can be cast in (mostly precious) metals, carved in stone, clothed, done in mosaic, in fresco, even printed on paper.

None the less, over time the icon became to represent non-religious symbols or images in a wide range of contexts. Secular objects or even persons can obtain iconic significance. The maple leaf, for example, is the symbol for Canada, the cedar tree, for Lebanon, etc. Political parties usually have their icons. A flag—that is a particularly patterned and/or colored piece of cloth—assumes the role of an icon inasmuch as it represents a political entity and its inhabitants. The iconic significance of the national flag is shown by legislation of most of the countries that punish the desecration of the flag. In fact, this is quite similar to religious punishments meted out for acts of iconoclasm.[1]

Even persons are called iconic. They are either politicians, religious leaders or—usually—popular artists whose ideas or behaviour influences or inspires a large number of people.[2]

This shows that the concept of icon has shifted from the religious to the secular domain. In our times the icon assumed the role of symbol.

Symbols represent something that exists either in reality or in beliefs or in concepts, respectively in ideas. They can be pictorial representations as in religious symbols or else, material signs such as a red octagon (a stop sign on the road).

All language consists of symbols. When we name an object, e.g. "house", it is just a symbol of the real object. When we say "chair" we apply a linguistic symbol for an existing object; one cannot sit on a combination of sounds. When we name a relationship, e.g. parents—children, we utter the words but these only symbolize an existing physical and emotional relation. It is the remarkable capability of the human mind that we can make an association between the symbol and the object or relation it represents.

Numerals are symbols of numbers that are symbols of relationships in the material world, however abstract they can be (e.g. negative numbers). Numbers, however, play an important role in religious mysticism, primarily but not exclusively in Jewish kabbalah or Christian Gnosticism. Actually, all high and everyday-life cultures and almost all religions have their secrets of numbers. This can be seen in the Maya and Aztec, in the ancient Egyptian, Indian and Chinese cultures as well. What is characteristic for Christianity is partly taken over from the astral cults of the Babylonians and much has been borrowed from the Pythagorean and kabbalistic way of thought. Pythagoras looked at numbers and their combinations as keys to understand and unlock the mystery of nature. The neo-Pythagoreans and neo-Platonians as well the Gnostics continued this tradition and developed a mysticism of numbers that is characterized by the following features:

1. The number influences the essence of things that are somewhat arranged in it.
2. The number will thus mediate between the divine and the this-worldly.
3. Therefore, if one carries out any kind operations with numbers, those operations act also on the things that are connected with the corresponding numbers.

Endres – Schimmel, 1985:30

Mysticism based on numerology (kabbalah, etc.) was partially made feasible by the fact that ancient Hebrew, Greek, etc. used letters to denote numerals. It made possible to interpret combination of numbers as expressing hidden meanings of nature, of thought, of relations. While in a lesser degree medieval Christianity contained gnostic trends and the practice of kabbalah did not entirely disappear in modern times, the rational and scientific thought became the prevalent form. Without wanting to draw far-reaching similarities, one has to remark that in the Renaissance art emphasis was put on depicting correctly body proportions. Leonardo da Vinci's sketches of the ideal proportions of the human body, respectively of its distorted forms also have symbolic significance.

The forms and interpretation of symbols depend on the historical and cultural conditions.[3] Also, the theoretical, ideological and religious or non-religious views of the interpreter influences the meaning one gives to a symbol. (Suffice it to evoke the contemporary controversies concerning the interpretation of what constitutes a family or a marriage.) In other words, they are embedded in society's fabric. Examples could be multiplied *ad nauseam*. Here I just want to refer to the well known symbol of the *swastika*. For the average Western person who grew up in the post World War II period, the swastika was the symbol of Hitler's Nazi Germany and his party, of the Third Reich, its government and of various Nazi institutions and organizations.

However, the swastika, as religious symbol, has a roughly 5,000 years old history. The term originates from the Sanskrit but archeological findings show swastika-shaped ornaments from the Neolithic period. Without going into details, one can remark that the swastika was used as motif in many cultures, including North American native ones, to symbolize—depending on the culture in which it has been embedded—either the cyclic round of existence (in Buddhism) or the male energy or the female generative power or the life force, solar power or the Supreme being (in Jainism) etc.

There are no societies that would not have symbols. This applies to the body symbolic as well. In the following part I shall suggest ways to consider the body as icon and symbol.

The body as icon and symbol

Human beings take their bodily existence for granted. This is quite natural. We also treat other humans as embodied beings. From times immemorial there were prohibitions to harm other members of your own group, clan, tribe and later, nation.

Nevertheless, in all societies there were rules for presentation of the body that spanned from defining the body scars and paintings through the permissible fabrics for dressing according to an individual's social position and the acceptable covering of the body (i.e. how much of the—mostly female—body can be shown in public). In post-modern society, however, these rules are increasingly being relativized. Individual preference often overtakes norms that used to define the socially accepted or preferred presentation of the body. Also, gender characteristics are often blurred, e.g. men wearing earrings or other ornaments that were previously used by women only.

What is then the iconic/symbolic nature of the body? To give an at least tentative answer to this question, one has to recur to an anthropological approach. Namely, since times immemorial the presentation of the body was connected to a person's social status. From tribal chiefs to kings or queens a certain presentation of the self was expected (and accepted) that was manifested in dress and behaviour.

A number of criteria ought to be mentioned in order to analyze the symbolic role of the body. First of all, the body constitutes the link between nature and society or, stating otherwise, between nature and culture. As such, it cannot be considered purely natural. Most developed societies do not allow public nakedness[4] although the degree a woman can show her body (cf. miniskirts, cleavage) has changed in post/modernity. This shows that the culture of a society usually puts limits on the display of the natural form of the body. However, depiction of the naked body was and is allowed in art.

The body—mind relation is also a combination of nature and culture. The human brain is part of our natural being-ness without which no thought, no speech would be possible. However, mind is a combination of natural factors (the brain) and the culture in which the concepts, attitudes, emotions, patterns of reasoning are developed and deployed. While there is a general pattern in influencing the working of the mind, specific cultural features do influence it as well as the perception of the world. Some languages are more prone to express the concrete varieties of nature (e.g. Inuit languages have a great variety for expressing different types of snow but are less adapted to generalized concepts); while other languages, such as German, are good in expressing abstract philosophical ideas.

Chris Shilling in his book *The Body in Culture, Technology and Society* (2005) explores the interaction between the body and society, respectively the impact of society—or the class position of an individual—among others, on such bodily functions as eating (taste, habits and consumerism like the proliferation of fast food outlets). In a broad overview he addresses the position of the body in many aspects: work, sports, impact of technology on the body and sociability. This latter is very important for our purpose because it involves several features that pertain to the relationship between body and society. While I am not considering that body is entirely determined by society (i.e. the theory of social construction of the body), it is undeniable that social forces influence the position of the body in different cultures and historical periods. One cannot disregard an individual's position in the stratificational hierarchy or in the division of labour. This was characteristic for all human societies; it is just the forms that have changed. External signs, such as regalia, allowable fabrics of clothing (e.g. in certain medieval societies silk was allowed only for nobility), uniforms indicating the wearer's rank in his or her unit, even the way individuals address each other (in many languages there is a difference in addressing the other by using either the familiar "thee" or the more polite or formal "you") and many other features indicate the symbolic aspects of the body in society and culture/s.

In his book, *Presentation of Self in Everyday Life* (1959) Erving Goffman uses a theatrical analogy for explaining people's behaviour in social interactions. This insight reveals the symbolic features in human behaviour. From our way of dressing, to our way of speaking, communicating our taste in food, in culture (high, low or middle brow), our desires (sexual and otherwise), ambitions, life-style choices, etc., etc.

The body, as Foucault argued, is also the means through which society controls the individuals. Indeed, government regulations and social control measures largely define our ways of presenting our bodies. Earlier I mentioned restrictions on how much of our natural bodies can be displayed in public. Dress codes—formal or expected as in business or social events—also have a symbolic meaning signifying the social position of an individual. Transgression is barely tolerated, mostly if it happens by a famous or popular person. However, President Kennedy made acceptable for men not to wear hats or the husband of the Dutch Queen initiated the abandonment of necktie.

Symbols of healthy—or desirable—body abound in contemporary societies. These are also culture bound; in many African societies the full female body is considered healthy, sexy while in modern Western societies the slim one is the model. It can, and does, lead to unhealthy practices, mainly amongst young females, such as anorexia nervosa or bulimia.

Religion plays an important role in regulating the body. In general, religions are concerned with the body—soul relation, and most (albeit not all) of them consider body inferior to soul. In the monotheistic theological thought there were two kinds of approaches to the problem: horizontal and vertical. The *vertical* scheme, advocated by Judaeo-Christian and Islamic religions as well as in Platonic thought, held that the spirit (soul) was superior to the body because the latter, by its weight, is pulled down to the Earth. Similarly, within the body, the lower parts were considered of lesser value: the head is nobler than the heart which is nobler than the stomach. There are, however, trends that regard body not as something that ties us to the sensual world but rather as a step to abortions involve the problem of a woman having a right over her body. Apart from the practical issues (e.g. legislation, regulations) the philosophical and moral problem centres around the question of what constitutes the beginning of life and the theological one is the sanctity liberating humans from it, thus serves as a springboard for the ascent from the sensual to the spiritual realm. The *horizontal* approach does not consider the human body as distinct from other life forms. It is characteristic to cultures that have close relationship with nature. Eastern religious-philosophical traditions do not acknowledge the inferiority of the body. For example, pre-Vedanta Hinduism considers the human body the seat of vital energy that links it to the cosmic energy.

The religious regulation of the body extends to sexuality as well. There are great variations in different religions, and in different historical periods, concerning the allowable or desirable sexual behaviour and practices, respectively attitudes toward monogamy or polygamy (rarely polyandry) and divorce. Lately the discussions and controversies about of life. As sexuality combines physical, emotional (and occasionally rational) elements, it also influences the presentation of the self. Male and female sex symbols do have an impact on many people of how to dress, how to behave, how to make oneself attractive to the desired person.

That leads to investigating into what I would call *commercial body*. It is linked to consumption. In spite of certain similarities there is a difference between the two. Consumption is necessary to maintain the body. As Marx already stated there is a dialectic between production and consumption. However, in post/modernity consumption has changed. The symbolic value and expression of consumption has been expanded and became increasingly subject to manipulation for commercial purposes. On of the most prominent expression of this trend is fashion.

Fashion was present in all societies even, as Ann Hollander, in her book *Seeing through Clothes* states:

> [T]hat peoples who do not wear garments nevertheless develop habits of self-adornment that seem, as Western clothing does, to be a necessary sign of full humanity: there are ways of clothing the human body in some completed concept of itself without actually concealing any portion of its surface or shape.
>
> Hollander, 1980:83

Modern or modernizing societies always had and have changing fashions in both men's and women's clothing. Moreover, dress codes—even though they became somewhat looser—often define the person's social position. For example, while they are not as compulsory as military, etc. uniforms, businessmen's suits are remarkably similar. Wearing tuxedos is almost compulsory in certain social occasions. However, fashion determines the type and colour of the fabric, the width of the lapel, the pattern of the necktie, the shape of the shoes, etc. Even the casual dresses have their codes, depending on the brand the logo of which is visibly displayed, and this indicates either the wearer's social status or financial means.

Women's clothing is more dependent on the rapidly changing fashion. The trend of displaying more or less of the body, at least in Western societies, has a long history. It is linked to the prevailing cultural atmosphere in a given stage of a society. Deep *décolleté* in the French Empire or modest cover of the body in, say, Victorian England or puritanical America show the importance of the overall cultural influence on the social display of the female body. In late modernity the rules have been more relaxed: the miniskirt or the bikini bathing suit is acceptable. But the fashion is constantly changing and the media's influence on what people want to wear cannot be neglected. The impact of consumerism on the presentation of the body is of paramount influence in modern societies. Here, again, one has to take into account the class differences in what is considered "in" or "cool" or "appropriate".

The image of the healthy or desirable body became also commercialized. True, there are medical indications against obesity and these ought to be taken seriously. But apart from those considerations, the images of a healthy, impeccable, thus desirable body are largely commercialized by advertizing various products that claim to eliminate blemishes of the skin, make teeth whiter, .keep the body slim, the hair shiny, etc., etc. Plastic surgery—mostly for cosmetic reasons—serves to render the body corresponding to the promoted ideal.

Contemporary societies accept body ornaments (tattoo or body painting) or scarification or body piercing that previously was unacceptable or linked to "primitive" societies. This is mostly attributable to opposition to the uniforming tendencies of post-modern, consumerist society that allows lesser opportunities to express and present an individual's own personality. "[P]racticioners of body alteration often explain their motives by referring to the body as the last remaining seat of control over one's own self." (Varga, 2005:229.) Similarly, it became a means for expressing oneself, distinguishing from others and at the same time creating a bond with people who share the same disposition. In that sense, the practice has a symbolic meaning and for people of similar inclination, it is an externalized icon.

In contemporary society the body cannot be perceived without a relation to technology, be it information technology or biotechnology. Information technology primarily affects the cognitive processes of the mind while biotechnology influences the function of the human body. In spite of the differences between the two types of technologies, their effects on the body are similar insofar as human invented technologies interfere with the natural processes of the body. And, as John O'Neill refers to Bourdieu's observation, "[n]o society seems content to leave the biophysical body outside the *symbolic system* whereby members communicate to one another their age, gender, marital status, sexual availability, social standing and the like." (O'Neill, 2004:8. My italics.)

The *technological bodies*, to use Shilling's expression (Shilling, 2005:173-197) comprise several facets. It includes the human use of technology that is increasing in modernity/post-modernity but also the symbiosis or even fusion of the body and technology (e.g. artificial limbs linked to the nervous system). But even in pre-modern times technologies interacted with the body; eye-glasses, magnifying glasses, telescopes, carved wooden legs—created by their contemporary technologies—performed the same functions, even if at a lower level, than the contemporary ones.

The rapid development of microelectronic technologies, their proliferation is affecting the minds, and eventually the bodies (as e.g. advice and directions in surgical procedures performed by non-specialists in remote communities) of a quickly increasing number of people. The positive effects, apart from the above mentioned ones, are the speed of operations as well as the vast and ever growing amount of stored information helps to acquire data, facts, opinions on issues of one's interest. However, over-reliance of ready data and opinions diminish the faculty of independent, critical thinking for many, perhaps most of, people, especially young ones. Even though we do not yet have sufficient evidence, one could find indications that certain negative effects are narrowing the scope of independent thinking. Programs, such as YouTube, while may broaden the circle people, mainly but not exclusively young ones, are contacting effect the bodily interactions amongst individuals and groups. We don't yet know the full psychological effects of the lack of person-to-person interactions but even the intimacy of transmitted images[5] is not protected.

In sum, one can state that the presentation of the body has had in all known societies an iconic and symbolic significance that was always shaped by socio-cultural conditions.

Notes

1. Iconoclasm was also a result of religious convictions. In Christianity there were periods when depictions of Christ and saints were forbidden by the Eastern Orthodox Church in the Bysantine period. Most of the Protestant reformers, e.g. Zwingli or Calvin, rejected depictions of of divinity and saints. The exception was Luther. In revolutions, such as the French one or the Chinese cultural revolution, destruction of religious (and even certain secular) imagery was widespread, encouraged, initiated or even organized by the central authority.
2. Even leaders whom we consider evil, such as Hitler, Stalin or Mao, do belong to this category because they shaped ideas and behaviour of masses. A recent example of iconic religious leaders is Ayatollah Khomeini.
3. Animals—especially those who live in proximity of, and interact with humans—can or do respond to symbols but this aspect is beyond the scope of our present considerations.
4. There are exemptions from this general attitude. Apart from nudist camps or beaches, until recently topless sunbathing was allowed in Paris; in Berlin public parks tolerate the same, and in Munich the largest city park has a nudist section, not sheltered from the public. In Canada, a court decision allowed women to go topless on city streets.

5. The practice of "sexting", i.e. sending provocative pictures of herself or himself by a teenaged individual via cellphones could be considered to be "child pornography" by some state jurisdictions in the USA.

Bibliography

Endres F. K. and Schimmel A. (1985), *Das Mysterium der Zahl. Zahlensymbolik im Kulturvergleich.* [The mystery of the Number. Symboly of Numbers in Cultural Comparision], Cologne: Diederich.

Goffman E. (1959), *The Presentation of Self on Everyday Life.* New York: Anchor Books.

Hollander A. (1980), *Seeing through Clothes.* New York: Avon books.

Honderich T. (Ed.) (2005), *The Oxford Companion to Philosophy.* Oxford, New York: Oxford University Press.

O'Neill J. (2004), *Five Bodies: Re-figuring Relationships.* London, Thousand Oaks, New Delhi: SAGE.

Shilling C. (2005), *The Body in Culture, Technology and Society.* London, Thousand Oaks, New Delhi: SAGE.

Tressider J. (General Editor) (2005), *The Complete Dictionary of Symbols.* San Francisco: Chronicle Books.

Varga I. (2005), *The Body–the New Sacred? The Body in Hypermodernity.* In: Bianca Maria Pirani, Ivan Varga (eds.), *Bodily Order, Mind, Emotion and Social Memory*, Current Sociology, Vol. 53, No. 2. Monographs. SAGE.

Chapter 6
Increasing diversity in a changing Europe

Umberto Melotti

Abstract

Since the Second World War European countries have experienced increasing diversity. This was due to many factors more or less closely interwoven. Certainly the most important one was the striking immigration from other European and non-European countries. Another important factor was the institution of the European Community (1957), with its subsequent enlargements and its later transformation into the European Union (1993), which was followed by other, and even greater, enlargements. A third, and perhaps even more deeply factor, was the process of globalization. All of these factors, in fact, have contributed to define the new forms of European diversity, which have developed upon the pre-existing remarkable diversities of cultural, ethnic, linguistic, religious and national character.

To face the problems involved by this impressive change, the various European countries have implemented different policies, according to their distinct political cultures. This is particularly evident in the case of the three main countries of Western Europe, France, Germany and the United Kingdom, which have also been, for a long period, the three main European countries of immigration. Yet, the relationships between the policy adopted and the prevailing political culture is quite clear also in the case of Italy, where immigration began much later than in the three countries above-mentioned, but developed much more rapidly.

The paper analyzes this process dwelling upon all main forms of difference involved in this situation.

Introduction

Since the Second World War European countries have experienced increasing diversity to an unprecedented extent. This was due to many factors more or less closely interwoven. Certainly the most important one was the striking immigration from other countries, both European and non-European. Another important factor was the institution of the European Community (1957), with its subsequent enlargements and its later transformation into the European Union (1993), which was followed by other, and even greater, enlargements. A third and even more deeply acting factor was the process of globalization, which brings people closer for better or worse. In fact, all of these factors have contributed to define the new forms of European diversity, which have developed upon the pre-existing remarkable diversities of cultural, ethnic, linguistic, religious and national character.

To face the problems involved by this impressive change, the various European countries have implemented different policies, according to their distinct political cultures. This is particularly evident in the case of the three main countries of Western Europe, France, Germany and the United Kingdom, which have also been, for a long period, the three main European countries of immigration. Yet, the relationships between the policy adopted and the prevailing political culture is quite clear also in the case of Italy, where immigration began much later than in three countries above-mentioned, but developed much more rapidly.

As for migration, since the Second World War between the various European countries there was an astonishing difference of policies. Some countries have long implemented a policy of cultural "assimilation" in order to limit the diversity and its risks. Others have worked to contain immigration itself, or at least its main consequences, by refusing the settlement of immigrants in their territory, even if they had to recruit foreign "guest workers" for a long time. Others have tried to manage diversity with a policy of uneven "pluralism" that accepted the presence of ethnic communities, but required their members to be law-abiding. Afterwards, under the influence of the experiences of some non-European States (in particular Canada and Australia and, to a lesser extent, the United States) almost everywhere there was a push toward some forms of "multiculturalism", though often in a rather rhetoric and non-realistic way. More recently, under the increasing problems of internal security due to immigration and the menace of Islamic terrorism, many countries have rediscovered the importance of implementing a sense of belonging to the country and have begun to promote "social cohesion".

The relationships between political cultures and the policies that the main European countries have adopted to manage the increasing diversity (and more particularly the diversity arising from migration) become quite evident if we define political culture—in contrast with the prevailing less comprehensive views—as the way each country conceives State, people and nation, the relationships between ethnicity, nationality and citizenship, and the citizenship rights and duties, included how it recognizes or attributes them to native and non-native people (Melotti, 2000a). In fact, political cultures, intended in this way, have exerted a major influence on both aspects of any migration politics, i.e. the "entry policies" (the norms and practices concerning the admission of foreigners) and the "policies for the immigrants" (the norms and practices meant to govern their presence).

To clarify this point, I will highlight the relationships between the political culture prevailing in France and its traditional policy of assimilation; the strong links between Germany's political culture and its long-standing tendency to keep immigrants in a precarious situation; and the connections between the political culture of the United Kingdom and its unequal form of pluralism. After summarizing these experiences, I will focus on the situation in Italy, which has rapidly become the fourth country of immigration in Europe and is now by yearly arrivals the first receiving country in Europe and the second in the world after the United States. Finally I will point out the increasing influence of EU institutions on the migration policies of the member States and the significant change that this has entailed in their policies to face and manage diversity.

Diversity, migration and political cultures in France, Britain and Germany

Diversity is a neutral term encompassing all types of differences between individuals and groups. It may have biological, social and cultural origins. Some types of diversity are "natural" or are perceived as such (for instance, race, gender and certain forms of ability or disability). But its forms are always ideologically "constructed" and this often entails forms of inequalities and structural discrimination that require counteracting actions (see Arnesen *et al.*, 2008).

Therefore, diversity has become an issue openly recognised and politically dealt with on a national and international level. Almost everywhere, economic, social and cultural differences based on ethnicity, class, gender, religion and disability are problems calling for social policies, which, of course, are based on the prevailing political culture. The following remarks will show this point clearly.

The case of France

France is a strong centralized State that considers itself a great homogeneous nation. Thus, migration has been always perceived there as an issue that calls for assimilation. This approach has appeared quite obvious in this country, where immigration began as far back as the first half of the nineteenth century and played a major role not only in coping with recurrent shortages of workers, but also with a chronic demographic crisis, which appeared to be dangerous even for military reasons, owing to the political unification and the rapid industrialisation of Germany (its traditional enemy since the early modern age).

Indeed, France, which before its great revolution (1789) was the country with the largest population in Europe, at the beginning of the nineteenth century experienced the demographic effects of the revolutionary and Napoleonic wars, which were also followed by a sharp fall in birth rate. So, soon after the start of its industrialisation (1820), the demand for labour could not be met by the internal supply. The demographic crisis has continued, with peaks and lows, to this day, owing, among other things, to the continuous toll in lives taken by the wars fought both in Europe and out of it (notably, in the nineteenth century, the Franco-Prussian war and the colonial wars in Africa and Asia and, in the twentieth century, the two great world wars and the repression wars in Vietnam and Algeria).

This situation has favoured continual immigration, both temporary and permanent. France has tried to integrate immigrants in the only way that was conceivable in a country deeply identified with a State that acknowledged no national minorities and ethnic groups and contested any claim for mediations between its institutions and its citizens.

In such a context immigration called for assimilation to French culture as shaped by the ideology of the *État-Nation*. Immigrants, far from being allowed to use their ethnic and cultural diversity as a strategic resource for a non-subordinate form of integration, were expected to shed it completely, in order to become "good Frenchmen": a process that, according to the prevailing political culture, implied assimilation as regards language, culture and, possibly, mentality and character too. In return, France granted them the same rights enjoyed by its native citizens through "naturalization" (how the attribution of citizenship to foreigners is traditionally defined, not only in France, with a revealing term). In addition, even those immigrants who did not want to acquire French citizenship, or could not obtain it, gave birth to French citizens, owing to the *jus soli*, there prevailing since 1889.

This project had long found a specular image in French colonial policy. Indeed, its ethnocentric universalism rested upon the principle that the *évolués* of all races and cultures could acquire the same rights as native Frenchmen thanks to their assimilation (though in practice only a small percentage of them achieved this). But just as that colonial policy, after significant results, ultimately failed to counter the desire for independence of the colonised, so too the national immigration policy, after undeniable successes, has entered a deep crisis and has appeared quite unable to cope with the problems entailed by new immigration.

For as long as possible, to favour assimilation, France preferred to draw immigrants from other Latin and Roman Catholic countries: in the nineteenth century from Belgium, which in its Walloon area was, moreover, inhabited by French-speaking people; then, from the beginnings of the twentieth century up to the 1970s, from Italy, Portugal and Spain. But the labour reserves of these countries have become exhausted and now most immigrants come from areas both geographically and culturally much more distant: the Maghreb countries, which speak Arabic and observe Islam; West Africa, another area where Islam prevails, and Southeast Asia, mainly characterized by Buddhist and Confucian traditions. Although many immigrants coming from the ex-colonies speak French as first foreign or vehicular language, the assimilationist project has collapsed in the face of the greater cultural diversity of these immigrants, their much more evident ethnic distinctiveness, their significant numbers and their organisation in family groups and ethnic communities that claim their own cultural identity and favour the preservation of links with their countries of origin. At the same time, assimilationism began to appear inherently less and less legitimate, as the old idea of France's civilising mission faded away and the respect for cultural diversity increased, together with awareness of the moral unfairness of subordinating the granting of many important rights to the acquisition of citizenship—all the more so in a context where this implies renouncing one's own cultural identity, the preservation of which is more and more regarded as an inalienable human right.

For these very reasons, sensitive circles have begun to underline the need of a more "secularised" view of citizenship, which in France, according to its powerful Republican tradition, was conflated with nationality even more than in other countries (for the difference between citizenship, intended as the status depending on the juridical belonging to a political unity, and nationality, intended as a sense of belonging to a nation, which is variously defined in the different historical, social and cultural contexts, see Melotti, 2000a).

Against this trend, the growing Arab and Muslim presence, which is the most sizeable element in today's immigration and is perceived as the hardest to assimilate, has generated a real "invasion syndrome" that feeds xenophobic sentiments and reactions. This situation has even worsened after the recent riots in the French *banlieues*, mainly due to young people with a migratory background (Melotti 2007a, 2008).

On the other hand, it is necessary to blame also the efforts of certain French leftist scholars, who have stigmatised not only racism and xenophobia, but also the defence of cultural diversity, which, according to them, would entail a sort of "differentialist racism" based upon culture instead of biology. In fact, for this very dubious reason, they have harshly criticized Claude Lévi-Strauss himself, in spite of his seminal critique of any kind of racism and ethnocentrism. These scholars, indeed, seem to be affected by a strong distrust against all attempts to confront the problems of immigration from outside the traditional assimilationist approach, which they usually call the "Republican model" of integration. Of course, this is partly a reaction to the faults of the New Right, which, using Lévi-Strauss's ideas in its own way, has theorised the "right to difference" in a way that seems to justify even ethnic segregation, completely ignoring Lévi-Strauss's complementary invitation to implement a dialogue between the different cultures.

At any rate, in France the problem of immigration is still prevailingly discussed in terms of "integration". But this word often seems to be hardly more than a euphemism for the old notion of "assimilation", though the stress is now on the social, rather than the cultural, aspects. This continues to reduce new settlers to France into raw materials, without history and culture, ready to enter the great assimilating machine of French society.

Since time, however, this machine has shown signs of grinding to a halt. Besides the increased resistance of the new immigrants themselves to assimilation, there is the crisis of the old socialising agencies (schools, army, factories, trade unions and political parties), as well as the difficulty of the Roman Catholic Church (which in this field has always played a certain role, in spite of the secular orientation of the country) of speaking to immigrants, who are now mostly Muslim.

Despite the evident ideological crisis, many institutions linger on by inertia, driven by old ideas. Suffice it to recall here that schools in France still fail to grant enough room for diversity, as have clearly shown the measures repeatedly taken against the Muslim girls guilty of wearing their headscarves, even before the explicit banishment of these clothes in 2004.

Nevertheless, the evident inadequacies and contradictions of the traditional model of integration have generated important debates on the way of managing diversity. Yet, what in practice still prevails is the old policy, so deeply rooted in French political culture, even if, as we will see over, an important change is now in course, under the influence of both globalisation and Europeanisation.

The case of the United Kingdom

The United Kingdom had a social policy quite different—as different was, and still is, its political culture. This, with well-known pragmatism, has always emphasized autonomy, decentralization and the role of the intermediate social formations (i.e., groups and organisations between primary groups and central State institutions). Abstract principles are regarded with suspicion and social regulation is largely left to local administrations.

However, also the British policy had a strong ethnocentric bias. Yet, this assumes quite specific form. If in France ethnocentrism manifested itself in a universalistic way, with the claim that all immigrants, regardless of their race and culture, could, and should, become "good Frenchmen", in Britain it manifested itself in a particularistic way. There was no expectation that immigrants could ever become good Britons (or, better, good Englishmen or Scotsmen or Welshmen). Their diversity was taken for granted and the main concern was to contain the damage that it might cause to the "British way of life", since it was implicitly assumed that the situation should remain under the control of the native Britons (who regard themselves as the "majority" in opposition to the so-called "ethnic minorities" of immigrants and their descendants).

Also this approach was in part a counterpoint to and in part and extension of the old colonial policy. But this policy, which in the French case was characterized by direct rule and an assimilationist project, in the British case was distinguished by indirect rule and a differentialist orientation. In other words, Britain generally allowed the colonized people to preserve their own culture, traditions and social and political organisations, provided that above them they recognized the authority of its representatives: the viceroy or the British governor. This tolerance of difference has subsequently become the basis for the Commonwealth, which has outlived decolonization, whereas the French Community, its untimely imitation, was soon obliged to declare its own failure.

In Britain immigration from distant regions started soon after the Second World War, typically from countries of the New Commonwealth (i.e., the former British colonies in Asia, Africa and the so-called West Indies), whose natives, until the Commonwealth Immigration Control Act of 1962, were allowed to freely enter the United Kingdom as "British subjects". The racially, ethnically and culturally distinct groups issued from this considerable population movement might form their own communities, whereas in France, until 1981, any kind of association between immigrants was practically forbidden. As a result, in Britain ethnic communities soon became very important reference points for public authorities, while in France, partly by choice and partly by necessity, immigrants were addressed only individually and directly.

In the United Kingdom juridical status itself, rather than drawing a clear-cut line between citizens and non-citizens, envisaged a significant range of intermediate positions, linked to the prior existence of a special category, the citizens of the Commonwealth, who were later distinguished according to their place of birth, their date of arrival and their possible British ancestry (so-called "patriality"). But all ex-Commonwealth immigrants domiciled in Britain enjoyed full citizenship rights, including the right to vote and stand for Parliament as well as in local elections, and this afforded them considerable political influence (all the more so since they were often concentrated in certain electoral districts), as the election of some of them to the representative assemblies clearly shows.

In effect, the British way of managing the diversity arising from immigration was highly flexible and long proved an enviable capability to cope with social and cultural change. Yet even this system has begun to reveal its own inadequacies and deficiencies.

The most important ethnic community push has been for a shift to a true multiculturalism, which would involve the abandonment of cultural hegemony by the natives. In particular, the second generation of immigrants has harshly criticised a system that recognizes the communities but tends to confine their members to a subordinate position, unduly emphasising their ethnic and racial "diversity".

We must recall that the debate on immigration, which in France has been mainly centred on the "integration of immigrants", in the United Kingdom has been focused upon "race and ethnic relations", and the problem most passionately discussed has long been that of the rights of "racial and ethnic minorities". Suffice it to say that at the beginning of the '90s a prominent British specialist could still define the condition of immigrants in terms of "segregated inequality" (Rex, 1990). Indeed, in spite of many important measures specifically designed to combat racial and ethnic discrimination and to promote equal opportunities for all groups, the social problems connected with immigration had become all too evident owing to the recurrent ethnic riots in many towns (London itself and the north-western towns of England that had been more deeply affected by the process of deindustrialization). This induced some observers to wonder whether in the future "Britain could become a successful multi-racial and multi-ethnic society or very much the opposite" (O'Donnell, 1991, p. 139).

In the last decade a lot of things have changed, but many problems have not yet been solved, while new alarming issues have appeared, as we will see over. Therefore, that question remains quite open.

The case of Germany

Germany has treated immigration and the related forms of diversity in a quite different way. Indeed, the German Federal Republic—in spite of being the European country with the largest number of immigrants (today about eight million)—for more than forty years even denied its long-standing character of country of immigration. Hence the phrase that German politicians used to reiterate uncritically: "Germany is not a land of immigration" (*Deutschland is kein Einwanderungsland*). In fact, foreign immigrants, even if targeted for a long time by massive recruiting campaigns, were considered "guest workers" (*Gastarbeiter*)—i.e., foreigners (*Ausländer*) admitted only temporarily and for working reasons alone.

However, the kind of migration that configured them as such vanished long ago, in 1973, when, at the beginning of the severe economic crisis due to the remarkable increase in the cost of oil, Germany closed its border to further labour immigration. But that stop to recruitment (*Anwerbestop*) entailed an unexpected consequence: a large part of the workers already present in the country decided to remain there and, with the families that soon joined them, eventually formed a real "population issued from migration", as demographers use to call it. But the persistent refusal by the German politicians to recognize this dramatic change prevented them from working out a clear project for that population. Indeed, the policy that they followed for a long period could more easily be defined in negative than positive terms: "Neither integration nor segregation" (Giordano, 1987).

In Germany, in fact, immigrants were destined to remain foreigners. Their economic contribution might even be appreciated, but their permanent settlement was in no way encouraged. They were allowed to live in the country for even lengthy periods, but this did not entail any significant change in their juridical status. Their acquisition of citizenship was not even envisaged and their children born in Germany, if they were not in a position to become naturalized, remained foreigners—in their own land—since *jus sanguinis* prevailed over *jus soli*. Far from favouring a "nationalization" of immigrants, Germans expected them to be ready to leave the country at any time, not only of their own free choice, but also as a consequence of an economic or political crisis or a government decision. Social policies did not aim at their assimilation, but rather at keeping them in a precarious condition that could favour their return to their own countries (which was even encouraged with some incentives, but with quite poor results). Education, too, in many *länder*, was oriented to give immigrants' children a basic knowledge of the language, history and culture of their parents' country, in order to favour their "return" there. Hence a deceptive respect of their real or alleged diversity.

This policy, too, was deeply rooted in a specific political culture. Germany was the last of the great western European countries to be constituted into a nation State and this was formed far later than the German nation. On the other hand, in Germany "belonging" to the nation—far from being conceived in subjective civic and political terms, as in France, where Renan (1882) could even define it as a sort of "daily plebiscite"—was always conceived in objective, ethnocultural terms, as a fact linked to blood and land (*Blut und Boden*), as well as to language, and founded upon the putatively irreducible distinctiveness of the German people.

Even after the Second World War, as a result of the humiliating division of Germany imposed by the winners, this notion of belonging continued to prevail over State membership. Thus, German refugees coming from Eastern Germany (the so-called *Übersiedler*) and even the descendants of the Germans who had settled many generations before in some eastern European countries (the so-called *Aussiedler*) have been regarded by the Basic Law of the German Federal Republic as potential citizens in their own right (and, therefore, have never been treated as "foreign" immigrants). More generally, this view has favoured a tendency to preserve the alleged homogeneity of the German people and to resist its dissolution (albeit implicitly rather than explicitly, owing to the memory of the Nazi theories, which thrived upon that myth with results that are all too well-known).

The influence of this idea on immigration politics could not be clearer. Its main concern was to distinguish between Germans and foreigners, as a renowned German scholar pointed out (Blaschke, 1993). In fact, for a long period, all existing norms encouraged the rotation of immigrants, to prevent them from putting down roots in the German land. The only integration conceived of was temporary (*Integration auf Zeit*), which—in spite of being an obvious contradiction in terms—was proclaimed in 1973 "to render the condition of immigrants more humane". But even this temporary integration was not really implemented, since the transformation in the nature of immigration due to the economic crisis of the '70s made the measures then foreseen quite inadequate.

Moreover, the complexity of immigration increased in the following years. The arrivals of clandestine and undocumented migrants rose, the rate of unemployment increased even among the legal immigrants, and masses of asylum-seekers (both genuine and fake) began to arrive from Third World countries. Later, at the end of the 1980s, the breakdown of the social and economic systems of Eastern Europe precipitated a flood of refugees, who poured into the Federal Republic of Germany in successive waves, unprecedented in peacetime: more than 1.5 million people between 1989 and 1990, before Germany's reunification, counting both the *Übersiedler* and the *Aussiedler*. After the reunification, in 1990, the arrivals of the latter did not stop and those of other refugees continued to increase, until new restrictive measures were introduced. Moreover, there was important internal migration from the new *länder* to the old ones, which, among other things, fostered competition between German and foreign immigrants. These factors favoured the spread of xenophobia and racism, and this entailed the explosion of many cruel riots all over the country. In 1993, after the death of five Turkish immigrants in an arson, the then president of the German Federal Republic himself, Richard von Weizsäcker, was obliged to intervene with clear words: "Do the extremists claiming 'Germany for Germans' want to change our Constitution? For it does not say Germans' dignity is inviolable; it says human dignity is inviolable [...]. As for the Turks [living in Germany], would it not be much better to begin to call them German citizens of Turkish origin?".

However, the change in this direction began only some years later, with the new citizenship law (1999) and the first immigration law (2004), which entered into force after a contrasted process. But, as we will see over, this change, which was largely due to the influence of the EU institutions, has not solved all problems.

The case of Italy

Italy, with its 4 million immigrants (3,600,000 documented and at least 400,000 undocumented in 2008), is now the fourth European country of immigration, after the three countries mentioned above, where immigration began far earlier. It is, moreover, the first country of immigration in the Mediterranean basin, the most crucial area for the current and the expected future migration towards Europe. And it is also the first EU country for the percentage of non-EU immigrants (about 90% in 2004, before the recent enlargements of the EU) as well as for the number of illegal immigrants, in spite of—or, rather, owing to—its long-standing tendency to regularize them periodically with general amnesties (five extraordinary regularizations in fifteen years, between 1986 and 2002, probably a world record, and the last, passed by a centre-right government, with its almost 700,000 regularizations, was even the largest ever carried out in Europe; besides, between 2006 and 2008, the subsequent centre-left government approved two *de facto* amnesties, disguised as an extraordinary increase of the normal entry quotas, which allowed the regularization of another million illegal immigrants present in the country).

The Italian attitude towards immigration is rather peculiar. This is due to various factors. First of all, Italy has only recently become a receiving country, after being for a long time the European sending country with the largest number of emigrants (the turning point was reached in the mid-1970s). Furthermore, in Italy immigration did not begin in a period of reconstruction and economic development (as occurred in the north-western European countries), but rather in a period of economic crisis, characterized by a remarkable increase in the unemployment rate, which reached 12.5% of the national labour force (double the EU average), even if later it decreased considerably, arriving at about 6.5% (but all of these data conceal a striking difference between the northern and the southern regions).

Therefore, in Italy immigration has always depended much more on the push factors in the sending countries than on the pull factors in the receiving one. This trait—which is common to other southern European countries, such as Spain, Portugal and Greece—became even more apparent during the 1990s, owing to the dramatic events of that period. Among them, we must recall the serious deterioration in the economic and social situation in many developing countries, the destructive wars in the former Italian colonies in the Horn of Africa, the implosion of the pretended socialist systems in Eastern Europe, the repeated crises in Albania and the breaking up of former Yugoslavia (two countries near the Italian borders), and the bloody and cruel conflicts that followed them. Briefly, in Italy immigration has been passively accepted rather than deliberately encouraged and has caused recurrent emergencies and a state of continuous precariousness. This situation has become even more worrying in the last decade, owing to the consequences of the enlargements of the European Union to some eastern European countries in deep economic crisis, such as Romania.

To all this we must add the effects of Italian political culture, a theme that up to now has been hardly treated. In effect, national identity itself in Italy has long remained a sort of taboo (due, among other things, to the memory of the Fascist regime, which exploited national feeling for its own aims). Italian scholars have preferred to investigate the so-called "political subcultures" (dwelling upon the distinction between the "red", which still characterizes the areas that have long been controlled by parties of communist and socialist orientation, and the "white", which still characterizes those under a prevailing influence of the Roman Catholic Church). More recently, they have focused their attention on the persistent contrasts between the northern, central and southern parts of the country and on the emergent differences between the north-western regions and the north-eastern ones.

As for the Italian political culture, the first point to stress is its weak and ambiguous idea of the nation. In effect, in Italy this idea has long fluctuated between the Romantic view (emphasizing the "objective" elements of ethnic and cultural character) and the quite different concept tracing back to the Enlightenment tradition (emphasizing the civic "subjective" factors).

In Italy the Romantic view inspired the first theorists of the Risorgimento and re-emerged later, in a deteriorated and degenerated form, in the ideological formulations of the Fascist regime (which eventually accepted even the racist theories of its Nazi allies). This approach has fomented a considerable distrust in foreigners, which had already been fed by the persistent memories of the long sequence of foreign rulers that the country had suffered after the collapse of the Roman Empire. After the barbarian invasions—which in some northern countries are euphemistically called "migrations of peoples" (*Völkereinwanderungen*), but in Italy still evoke the figure of Attila and the destructive raids of the Vandals—there were the dominations by the Arabs and the Normans in southern Italy and later by the Spanish, the French and the Austrians in many parts of the country, not to mention the recurrent Saracen incursions along its coasts and the expeditions and interventions of the German emperors (some of which were confronted and defeated by the Lombard League to the cry "Out with the barbarians!"). Anyhow, the Italian national anthem (written by a young patriot who died in Rome while fighting against the temporal power of the Church, supported by the French army) harshly thunders "Get out of Italy, get out, foreigners!". Similarly the most important song of the First World War boldly proclaims "No entry for foreigners!".

On the other hand, the Enlightenment concept of the nation inspired a sort of cosmopolitan wishful thinking, which recently, under other influences (coming, mainly, from some English-speaking countries), has nourished a superficial and rhetorical form of multiculturalism (see Melotti, 2000b). As a typical example, we may quote the first book on the "multicultural society" in Italy, published by the Department of information of the Italian government, which, completely forgetting the harsh problems entailed by immigration in Italy, portrays a smiling world made up of "oriental rhythms and *lambadas*, couscous and Eritrean *zighinis*, African textiles and Arab-style trousers, kink hair plaits and gorgeous carpets in the underground stations, Muslim prayers and Ramadan, Brazilian *viados* and Nigerian prostitutes, Egyptian bakers and Polish and Bengali windscreen washers, Cape Verdian maids and Filipina nurses, Andean flutes and greetings with joined hands" (Ghirelli, 1991, p. 78).

Moreover, in Italy the national sentiment, according to the latest research, is far weaker than in most European countries. Among the main reasons, we must recall that, since its origins, this sentiment has been tempered by the cosmopolitan heritage of both the Roman Empire and the Roman Catholic Church. Later, owing to the effects of internal conflicts and foreign rule, it was contradicted by a strong attitude to mind one's own affairs rather than the public interest, as was even theorized by some important authors in the Renaissance age, which in Italy was a period of political decline, in spite of the luxuriant development of arts and literature. Therefore, Italians, who had to live for centuries under rapacious foreign oppressors and inefficient domestic sovereigns, to survive were obliged to develop private virtues and public vices (Barzini, 1964).

More recently "civil religion", to use Rousseau's well-known expression, or the "sense of State", to use the old liberal definition, or rather "civicness", as modern sociologists prefer to say, has undergone the attacks of the two subcultures above mentioned: the one characterized by the ecumenical orientation of the Roman Catholic Church and the other marked by the internationalist stance of the leftist parties and movements. Italy had the largest communist party in Western Europe and many of its exponents are still active in Italian political life, though under new labels that enabled them to take part in the leftist governments of 1996-2001 and 2006-2008, overcoming a semi-secular *conventio ad excludendum*.

In the same period there was another important development: a widespread reaction against bad government, centralism, corruption and excessive taxation, not to mention the notorious ties between some members of the central government and the criminal organizations based in the southern regions. This fostered the expansion in northern Italy of an autonomist movement, the Northern League, characterized by both anti-national feelings and strong xenophobic overtones.

Moreover, Italy is affected by its very special historical legacy. The Italian national State was formed comparatively late (after all the others in Western Europe except for Germany, if we consider the date of its formal proclamation, but even after it, if we do not forget that the important regions of Trent and Trieste could join the Italian kingdom only after the First World War). Besides, the formation of the national State was not due to the initiative of the core part of the country, as occurred in France, Britain and Germany (and also in Spain, Portugal and Greece), which offer important examples of the "ethnic origins of nations" (Smith, 1986). In Italy the first steps of unification were taken by a small, peripheral and rather backward kingdom, the then still ethnically and linguistically composite Piedmont, whose ruling class could hardly speak Italian.

Last but not least, the Italian national State was formed without (and even against) the masses, which were largely pro-Bourbon in the southern regions and Catholic and socialist in the central and northern ones (not to mention those of anarchical orientation, less numerous, but widespread all over the country). The masses did not take part in the building of the nation, despite the idealized picture of the Risorgimento which was later "invented" for ideological reasons. Their "nationalisation" (Mosse, 1974) began only during the First World War (and certainly not in a painless way) and continued during the Fascist regime, but in a contradictory and counterproductive way that eventually increased the pre-existing distrust towards the State. The Resistance to both Fascism and Nazism was important, especially in the last phase of the Second World War, but could hardly change this situation, since, in contrast with what occurred in France and other countries, it had partisan social and political aims rather than national ones.

At a theoretical level we could discuss whether a weak concept of the nation represents an advantage or a disadvantage for the social and cultural integration of immigrants, and our conclusion would also depend on our idea of "integration", a term that may mean many different things (see Niessen *et al.*, 2007). As for me, I would define it as a condition of normal social relationships between individuals of different origin, race, class and culture, which allows civilised living in a relatively cohesive context. In fact, in Italy, the lack of an organic policy for immigrants has given much scope to private organizations (in particular, those controlled by the Roman Catholic Church, which, by using State funds with little or no accountability, have almost replaced public institutions in charity and other important activities).

Furthermore, the State, as far as immigration is concerned, has been quite unable to enforce its own laws. Hence a series of problems—both of a social nature and of public order—which make it very difficult to face the situation in a rational and systematic way.

Unfortunately, this has led to two extreme positions. Some advocate the mass expulsion of all illegal immigrants, adopting an essentially xenophobic stance. Others, in a rather irrational way, regard the existing situation as a marvellous opportunity for the development of a multicultural society. In effect, in Italy these widespread opposite stands seem to be even more radical than in other European countries. Hence the tendency—much stronger than in any other EU country—to leave immigration policies to the EU institutions, with the Italian government and Parliament shirking their own responsibilities. In effect, many Italian politicians have long preferred to avoid this problem, which deeply divides public opinion and the electorate.

As for diversity, we must underline that an important trait of immigration to Italy has been the presence of a very large variety of national and ethnic groups since its very early phases, while in France, the United Kingdom and Germany for a long period the large majority of non-European immigrants came from a much more limited number of countries: Algeria and Tunisia as for France; India, Pakistan and the West Indies as for the United Kingdom; and Turkey as for Germany.

The first immigrants to Italy were Chinese. Some small groups of them arrived between the two world wars, mainly from France, where they had been recruited during the First World War to substitute in agriculture and industry the men at the front. This arrival did not remain unnoticed, but it was not regarded as a migratory event, though many of them established themselves in Milan (where they created the first Chinese district in Italy, which has recently become the second largest Chinatown in Europe after that in Paris). In the beginning, the Chinese worked mainly as street vendors of silk ties; then they developed leather goods workshops and later, in the mid-1960s, they opened the first Chinese restaurants in Italy, which have become very numerous and have spread all over the country. More recently, they have developed other kinds of workshops (mainly in Tuscany and in the area of Naples) and various commercial activities, some of which of clearly illegal character (in particular, in Rome, Milan, Naples and Catania).

The start of the 1960s witnessed the first arrivals of foreign women intending to work in Italy as maid. Most of them came from the Italian former colonies in

East Africa, accompanying the Italian families who left Somalia, when the Italian trusteeship came to an end (1960), and Eritrea, when it was annexed by Ethiopia (1962). These women soon called for their sisters, cousins and friends, and many of them came willingly, all the more so as the economic and political situation in their countries had rapidly worsened.

Between the end of the 1960s and the mid-1970s, other women arrived with a similar project from other countries: Cape Verde, the Seychelles, Mauritius, El Salvador, the Dominican Republic, Peru, the Philippines and India (specifically Kerala). Most of them were Roman Catholic and were helped to immigrate by the Catholic organisations working in their countries. Because of the high demand for maids, these flows rapidly increased and women soon came to make up 50% of all foreigners in Italy. This proportion—which was mainly due to the scarce demand for male labour—remained almost unchanged until the second half of the 1980s, when male mass migration from North and West Africa considerably changed the situation. This high percentage of women, quite unusual in the initial phases of immigration, was for many years one of the most remarkable traits of immigration to Italy.

During the 1970s, two other important groups arrived in Italy: the Tunisians in Sicily, called to work in agriculture, quarries and fishing, and the Yugoslavs in Frioul, called to work in the building industry after the earthquake which struck that border region in 1976. The same period also saw the arrival of the first Egyptians. They settled in the main towns, particularly Milan and Rome, where they could find jobs in the tertiary sector, and to a lesser extent in some other areas, where they were inserted into industrial activities, mainly foundries and pottery workshops. But employment in industry remained quite exceptional for all immigrants until the end of the 1980s.

In fact, throughout the 1970s the most important flow was that of political refugees, coming from a great number of Third World countries: Brazil, Chile, Argentina, Uruguay, El Salvador and Cuba in Latin America; Algeria and Libya (another Italian former colony) in Northern Africa; Palestine, Iran and Turkey (Kurds) in the Middle East; Somalia and Ethiopia (especially Eritreans) in East Africa; Sri Lanka (Tamil) and Vietnam (the so-called "boat-people") in Asia. Italy did not grant them the refugee status foreseen by the Geneva Convention of 1951, since like some other countries it had signed that convention with a "geographical clause" that limited its application to people coming from other European countries (actually the few refugees that were able to escape from the communist regimes of Eastern Europe). Nevertheless, all refugees could remain in Italy, even if not officially recognized as such, with substantial Italian financial support and under the protection of the United Nations High Commissioner for Refugees, which had signed a specific agreement with the Italian government. The "geographical clause" was abolished in 1990, by the second Italian law on immigration.

In the 1980s there was the arrival of masses of North Africans (mainly from Morocco and Tunisia), West Africans (mainly from Senegal, Nigeria and Ghana) and South and East Asians (mainly from Pakistan, Sri Lanka and the Philippines).

The same period also witnessed the arrival of many East Europeans (first Poles, attracted, among other things, by the presence of a Polish pope in Rome; then Romanians, Bulgarians and Russians).

In the next decade there was the shock arrival of three unprecedented waves of Albanians (1991 and 1997) and repeated arrivals of refugees from Somalia and the former Yugoslavia, fleeing violent civil wars. From Eastern Europe (especially Ukraine, Romania and Moldavia) and Latin America (especially Ecuador and Peru) also came hundreds of thousands of women to work as nurses for children and old people. The arrivals from Romania considerably increased after its entry into the European Union (2007), and among its citizens there were many thousands of gypsies (especially Roma), attracted by the facility of remaining in Italy even if undocumented and involved in hideous crimes, owing to the notorious laxity of its laws, the inefficiency of its police and the unbelievable mercy of its judges, who are used to set free even the perpetrators of heavy crimes caught in the very act (as was denounced even by the minister of Interior of the last centre of left government Giuliano Amato, who attributed this practice to the bad influence of the extreme left, inside and outside that government).

This large variety of national and ethnic groups in part favours and in part impedes immigrant integration. However, this also depends on the kind of laws in force. For many years (until 1986) the political response to immigration was a sort of *laissez-faire*, i.e. no response at all. Then, for some years (1987-1990), there was a policy of unconditional regularisation, inspired by Christian and socialist feelings of solidarity. Later (1990-1997), when owing in part to it, the presence of unemployed immigrants (both legal and illegal) reached a worrying level, there was a shift to a policy of control and restrictions, though poorly implemented and scarcely effective owing to the pressure of the leftist parties that formed part of the then political majority. In 1998 another left majority approved an ambitious, but unrealistic law, which completely failed to control both immigration and the crimes connected to it. In 2002 the new centre of right majority approved a new law that, in spite of its declared desire of better controlling these two phenomena, could not change the situation too much, owing to the strong resistance of the trade unions, the religious organizations and the leftist movements, which were able to contrive pretexts not to respect it.

However, as for diversity, the Italian laws on immigration are not too bad. In fact, they outline immigrants' rights clearly and define a complex of measures aiming at their integration. They guarantee all non-EU workers legally resident in Italy and their families complete equality of rights and treatment with Italian and EU workers. They recognize their right to obtain houses, organise associations and maintain their cultural identities. The right to education is assured also to non-legal immigrants' children and the right to health is recognized, at least to some extent, to non-legal immigrants themselves. The regional and local authorities have been entrusted to implement measures dealing with the initial reception of immigrants, to introduce literacy and vocational courses and to promote the constitution of immigrants' associations and consultative bodies composed of representatives of their "communities".

This legislation tends to endow immigrants with the same civil, social and economic rights enjoyed by Italians. However, the right to vote is not yet granted even at local level (except for the EU citizens). By contrast, the right to sustain their national and religious identities is strongly affirmed. On the other hand, the period of regular stay in Italy required for the acquisition of Italian citizenship, which before 1992 was five years for all aliens, has been reduced to four years for EU citizens, but prolonged to ten years for the others.

The communitarization of the migration policies and the management of diversity

The situation presented above has somewhat changed in recent years and will change further in the near future. The main factor in this change has probably been the process of globalisation, which has considerably affected both the migratory flows and the national political cultures (see Melotti, 2004). But also the increasing communitarization of the European migration policies has played an important role.

It is worth recalling that already in the 1950s the German Federal Republic, France, Italy, Belgium, the Netherlands and Luxemburg gave life to the first European communities, whose officers were later merged (1965). These communities, progressively extended their borders owing to the entry of the United Kingdom, Ireland and Denmark (1973), Greece (1981), Spain and Portugal (1986), Sweden, Austria and Finland (1995). The main result—beyond an unprecedented period of peace and prosperity—was the constitution in Western Europe of a great "common market", which, among its basic "freedoms", envisaged the free circulation of the workers of its member States.

In 1985 five of the six initial member States (the German Federal Republic, France, Belgium, the Netherlands and Luxemburg) signed the Schengen agreement, to which later, between 1990 and 1996, all the others joined, except the United Kingdom and Ireland. This agreement was not part of the community framework, but had a clear communitarian vocation, being open to all of its members and only to them. Yet, when Denmark, Finland and Sweden joined, its main effects were extended to Norway and Iceland, which were associated with them in the so-called Nordic Union.

The Schengen agreement envisaged the dismantling of all border controls between the member countries and their substitution with stricter controls at their borders with other countries. However, it also concerned the aspects of immigration and asylum policies that were more closely connected with security and public order. This agreement came into force in 1995 for seven countries and some years later, at different times, for the others.

The Treaty of Maastricht (which was signed in 1992 and came into force in 1993) marked a turning point in the process of European integration. It established the European Union, an international organisation *sui generis* with a clearer political orientation and wider competence than the previous communities. Yet, these were not suppressed, but were absorbed into one organisation, the European Community, which, in turn, was inserted into the new body. Besides, the Treaty of Maastricht provided for an economic and monetary union, extended the functions of the European Parliament and attributed a new "European citizenship" to all citizens of the member States. On the other hand, it affirmed the principles of "subsidiarity" and "proportionality", to limit the interventions of the Union outside its areas of exclusive competence.

The European Union is based on three "pillars". The first, which comprehends the matters that were already the competence of the old communities, uses the "supranational method", which leaves only a secondary role to the member States. The other two, which comprehend the matters that were previously the exclusive competence of the States (namely international politics and common security, attributed to the second pillar, and justice and home affairs, attributed to the third), use the "intergovernmental method", which leaves the last word to the Council of Ministers, i.e. to the States. Migration, in spite of being recognized as an issue of "common interest", was inserted into the third pillar.

However, five years later, the Treaty of Amsterdam (which was signed in 1997 and came into force in 1999) transferred co-operation for justice and home affairs (including common action for immigration, asylum, visas and circulation of persons) to the first pillar. In order to facilitate this passage from the intergovernmental approach to the communitarian one, the treaty introduced a period of transition of five years from its coming into force, which expired on 30 April 2004.

One of the protocols attached to the Treaty of Amsterdam incorporated the Schengen *acquis* into the European Union, though by dividing it between the first and the third pillar. Only three countries failed to sign this protocol: the United Kingdom and Ireland, which had remained outside the Schengen Agreement itself, and Denmark, which preferred to remain free to choose whether or not to accept any new decisions.

Therefore, the "communitarization" of the migratory policy envisaged by the treaty of Amsterdam has remained defective and has operated variably in different geographical areas. Yet, it opened a new phase, marked by a much more active role of the European institutions and a final overcoming of the prevailing "defensive" character of the Schengen *acquis*.

In this perspective the European Council of Cardiff (15-16 June 1998) requested the competent institutions of the Union to prepare an action plan for implementing the "area of freedom, justice and security" foreseen by the Treaty of Amsterdam. This plan was approved by the European Council of Vienna (11-12 December 1998).

In the following year the European Council of Tampere (15-16 October 1999) gave an important impetus to the homogenisation of the migratory policies of the UE countries. It explicitly recognized the need for a common policy for asylum and immigration and solicited an approximation of the legislations of the member States on the conditions for admission and stay of non-EU nationals. According to the Council, this development had to take into account both the demographic and economic conditions of the EU countries and the needs of the immigrants' countries, without neglecting the historical and cultural links that might exist between them. In other words, it underlined the necessity of balancing the interests of the sending and the receiving countries. Therefore, to contrast illegal immigration, it advised the member States to favour legal immigration and use the partnership with the countries of origin and transit. It also stressed the opportunity of a progressive improvement in the immigrants' status, in order to grant long-term legal residents rights and duties "comparable" to those of EU citizens. In addition, it decided to introduce common measures to fight racism, xenophobia and all forms of economic, social and cultural discrimination. As for asylum, after reminding the obligations deriving from the Geneva Convention, it affirmed that some form of protection had to be granted to all persons who needed it, even if ineligible as refugees. It also emphasized the opportunity of defining a common procedure and a uniform status for asylum seekers.

After that Council the European Commission sent two important "communications" on a common policy for asylum and immigration to the European Council and the European Parliament (22 November 2000). As for asylum, the commission underlined the importance of reconciling respect for humanitarian reasons and the need to contrast illegal immigration. On the other hand, it supported the proposal of a common procedure and a uniform status in all EU countries, to prevent unnecessary secondary flows between them. As for immigration, after recognizing the right of any member State to limit the admission of non-EU citizens, it advanced the proposal to insert the permit of stay and the permit of work into one document, which could become permanent after some years; in addition, it asked for a common status, independent of any national ruling, for long-term residents. It also advocated a rapid homogenisation of the norms for family reunification, to prevent any differences that could induce immigrants to prefer some countries.

According to the Commission, the integration of legal immigrants was a priority for the European Union: "a pluralistic society by its very nature, enriched by a variety of cultural and social traditions", which were expected to increase in the future. Therefore, it was necessary to respect such traditions as well as the EU fundamental principles, described in terms of assertion of human rights and dignity, appreciation of pluralism and recognition that membership of a society implies rights and duties for all of its members, be they nationals or not. However, owing to the length of time required for a real integration, special attention had to be paid to immigrants' children, including those born in the EU, to prevent their social exclusion and their possible involvement in crime. As for the fight against illegal immigration, according to the Commission, it was essential to facilitate both legal immigration and rejection, or rapid expulsion, of illegal migrants and bogus asylum seekers. In order to implement these policies, EU member countries were invited to set up talks and partnerships with immigrants' countries of origin and transit.

This "comprehensive strategy" regarded immigration as a broad long-term process, which required different kinds of measures, including co-operation for the economic and social development of the sending countries, in order to control the push factors, which in this phase appear to be far more important than the pull factors, which, however, exist and must be taken into account.

The European Council of Laeken (14-15 December 2001) confirmed this trend towards a common policy of asylum and immigration, in spite of the difficulties due to the resistance of some national administrations. But each member State had its own concerns. Germany and Austria, particularly exposed to mass migration from eastern European countries, did not want to facilitate the granting of visas to their citizens. France, very suspicious against any limitations of its sovereignty, insisted on the principle of subsidiarity. Spain, under the double threat of mass migration from the Maghreb countries in the South and Basque terrorism in the North, asked the other European countries to admit more non-EU migrants and to refuse asylum to EU citizens, to prevent the establishment of terrorist bases near its borders. And Greece, while calling for a more open policy for legal migrants, resisted the proposal of recognizing more rights to the illegal ones.

Other differences emerged in the European Council of Seville (22 June 2002). Spain, which had the presidency of the Union, proposed to cut economic aid to the countries unwilling to contrast illegal migration. The United Kingdom, Italy, the Netherlands and Denmark supported this suggestion; France, Sweden and Luxembourg opposed it for both political and humanitarian reasons; and Germany, owing to its internal disputes on its recent migration law, did not take a clear stand. The final compromise rewarded the more co-operative States without damaging the others. However, it was decided that all future agreements on co-operation with developing countries should include norms aiming at committing them to control the flows and to re-admit their illegal migrants. The Council also asked for the definition of new criteria in order to decide which State would be competent to process the applications for asylum presented in other member States.

In this period, however, Europe had to face a new problem: the threat of Islamic terrorism, which became alarming after the tragic attacks on the Twin Towers in New York and the Pentagon in Washington D.C. (11 September 2001) and the subsequent US intervention in Afghanistan and Iraq, supported by three EU member countries (the United Kingdom, Italy and Spain) and a then candidate country (Poland). The seriousness of this threat was proved by the terrorist outrages committed in Madrid on 11 March 2004 and in London on 7 July 2005 and the attempts against British and Italian objectives in various parts of the world. This global crisis obliged the European institutions to re-examine the policy of immigration in the light of common security, however avoiding any undue penalization of migrants coming from Islamic countries. Therefore, in recent years a great deal of attention has been paid to border control. This clearly emerged during the semester of Italian presidency (July-December 2003), when, among other things, the member States agreed upon the establishment of a European Agency for the management of operational co-operation at their external borders (which became effective some years later).

In the following years the EU has become even larger, owing to the unprecedented entry of ten countries at the same time (Poland, Hungary, the Czechia,

Slovakia, Slovenia, Estonia, Latvia, Lithuania, Malta and Cyprus on 1 May 2004) and two others afterwards (Romania and Bulgaria on 1 January 2007).

During this process the old and new EU member countries signed a Constitutional Treaty (29 October 2004) that paid much attention to diversity itself as well as migration. This document proclaimed the pluralistic character of Europe, "united in its diversity" and based upon "the values of human dignity, liberty, democracy, equality, the rule of law and human rights". It also emphasized "pluralism, tolerance, justice, solidarity and non-discrimination" and the respect of "cultural, religious and linguistic diversity", i.e. all main forms of diversity present in Europe. It explicitly committed the Union to prevent and combat "crime, racism and xenophobia" and declared the willingness to implement "a common policy for asylum, immigration and the control of the external frontiers", which should be characterized by solidarity between the member States and fairness towards non-EU citizens.

The treaty could not enter into force, since it was rejected in two countries (France and the Netherlands) where it had been submitted to referendum. Yet, all preparatory and subsequent works have exerted an important influence on the orientation of the member States.

According to that treaty, the Union had to develop a common immigration policy in order to ensure an efficient management of migration flows. To re-launch the debate on this theme, in January 2005 the European Commission issued a Green Paper on a possible EU approach to manage economic migration. This document was conceived as a basis for a policy plan on legal migration, with admission procedures capable of responding promptly to fluctuating demand for migrant labour. The basic idea was that, owing to their demographic ageing, EU countries required more sustained immigration flows to meet the needs of their labour market. Therefore, EU had to attract economic migrants by assuring them a secure legal status with a clear set of rights. The Commission recognised that the number of economic migrants to admit was to remain competence of the member States, but underlined that the admission of third country nationals in one member State could affect the others in many ways (due to their indirect impact on the EU labour market and to their right to travel within the Schengen area, to deliver services in other member States and even to move to them after acquiring a long-term resident status). Therefore, according to the Commission, it was necessary to agree transparent and harmonised common rules for admitting such migrants.

Some months later (10 May 2005) the Commission launched a 5 year Action Plan for Freedom, Justice and Security that detailed proposals for EU action on terrorism, migration management, visa policies, asylum, security, the fight against organised crime and criminal justice. This plan (to be known as the Hague Programme) was presented as a cornerstone of the Commission's Strategic Objectives for 2010, built upon the basic ideas of prosperity, solidarity and security.

The plan singled out ten priorities for the next five years. As for migration and related issues, it defined a balanced approach envisaging a common immigration policy at the Union level and a stronger fight against illegal migration and trafficking in human beings (notably women and children). Moreover, it further developed an integrated management of external borders and a common visa policy and implemented the work to establish a common asylum area.

To fulfil the mandate included in that Programme, before the end of the year (21 December 2005) the Commission adopted a Policy Plan, primarily focused on economic immigration. This document defined a road-map for the remaining period of the Hague Programme. It listed the legislative initiatives and the other actions to take in order to pursue the development of the EU legal migration policy. The results of these measures have been object of the subsequent Reports on Migration and Integration, annually published by the EU Commission.

After a long period of reflexion and discussion, the member States signed the Treaty of Lisbon (13 December 2007). This treaty, though much less ambitious than the previous Constitutional Treaty, as for migration and diversity confirmed the same basic orientation. However, this treaty, too, which had to enter into force on 1 January 2009, has met its difficulties, since it was not ratified by Ireland, the only member State where it was submitted to referendum.

The convergence of migration policies in EU countries

In spite of these difficulties, the process of integration summarized above clearly explains the convergence of the policies of EU countries as for migration and the management of diversity. For sake of brevity, I'll mention here only some of the most important changes that have occurred in France, Germany and the United Kingdom, with a final reference to Italy.

France, though without renouncing its prevalent individual approach, has come to recognize the representatives of the ethnic communities as proper partners of its policies and no longer considers the acknowledgement of their specific traits as an unacceptable form of *ghettoïsation à l'americaine*. To meet with the particular difficulties of immigrants, special measures have been approved, especially for housing problems. Also some elements of an intercultural approach have been introduced, both in education and in everyday life. Some well-known *maîtres à penser* (such as Amselle, 1996, Viewiorka, 1996 and Touraine, 1997) have even called for a French form of multiculturalism: a sort of compromise between the traditional "Republican" values and the religious and cultural claims of immigrant communities. Afterwards a similar idea was expressed by M. Nicolas Sarkozy (2004) himself, who was then the leader of the right-wind majority party and later became the president of the French Republic.

An important step in this direction was the new law on secularism (*laïcité*), approved in March 2004. Its preparatory commission (which included many prominent personalities of various cultural and political orientation under the chairmanship of Bernard Stasi, the "mediator of the Republic") supported an explicit prohibition of entering State schools with "conspicuous" religious symbols (such as large Christian crosses, Islamic veils, Jewish kippahs and Sikh turbans), but suggested allowing minor signs of faith and ethnic origin. Moreover, it proposed to recognize as national holidays the Jewish feast of Yom Kippur (the day of expiation) and the Muslim feast of Aid-el-Kebir (the day of sacrifice): an important sign of change in the political culture of the country (though this idea was rejected by then president Jacques Chirac). Similarly, the commission confirmed the prevailing orientation for the religious and political neutrality of public institutions (according to the pre-existent 1905 law on secularism), but asked them to respect all religious and cultural sensibilities whenever possible, with explicit reference to schools, barracks, cemeteries and hospitals (it is worth recalling, in particular, that many Muslim women, of their own will or under the pressure from their fathers, husbands and brothers, refuse the services of male doctors and nurses).

In fact, "owing to the change in the spiritual situation which took place in the last century", that commission interpreted secularism as an important means of favouring the cohabitation of people of different origins, cultures and religions (the Muslims now number between 4 and 5 million and the Jews are about 600,000). Therefore, secularism is now regarded as a factor which could favour the integration of immigrants. In effect, this approach (quite different from the old "Republican" formula, which had been worked out to face the then reactionary claims of the Roman Catholic Church) could foster the development of a new culture, which, among other things, could improve the condition of women and promote sexual equality without pretending an immediate assimilation of uses and customs that even in Europe have emerged quite recently.

The United Kingdom considers immigration an important resource, in spite of the many economic and social difficulties that have caused strong resistance among the natives and violent reactions among the immigrants. Even the long-term consequences of immigration are substantially accepted, though some scholars and some politicians have underlined the need to "reclaim Britishness", to quote the controversial book published by an influential think tank of New Labour (Leonard and Griffith, 2002). Both the past premier Tony Blair and the present premier Gordon Brown have agreed upon this suggestion. On the other hand Prince Charles—who, if and when he ascends the throne, will inherit the function of *defensor fidei*, i.e. Defender of the Faith—has promised to act in favour not only of the Church of England (of which he would become the head), but of all religions present in Britain.

As for migration policy, while legal immigration for working reasons has been facilitated, illegal immigration (including that which adduces dubious political reasons) has been more strictly controlled. The new law on asylum, immigration and citizenship (2002) has limited the flow of asylum seekers, who had become quite numerous (more than 100.000 in 2002). Among other things, the new law provides for the biometric identification of the immigrants: an important step in a country that since the end of the Second World War had become used to live without identity cards (see Bennet, 2008). Now this situation, which made the stay of illegal immigrants much easier, has finished, owing to this "time of global uncertainty with an increased threat from international terrorism and organized crime", to quote Queen Elisabeth herself. The law makes both long-term stay and the acquisition of citizenship more difficult and requires foreigners eager to settle in Britain to learn one of the British languages. But it also introduces important measures aiming at their integration.

However, something has changed after the terrible outrage committed in London on 7 July 2005 by some young British citizens of Pakistani origin and Muslim religion, which was followed by other similar events (in August 2006 other 24 young British citizens of Pakistani origin and Muslim religion were arrested for planning an attack to six airliners during their flight to the United States and in June 2007 two other attempts of similar inspiration were foiled in London while another one partially succeeded in Glasgow). Indeed, these facts have alimented increasing doubts on the British pluricultural model, which was openly blamed for favouring the formation of real "Londonistans" in a country with almost 2 million Muslims and 1,500 mosques. In fact, it was Britain that reared those "robots of death", ready to sacrifice themselves in order to strike the West they hate.

Germany, which, as we have already mentioned, has abandoned its semi-secular denial of being a country of immigration, has also dismissed the idea of a merely temporary integration of immigrants. Suffice it to say that now immigrants meet much less difficulties to become naturalized and their children born in Germany can acquire its citizenship easily when they come of age. Also the later request to immigrants to learn German seems to be due to the new acceptance of their settlement rather than to a persistent resistance to it.

Of course, the new immigration law, which has become effective on 1st January 2005, contains new provisions related to internal security, such as the easier expulsion of aliens suspected of terrorist activities or involved in preaching hatred. But it also provides important opportunities for those willing to integrate and grants the refugee status for reasons not considered before, such as persecution by non-governmental groups, discrimination for sexual orientation and risk of genital mutilation.

As we can see, beyond their persistent difference, the migratory policies of these three countries seem to converge towards a social integration of immigrants with respect for their cultural identity: the new "common policy" that had been proposed by the European Union.

Paradoxically, the first country to move in this direction, even anticipating any EU soft laws on this theme, was Italy, which, as we have already seen, became a country of immigration only in the 1970s. In fact, the idea of the social integration of immigrants with respect for their cultural diversity already inspired its first immigration law, approved in 1986 by a centre-left government, and this approach was confirmed by all of its three subsequent laws on migration: both those proposed in 1990 and 1998 by other centre-left governments and that proposed in 2002 by a centre-right government, whose main care was security and public order.

This paradox may be easily explained. In Italy immigration began only in the second phase of post-war migration in Europe (1973-1982) and has developed only in its subsequent phases, under the influence of globalisation and Europeanization (Melotti, 2004). Besides, in many ways, the Italian political culture is a sort of compromise between the French and the German types. Another important factor was the "universalistic" approach inspired by the Roman Catholic Church (suffice it to recall that after the Second World War, for more than forty years, Italy was governed by a Christian democratic party). To this, we must add the "proletarian internationalism" professed by the leftist movements, the main trade unions and some of the main parties (in Italy the Communist party governed many important regions, and its epigones continue to govern them). Moreover, strong and widespread opinion movements in favour of European integration have made Italy very sensitive to the positions of European institutions, even before their formal declaration, while the growing crisis of the immigration policies of the other European countries has suggested the need of a new approach.

On the other hand, as already mentioned, globalisation has deeply affected the patterns of immigration themselves, which have become more and more similar almost everywhere. And this has facilitated both the work of the EU institutions and the acceptance of their decisions and proposals by the member States.
However, we are still at the beginning of a very long road. It is the duty of the enlarged Union to move ahead, overcoming also the new difficulties arising from its increased diversity in history, traditions, conditions and political cultures. But diversity is also an important resource, and we must not forget it, apart from any rhetorical statement on this subject.

Bibliography

Akgündüz A. (2008), *Labour migration from Turkey to Western Europe 1960-1974*. Ashgate, Aldershot.

Amselle J.-L. (1996), *Vers un multiculturalisme français*. Aubier, Paris.

Arnesen A.-L., Bîrzea C., Dumont B., Essomba M. Á, Furch E, Vallianatos A. and Ferrer F. (2008), *Policies and practices for teaching sociocultural diversity*. Council of Europe Publishing, Strasbourg.

Bagguley P. and Hussain Y. (2008), *Riotous citizens. Ethnic conflict in multicultural Britain*. Ashgate, Aldershot.

Bazin L., Gibb R. and Selim M., eds. (2007), *Identités nationales d'État*. Journal des Anthropologues, special monographic issue, Fondation Maison de Sciences de l'Homme, Paris.

Barzini L. (1964), *The Italians*. Atheneum, New York.

Bennet C. J. and Lyon D., eds. (2008), *Playing the identity card*. Routledge, London - New York.

Blaschke J. (1993), *Tendenze delle migrazioni e relazioni etniche nella Repubblica Federale Tedesca*. In: Marcella Delle Donne and Umberto Melotti, eds., *Immigrazione in Europa: solidarietà e conflitto*, Dipartimento di Sociologia, Sapienza Università di Roma, Pg. 143–156.

Cerroni U. (1996), *L'identità civile degli italiani*. Manni, Lecce.

Ghirelli M. (1991), *La società multiculturale in Italia*. Presidenza del Consiglio dei Ministri, Dipartimento per l'informazione e l'editoria, Roma.

European Commission (2000), *Communication from the Commission to the Council and the European Parliament on a community immigration policy* [com 757 final], Bruxelles.

Giordano C. (1987), *Né integrazione, né segregazione. Il contesto migratorio nella Repubblica Federale Tedesca.* In: Giuseppe Giordano, ed., *L'immigrazione dal Terzo Mondo verso l'Europa: un fatto umano e un problema sociale destinato a crescere*, La Quercia, Genova, Pg. 61–71.

Leonard M. and Griffith P., eds. (2002), *Reclaiming Britishness.* Foreign Policy Centre, London.

Melotti U., ed. (2000a), *Etnicità, nazionalità e cittadinanza.* Seam, Roma.

Melotti U., ed. (2000b), *L'abbaglio multiculturale.* Seam, Roma.

Melotti U. (2004), *Migrazioni internazionali, globalizzazione e culture politiche.* Bruno Mondadori, Milano.

Melotti U. (2007a), *Le banlieues. Immigrazione e conflitti urbani in Europa.* Meltemi, Roma.

Melotti U. (2007b), *Cambiamenti politici, immigrazione e nuovo pluralismo culturale in Europa.* In: Roberto Segatori and Giovanni Barbieri, eds., *Mutamenti della politica nell'Italia contemporanea*, Rubbettino, Soveria Mannelli, Vol. 1, Pg. 265–300.

Melotti U. (2008), *Migration, political cultures and urban conflict in Europe.* In: Bianca Maria Pirani and Ivan Varga, eds., *The New Boundaries between Bodies and Technologies*, Cambridge Scholar Press, Cambridge, Pg. 106–123.

Mentasti L. and Ottaviano C. (2008), *Cento cieli in classe. Pratiche, segni e simboli religiosi nella scuola multiculturale.* Unicopli, Milano.

Niessen J., Huddleston T. and Citron L. (2007), *Migrant integration policy.* British Council, Bruxelles.

Mosse G. L. (1974), *The Nationalisation of the Masses.* Howard Fertig, New York.

O'Donnell M. (1991), *Race and Ethnicity.* Longman, New York.

Renan E. (1882), *Qu'est-ce qu'une nation?.* In: *Oeuvres complètes*, Calmann-Lévy, Paris, 1947, Pg. 887–906.

Rex J. (1990), *L'atteggiamento verso gli immigrati in Gran Bretagna.* In: Albert Bastenier *et al.*, *Italia, Europa e nuove immigrazioni*, Edizioni della Fondazione Agnelli, Torino, Pg. 67–85.

Sarkozy N. (2004), *La République, les religions, l'espérance.* Éditions du Cerf, Paris.

Skenderovic D. and D'Amato G. (2008), *Mit dem Fremden politisieren.* Chronos, Zürich.

Smith A. D. (1986), *The Ethnic Origins of Nations.* Blackwell, Oxford.

Touraine A. (1997), *Pourrons-nous vivre ensemble? Égaux et différents.* Fayard, Paris.

Wiewiorka M. (1996), *Une société fragmentée? Le multiculturalisme en débat.* Paris, La Découverte.

Wiewiorka M. (2002), *La différence.* Balland, Paris.

Weizsäcker (von) R. (1993), *Ansprache bei der Trauerfeier für die fünf Opfer des Anschlags von Solingen in der Hauptmoschee in Köln am 3. Juni 1993.* Presse und Informationsamt der Bundesregierung, Bonn.

Second Part

Social Icons of the Body: Case-Studies

Chapter 7
The play actors: fleeing bodies, remaining prudishness or... body under control?

Chan Langaret

Abstract

Unlike many readymade ideas, the release of the body is not so obvious in the everyday life. In the world of entertainment, the evocation of body matters, whether with players or spectators, often remains rather vague, between hesitations, reluctance and reserve. A path, to study these "rearguard" phenomena, is to examine a possible "endoreïc" dimension, according to the concept proposed by Pierre Bouvier.[1] Thus, because of a simple desire to keep secret "what is going on backstage" or because of a prudish tradition, discretion seems to dominate when the actors face the question of the body. Only a few themes diffusely relating to the bodyness are accessible with the respondents.

We see here that the body is often kept distant and controlled, whether through techniques of acting or consciousness of the presence on stage. The body is an instrument of the artist but it is also limited through the role to be performed. Moreover, this self-consciousness also refers to the links between the actor's personal life and ways to enter "the skin of the character", which he must also fulfill.

The method used in this work, carried out by the socioanthropologic approach, consisted in conducting qualitative interviews in a strengthened non directive way, to respect a necessary distant first approach. The use of this methodology of absolute non directive way may partly explain the scarcity of the body evocation (no items on the body as such has been raised) but as such, underlines the actors' reluctance, to let the body as a conversational subject arise spontaneously, otherwise than with detachment or control.

The testimony of Pippo Delbono,[2] actor and producer, in a set of interviews with the cultural journalist and theatre critic Hervé Pons, reveals, through it's Italian cultural background, a possible anthropological distance of the French culture regarding corporality. We are going to examine, in both a sociological and anthropological perspective, the meanings of this partial emergence of a diffuse body.

Body kept distant and under control

The social sciences as well as the society on the whole seem concerned by the issue of the body, as in this title of a recent number of the journal *Sciences Humaines*: "Le corps sous contrôle".[3] Besides, we can observe that the body of the actors stays a much "segmented", differentiated issue, according to their primary culture. The specialized publications of the world of the entertainment, for example *Les Cahiers du Cinéma*,[4] often underline the fact that French actors are less "physical", and play less physically than, for instance, American actors. The special number of the TV journal *Télérama*,[5] in August, 2008, also refers to this notion with this title: "Le corps des acteurs: zero en physique" ("the body of the actors: bad marks in physics"), asking the question of the use and exploitation of the body by the French actors:

> "The last outlet offered to the adventurers of the body would be the biography of famous characters, as Piaf for Marion Cotillard (La Môme) or Sagan for Sylvie Testud, and soon Coluche (september), Gainsbourg or Coco Chanel (next year). But in these often stunning exercises of resemblance, what is really left for the body, autonomous entity which escapes, betrays and reveals? Isn't the complete imitation, postures and gestures included, a waterproof gangue, an obstacle to the appearance of life? Tool of the actor, does the body have to become the object of an absolute control?"[6]

In France, the body would be so much perceived as a belonging of the artist that it would finally be taken as an instrument requiring, at least implicitly, a technical training to acquire the master's degree, in the way Mauss presents the systematic use of the body in numerous techniques during the exercise of common activities.[7]

Training and techniques of acting

The apprenticeship of the techniques of acting is a crucial key of the management of the body on the scene. There, through them, the actor rules and animates his body. These techniques, learnt during a singular training for each artist, seem to play a role in this relationship between the physical dimension of his manner of acting and his presence on stage; Emmanuelle Devos, former student of the "Florent School" (one of the most famous training course for young actors), exposes this aspect in the pages of *Télérama*:

> "The typical timetable of the apprentice actor consisted in working the scenes of the directory and going to the pub. We were actually proposed a lot of others courses, such as fencing, or even equitation, but after three months, nobody was coming. Besides, for the Americans, and the Russians, the physical language is full part of the apprenticeship. The fact that the American method of Strasberg and the Actors Studio refer to the Russian method of Stanislavski isn't a random choice."[8]

The characteristics of the training make sense for the actor, in the way he will approach his relationship with the role, as Catherine Paradeise explains:

> "The work on the role consists in building the qualification of the role by elaborating the agreement between the characteristics of a workstation (negotiated with the author and the producer) and the characteristics of a person (arisen from the apprenticeship of the actor)."[9]

The apprenticeship, there, expresses a plurality of approaches, among which the actor concentrates on the connections to the character and more widely, the place of the technique in the manner of acting:

> "From a school to another, the means change: one bases the success of the performance on the dissociation between the intuitive being and the technician; the other looks for the communion with the public through the fusion of the actor and the character."[10]

The plurality of these approaches, more "detached" for some, more physical and instinctive for the others, reveals the accent put on one or another aspect of the relationship between an actor and acting. These differences of training and in the way an actor follows and preserves this apprenticeship or not, appear in the narrative made by the respondents about their relationship with their training. Whatever the school - even the tradition - they refer to, the expression of their relation with the scene clearly shows the variety of their postures regarding Method. Marc, self-taught actor, embodies the "intuitive" posture by placing the technique behind the experience, the basic experience built through the contact of the public which drives its acting more than any directive:

> Marc: "when I started, that was the theater workshop: it had nothing to do with acting for family and friends. That allowed me to develop myself, to learn on myself, on actor's profession and above all, to come into contact with, maybe, the best professor of all: the public."

In another perspective, Sophie, trained in a Parisian school, places the method at another level. In a more "pragmatic" posture, she mentions the ins and outs in the implementation of a collective project:

> Sophie: "There was a project, I knew two actors in the project. The producer, he had done the same school as me, I had not met him, he was in another year. Well, a priori we were of the same school, so... But I spent a whole week asking a great deal of questions, I said *I don't understand*, - you see - I did not really know... When I work with William, I do not ask any question, I know exactly - you see - how to do."

Her approach marks a bigger distance with the acting and emphasizes the importance of these techniques in the "compromise with the producer" mentioned by Catherine Paradeise. The way the protagonists can define the characteristics of the job establishes between the actor and his partners a real practical-heuristics construction,[11] which is necessary for the coherence of the group. The teamwork is made possible by defining collectively a "way of doing", connected to the training and by adjusting constantly the common knowledge and the practices they support.

As for Pippo Delbono, he insists on the physical dimension, which is leading his voluntarily very physical approach of the theatrical act:

> "The step of samurai, for example, which I use in Henri the fifth's staging. It is the way the kabuki actor enters on stage. There is three steps which are

> based on the ambivalence male-female. The upper part of the body must be totally relaxed; all the strength concentrates in the stomach. It's three steps very simple and very complex at the same time, because all the body is engaged, pulled between two opposite strengths, the one towards the sky, other towards the ground."[12]

We find here an ascendancy of the body and its expression in the scenic act, before joining a reflective and conscious process. The reference to the kabuki actor (epic shape of the Japanese traditional theater) brings to light the cultural dimension of the relation with the body. Pippo Delbono concentrates his efforts and his steps on an eminently concrete aspect, connected with the physical reality of the actor. This approach joins the concern for nature which Edgar Morin mentions:

> "Besides, the acting doesn't lack sense and interest in the cinema, but it is implanted on a particular dialectic. 'Be natural', as it is asked to the actors. So the nature somehow becomes the only technique which is taught them."[13]

The relation to the technique of acting is more or less methodical, instinctive, the actor approaching the activity with the knowledge based on its particular learning. This knowledge is the basis of the relationship between the actor, the acting, his partners and the stage.

Awareness of the presence on stage and relation to the technique

What is the fugitive relationship between the actor and his own character, in the way he approaches the role, when he is, physically, on stage? The consciousness of this presence gets involved in a plurality of technical constraints imposing, on stage, an alternation between acting and consciousness of acting:

> Benoît: "It is necessary to keep moments of consciousness as we play; I think of the cinema, in particular where we actually have to play a character, but there are technical constraints. (...) It is necessary to keep this consciousness, and to get totally involved in acting, but to keep it a little buried. There is a moment when carelessness is necessary and then, afterward it is necessary to re-frame."

If the interpretation demands a kind of carelessness, a fluidity of acting, to offer the character coherence, the actor also has to keep in mind the frame of the stage setting, to fit the contingencies. These necessary comings and goings between the interpretation and its awareness can, at this level, lean on the technique, as underlined by Paradeise:

> "The technique is a necessary condition of the work of interpretation because it canalizes the sensibility by offering codes of memorization and expression stabilized by the exercise."[14]

We conceive then that the acting "genuinely" must be directed so that it stays in the frame and a very constructed or elaborated manner of acting has to keep spaces of free expression to give the character his measure. Paradoxically, the constraints can also help to contain the expression of the feelings.

> Sophie: "In theater, due to the technical constraints, you have to push up the feelings, the emotions, you need to have a bigger control (...) The actors of cinema don't have to rehearse too much, otherwise they lose spontaneity; in theater, to be spontaneous on stage, it is necessary to rehearse, to rehearse, to rehearse and to work enormously."

These differences make the actor's relation with its acting strongly dependent on the requirements of the scenic devices, as Morin underlines it:

> "For an actor of theater, the acting is determined by certain practical necessities. The distance which separates the scene and the spectators calls for a gestural and vocal exaggeration. As Dullin says, the actor of theater has to enlarge the emotion. (...) [In cinema,] the ostentation of the camera and editing replaces the ostentatious art of the actor. The semi-close-up, the mid shot, the close-up destroy the distance which separates, in theater, the actor and the spectator, and make the ostentation of the gesture or the mime unnecessary. 'An actor of theater, it is a small head in a big room, an actor of cinema a big head in a small room' (Malraux). The expression of the "big head" supplants the expression of the gesture. It makes even the mime of the face useless: a quiver of lips, a fluttering of eyelashes are visible, thus legible, thus eloquent. The image of the close-up exaggerates."[15]

Thus, we perceive here the plurality of elements having an influence on the way the actor animates his own body on stage, between auto-control (which reminds of the increasing process of the modern societies described by Norbert Elias[16]) and the actor's physical and immediate relation to the material reality of his spatial and human environment (as analyzed by E. T. Hall[17] with the concept of kinesthesia). It would also be advisable here to study the physical relationship of exchange between bodies, as the actor very often interacts on stage with the other actors but also with the technicians, sometimes, the prompter. Bodily questions which open new tracks of study.

Body instrument of the artist - physical limits

All these "gesticulations" of the marionette impose to the actor, a relative, if not complete control of his body. And so the actor builds a singular relation to his body, as a craftsman to his tool, an artist to his instrument:

> Benoît: "This profession, we could compare it with the musician's one, where the instrument does not come from outside but is our own body, our own inside, the instrument works this way; to enrich our tact, our art, we need to explore ourselves."

The body-tool has its own characteristics, and thus its own autonomy, and so when the musician plays his partition, the body of the instrument also "rings". The actor makes his instrument ring, for Pippo Delbono the body even gives the acting its partition;

> "The body is my actor's tool, the movement precedes the thought. (..) The body thinks in a autonomous way, without the head. It is what we look for in the training, when our entire being stands in the pelvis. To avoid aesthetics."[18]

This presence of the body, which imposes a shape on the acting and on the character, is also added by the physical appearance, which can limit the possibilities of acting:

> Marc: "A 20-years-old young person, even in a suit, will never be taken for someone in his forties (...) the same matter for the size, and it is not a surprise. We will never see a small person playing Michael Jordan's biography, even if he is black. That also fits in with the fact that an actor cannot play everything. We can, nevertheless, mystify the spectator through appearances."

The physical appearance can almost be modified at leisure; this "game of appearances" helps the actor to enter into the role. For the actor, the physical modification can carry an important part of the character. From that angle, the costume often plays a central role:

> Benoit: "We cannot speak the same way wearing a suit and, like in pair of shorts! There are clothes which impose different dresses, manners, and we can very well approach a character simply by its physical aspect. (...) I had played a role in d'Artagnan, an American film; I had 4 hours of make-up there and a wig, a mustache (...) when you get up after the make-up it's amazing, the character is already there, it's incredible. All the outside, the entire outline, everything seemingly derisory, superficial, is actually useful; useful to build a character. A character, that needs to be built."

The body, as an instrument, decked out in hairpieces and "packaging", offers the actor a range of possibilities. But if this formal aspect (in Georg Simmel's sense) of the actor's physical dimension offers so many possibilities, it also limits the potentialities of acting. For some, to obtain a role, it seems necessary "to look the part" to get into the skin of a character:

> Marc: "I made short films, often after an announcement or a call of the director who was looking for an actor with a very specific physical appearance: 5'6 feet, brown hair, blue eyes."

If the actor can benefit from his outside aspect, he can suffer because of it as well, when too far from the demanded standard. Character parts, which can reveal the value of the performance, stay a risk which is not always taken by the director:

> "In cinema, the actor has to stick so strictly on his role that he is chosen according to his type, that is, to the immediate natural meaning and expression of his face and body. (...) The facial type, the dominant and characteristic expression of the face, takes such an importance that the director fetches faces in the street and uses faces of the street. Henceforth, it is less important to play with one's face than to have a face - a mask -. The physical type tends to be even more important than the traditional manner of acting, or the art of Playing. Thus, the actors who can play character parts are less and less present on cinema. Here again, there is a development of the character, the character who is the same in the real life and on the screen, and at the same time, a devaluation of the acting."[19]

As a way out, or to reduce the preparation work or simply to seduce immediately the public, ready-made roles are coming back. The mask, immediate perception of the role, supplants the character, profoundly personified by the actor. So, the physical appearance of the actor sometimes defines the limits of his performance. The physical limits are, besides, dependant on the "limits of the body", physical ability to stand certain wild rhythms:

> Sophie: "there were years, when I worked, acted, enormously. Almost too much, and as always, excess does no good either. There were periods when I was rehearsing a show all day long, and I acted another one in the evening, and so, during 4-5 months and there is a moment, an evening, on stage, when you cannot go further, you're exhausted. (...) Antigone, we played it 80 times, well, we had to walk there also[20], but I was bored, finally I was also physically tired, I needed to be in another context."

These steady, physically hard rhythms, impose a necessary alternation, in actor's profession, between activity and inactivity. This profession often stands in a system where irregularity is taken into account. It is, first of all, because short-lives projects give way inevitably to temporary contracts, but this necessary alternation between worked and unemployed periods can also constitute a working mode. Indeed, we will see that the actors need these moments of peace to face storms and most of all to incarnate and then "give substance" to a character.

The links with the personal life - the manners to enter the skin of the character

The "Character"

We observe several configurations in the processes of construction of the character. A first figure is composed by the *Character*. Some actors, defending the same fictitious identity throughout their career, practically stick with their character and embody these ready-made roles they created themselves. Besides the interest of this type of interpreter from the angle of the definition of an artistic project, with a producer relying on a fictitious figure previously defined, the investment in this kind of character, recognized by the public, can also turn out to be profitable for the production:

> Benoît: "some actors are chosen because they are stars, because they are 'bankable' as we say, because they give a high return, but also because, due to their star's status, they have a character whom we respect and we want to see. (...) Actors are taken because they represent a character; Charlot for example, Charlie Chaplin, he found his own character."

This "starisation" allows the artist to work more regularly, granting regular incomes. The fame is considered as the way to exercise the profession without anxiety. As such, the actor relies on the networks he has integrated to realize successful projects which can engender popular recognition, through media:

> Sophie: "That corresponds to encounters, it turns out that I had this meeting with [the theater of] the Etreinte, we began the theater together. It turns out that if tomorrow Philippe becomes a very known film-maker, he will engage me in his project, and perhaps there will be a press and all this. You see the most important in this profession, it is the encounters, you see. Later, if fame happens - with the cinema because of the media - so much the better, but it has to correspond to something."

On the other hand, for the actor who is known with a successful figure, this fame can also bring the risk of locking himself into the so created character, whose autonomy passes the creator. Thus the actor has to oscillate between fiction and reality to preserve his interpreter's identity:

> Benoît: "Can we, despite ourselves, become the character we interpret? Of course, we can; there are persons who don't make the distinction between fiction and reality and who continue to play a role in the real life, somewhere. The problem is to know if it remains the exercise of a profession or if it hangs on the personal psychology."

Without questioning the psychology of the actors, the character, as an autonomous entity, sometimes makes a character of the actor. He is then automatically identified by his function. This movement of the centre of gravity of the artist's identity can constitute a frightening spectre for the actor:

> "I would not like to become a seventy-year-old person, whom everybody realizes, as soon as he enters a restaurant, that he is an actor. If we can say about somebody that he is an actor, it means that he has lost his human dimension and he has become a character: a special being with a certain way of laughing, moving, sitting down... In that case, the actors become monsters. In this way, the autonomy of the actor we spoke about previously is important, because, otherwise, the film directors treat them as madmen, patients. It is necessary to be strong in front of a director. An actor must never abandon his own identity. We think of Brecht, of course; I don't want to see on the scene people who say: personally, I am such character... Especially in our current world where everything is only fiction, the politics, the society, the fashion, the television... We henceforth search life in theater."[21]

This halving between the person and the character, identified by Morin, constitutes a system going from fiction to reality and from reality to fiction, with a very perceptible support, which is the figure of the actor, the actor "personally":

> "We find again the dialectic, from which the star is born, which goes from the real person to the cinema character and vice versa. Let us remind that, if the star plays his myth in the life, (cinema character) this myth is inscribed in his type - his face and body."[22]

But these effects of superimposition between the person and the character, based on physical and expressive characteristics, only concern actors benefiting from the public recognition, through the immediate identification of their "character". For all the actors, the construction of their characters happens through a personal creative process.

The instigator of the vocation: sensation, pleasure, feelings

The basis of this creative process, the "fuel" which seems to be the raw material of the acting, appears to be the pleasure:

> Benoît: "There is also a dimension of pleasure; there is a dimension of pleasure in the fact of wanting to interpret a character... This pleasure has to be felt. If it is not felt, something can be missing; this notion of enjoyment is necessary to transport oneself in the other, in the character. It is tremendously important, this notion of pleasure, really. We do stay in an artistic domain where we have fun, in a sense, we get pleasure."

This very physical notion of pleasure, of feelings, stresses the actor's visceral attachment to his activity, which overruns the material or spiritual constraints, when he wonders about the deep reality of a sometimes manhandled vocation:

> Gaëlle: "I miss it, because even if it is a lot of time, of work, acting really brings pleasure and the feeling of being on stage..."

This lack also goes beyond the material dimension of a job that pays the rent to concern a more fundamental aspect of artist's life: necessity. The sensation of lack or need marks out the testimonies of the actors:

> Sophie: "I don't do that to earn money; I do my job because I need it. (...) I don't know any actor who does not feel like acting."

This aspect of the physical, intimate relationship imposed to the actor, underlines the specificity of the activity. The phenomenon of embodiment of the characters by the actors asks from them a total investment in their activity as well as a kind of detachment regarding the acting. This embodiment brings to light the specificity of a "total" relationship between the body and the actor.

Embody: give substance to the character.

We saw the psychological dimension of the relationship to the character mentioned by certain actors. Actually, actor's acting appeals to this duality where the actor has to control himself and also get taken.

> Sophie: "you persuade yourself that you are this person (. . .) On stage, you can go into a certain madness, an excessiveness, a controlled excessiveness, we are not schizophrenic either. It is a totally controlled, chosen schizophrenia."

This "going out of oneself", fleeting phenomenon which let see someone else than oneself in the space of the scene and during the time of the play, also takes root in a long term halving, engaged throughout the work on the role. One must learn to discover, to know and, in the end, to incarnate this other one:

> Benoît: "By going towards the character, interpreting, skipping into the skin of the others, the possibility is given to go out of oneself; to go out of oneself or to allow oneself to do things normal life does not allow. In that way, we can say, indeed, that there is a. . . We go out of ourselves to go embodying a character. That is, there, there's a measure to transcend, of transcendence. (. . .)We usually speak about interpretation, about going into the skin of a character, we are really below and we inhabit it."

Duvignaud analyzes, by underlining the anthropological dimension, this halving, this transcendence, so as to enlighten and determine the causes and effects, the social function of the phenomenon:

> "The actor is depersonalized, divided into halves, possessed, but it is to conquer a supplement of existence and realize a concrete experience."[23]

Here, the author compares the theatrical act with other social facts observed by the ethnologists, such as certain rites of possession[24]. During these rites, the protagonists embody symbolic figures which "come down" on them, like certain actors who say they are "possessed" by their character:

> Marc: "On stage, the body doesn't belong to the actor anymore, the body is inhabited. The actor isn't there anymore because he lends his body to his character; as an evil spirit would inhabit the body of a possessed man. The character is a kind of mask behind which the actor pulls the strings."

Nonetheless, the actor, by staying on the manipulator of the puppet's side, shows his control, his construction of the character who will take a temporary control over the body of the creator afterwards, during the time of the play.

Fill the skin of the character

In the character's construction process, when the actor gives gradually existence to the imaginary entity, we can wonder how the actors fill the skin of their characters and how they can "give flesh" to the character. Benoît has his point of view on this issue:

Benoît: "But I think that it needs a dimension of actual experience; there are characters we cannot interpret because we don't stand back from life enough to carry the contents of these characters. (...) There are very heavy things, very ... If we enter the skin of a character but if have nothing to propose, it might be a little tenuous. It's better if there is matter inside. "

And this matter is acquired little by little in the conscience of the actor. Catherine Paradeise analyzes these contents, the "supplement of soul" the actor can show by making it sensitive through his performance:

"The sensibility is also developed by reading texts, but is not only accessible by the technical training. It goes richer thanks to the experience lived by the actor, in his own life as well as on stage."[25]

The actor absorbs "like a sponge" things in his life, fills up with experiences, becomes impregnated with feelings, with perceptions, which will be as many sources of inspiration to create the substance of the characters he will play.

Sophie: "There are also periods with nothing, you do not act during 3-4 months and, well, there is *not* nothing. Finally you grow up. (...) If there is nothing in the real life, what are you going to give on stage? In the real life, it's also necessary to walk, make shopping, see friends, go to the cinema, see an exhibition (...) because such things opens senses when you work on a character, we necessarily have references. If you only act, rehearse, act, rehearse, and if you don't live, after some time you are completely dry, you can't give anything anymore."

These moments of break allow the actor to stay in sync with reality and to give the characters this famous "naturalness", this realism which takes root in real life. Through this nourishment of the character, the link between the person of the actor and the character he embodies allows bridges disturbing the strict separations between fiction and reality.

Sophie: "It is violent, it's violent but it is not suffering, it is a violence we inflict to ourselves, which is sometimes difficult but which must also be happy. That doesn't mean that we cannot be moved on stage; sometimes, there are moments when you are so relaxed, where things take place so... that, sometimes, the text you are saying on stage catches up things that perhaps, you are living at the same time in your life, and you can be suddenly totally moved, you can have a true emotion, you see, you can begin to cry because you are in a state of availability. I believe that an actor is also a sort of receptacle, you see. (...) You let yourself be passed through by things."

Actor's susceptibility to a foreign personality, a character, is for Jean Duvignaud the demonstration of a kind of alienation:

"Alienation, very peculiar possession: the actor let himself be inhabited by a coordinated system of signs appointing an imaginary character whose current display exists only through him (...) When we use here the term of alienation, we point out this movement of realization of the being by the representation of an imaginary figure which appoints especially the fulfillment of the collective activity, through aesthetics."[26]

There, Duvignaud describes the vector, realized by the actor, between the collective sense and its singular demonstration. Thus we perceive the emergence of the sense of the collectiveness in the individual order, a demonstration of the social ties in the personal expression of codes and significant signs, expression of a societal sense in a singular will and a surge of sometimes diffuse transmission through the artistic act. We can then speak of a phenomenon of "endoréism" at an individual level, that is, in the act of the playing, a resurgence of incorporated structures which are as many personalized collective representations. When these structures and representations make sense for the public, the transmission is then effective; "we believe in it". Taken between these collective representations, to which he gives reality, and his personal perception of the activity, for which he has to build an engaged posture, the actor has to look after the individual so that he is vector of the collective:

> "Art does offer us the occasion to live extraordinary moments, but we permanently have to put ourselves in a state of crisis and personal questioning. I don't want to fall in the cliché of the artist and to play the artist. The actor's way can be a mystic way. Certainly, the theater invents and builds itself for the public; however the actor must be autonomous, dedicate himself to himself, be in permanent revolution, not with the world, but with himself. As the priests! Otherwise he becomes an empty, plastic, gossiping, smooth, formatted and unbearable shell. There are many things in the world; war, poverty, violence... Art makes sense only if people who practice it are engaged in a revolutionary process. Otherwise, so what? Let's stop!"[27]

Thus the person of the actor seems "multifaceted", in the way that he embodies a plurality of roles and gives reality to a variety of senses collectively recognized, as references for the group, translating the "spirit of the times". The specificity of the actor is that he seems to be conscious, sometimes diffusely, of being a part of the collective and also of his place as an individual. As a conscious division of the collective, the actor plays his role of interface, revealer, by restoring the sense perceived, and he gives to think, by a kind of mirror effect. The search for sense, in a context often considered, by many aspects, as much absurd, nourishes the creations as well as the actors who embody them.

Besides, the distinction individual / collective underlines issues regarding the modesty of the persons, but the question of an endoréism of the prudishness at the global level of the whole society still remains. We can only confirm that some actors keep the body at a distance, separating the body and what moves it, whether a conscious "technique" or a kind of "evil spirit", foreign to the actor. In the individualistic dimension of our modern societies, this distinction reminds that the search for sense, the focal of the understanding, concentrates at first on the very individual, who finds his place in the collective only when he conceives his place according to himself. Thus the relationship with the body, modesty which redefines each and everyone's secret garden, can take refuge behind the individual barrier. Nevertheless this border is not unambiguous, some manifestly placing the barrier of the personal garden *behind* the physical exhibition. Would this border be bound to the individual consciousness of the collective, the individual considering his (individual) body all the more prudishly or modestly that he strictly distinguishes it from the collectiveness?

Notes

1. In particular: Bouvier 2000.
2. Delbono 2004.
3. *Sciences Humaines* n° 195, Juillet 2008.
4. *Les cahiers du cinéma* is a French monthly magazine about cinema.
5. *Télérama* is a French weekly magazine about culture.
6. Mathilde Blottière et Louis Guichard, in *Télérama* n° 3057, 6 août 2008, Pg. 22.
7. Mauss 1936.
8. Emmanuelle Devos, *Comment reported by* Mathilde Blottière et Louis Guichard, in *Télérama* n° 3057, 6 août 2008, Pg. 21.
9. Paradeise 1998, Pg. 128.
10. Ibid., Pg. 132.
11. Bouvier 2005.
12. Delbono, op. cit., Pg. 32.
13. Morin 1972, Pg. 113–114.
14. Paradeise, op. cit., Pg. 129.
15. Morin, op. cit., Pg. 103–104.
16. Elias 1993.
17. Hall 1966.
18. Delbono, op. cit., Pg. 50.
19. Morin, op. cit., Pg. 113–114.
20. Itinerant show, the troop reached the places of the shows by foot, followed by vehicles loaded with the equipment...
21. Delbono, op. cit., Pg. 80.
22. Morin, op. cit., Pg. 117 (43–163)
23. Duvignaud 1965, Pg. 243.
24. In particular, this rite related by Jean Rouch in the documentary film "Les maîtres fous", where every interpreter embody a divinity, who comes to live in them and makes them act, during the time of the ceremony.
25. Paradeise, op. cit., Pg. 130.
26. Duvignaud, op. cit., Pg. 242.
27. Delbono, op. cit., Pg. 56.

Bibliography

Bouvier P. (2000), *La socio-anthropologie*. Armand Colin, Paris.

Bouvier P. (2005), *Le lien social*. Gallimard, Paris.

Delbono P. (2004), *Le corps de l'acteur ou la nécessité de trouver un autre langage*. Ed. Les Solitaires Intempestifs, Besançon.

Duvignaud J. (1965), *L'acteur, Esquisse d'une sociologie du comédien*. Gallimard, Paris.

Elias N. (1993), *Engagement et distanciation*. Fayard, Paris, 1993. (tiré de: *Problems of involvment and detachment*. The British Journal of Sociology, 1956.

Hall E. T. (1971), *La dimension cachée*. Seuil, Paris. (*The Hidden Dimension*. Doubleday & C°, New York, 1966.)

Mauss M. (1936), *Les techniques du corps*. Journal de psychologie, repris dans *Sociologie et Anthropologie*, Quadrige/PUF, Paris, 1950, "avec une introduction à l'œuvre de M. M." par Claude Lévi-Strauss.

Morin E. (1972), *Les stars*. Seuil, Paris.

Paradeise C. (1998), *Les comédiens*. Profession et marchés du travail, P.U.F., Paris.

Special thanks to Frédérique Chave and Aude Napoli for their help and their patience.

Chapter 8
Body cult and consumption: fashion and beauty in the construction of lifestyles

Ana Lúcia de Castro

Abstract

The aim of this text is to supply elements that are the basis for a discussion on body cult[1] in contemporary culture. It needs to be clarified that, in dealing with the material dimension of a behavioral phenomenon, like body cult, this study draws most of its inspiration from the proposals of Raymond Williams on the understanding of culture. This English author, looking at the issue from a largely anthropological perspective, proposes the notion of *common* and *ordinary culture*, in order to refer to the fact that the term involves the *everyday experience*, as well as the *sharing* of it, bringing together those individuals in society who are united by the common experiences with which they are actively involved. Furthermore, the author points to the idea that the term is concerned with the production of meanings and values, involving relationships between socially diverse structures and institutions.

Along these same lines, culture is seen not just as a product, but as material production and therefore as a *way of life*, in other words, social practice that articulates institutions, has political implications, drives finances and promotes and challenges values and ideas. An approach to the different types of institution and formation by which culture is produced, distributed and divulged is very promising, when it provides an analysis of the relationships between material processes and the preparation of symbolic meanings.

Starting with these assumptions and seeking to contribute to an understanding of the relationship between economics and culture this reflection seeks to present the material dimension of one of the most significant features of contemporary culture: the present-day obsession with constructing appearances, as expressed in the search for a socially established body standard. To do so we shall first of all discuss the body/society relationship, indicating the role the body plays in supporting culture. Then we shall deal with the evolution of the financial performance of the economic sectors involved with the production of appearance, above all the cosmetics and fashion sectors. The hypothesis orienting this reflection relates the obsession with constructing appearance with the complexification, fragmentation and instability that mark the contemporary condition.

Body: both a product and producer of culture

There are various studies in the social sciences that show the way in which the body is configured as a symbol of a culture, a space onto which identity and alterity codes are projected, the uses that are made of it being associated with the clothes, ornaments and body painting that indicate the symbolic universes that are capable of helping us better understand the world that surrounds us.

First of all it is worth recalling the reflections of Durkheim (1984) and Mauss (2003), who declare the prominence of the social over the individual, by claiming the social origin of every classificatory act and discarding psychlogizing explanations, which start with the idea that man classifies things and the world out of an internal need for individual understanding. For these particular authors this is a matter of finding a place from which the classificatory systems originate and, at the same time, explaining the internal logic that presides over their formation, organization, processing and up-dating. Therefore, what characterizes these classifications is that ideas are organized within them in accordance with the model that society supplies.

Mary Douglas's contribution helps with the reflection here proposed when she demonstrates the evidence of social symbolism that exists in the human body. According to this author the public rituals involving the body that she studied are evidence of collective and not personal interests, because if the body in question belongs to the individual who is taking part in the ritual, what is being engraved on the human flesh is the image of society. (1976:143). There is a type of stock of socially created symbols that appear in rituals and that in turn represent the types of social relationship that allow individuals to understand their own society. In the author's own words:

> *Just as it is true that everything symbolizes the body then it is also true that the body symbolizes all the rest. From this symbolism, which from layer to layer of interior meaning refers to the experience of self with its body, the sociologist finds justification for removing some samples of the insight into the experience of self in society.* (Douglas, 1976:150).

Along the same lines we have Marcel Mauss, who when he coins the concept of corporal techniques, defines them as the *ways in which men, in society after society and in a traditional way, know how to make use of their bodies* (2003:401). Drawing attention to the understanding of bodily practices and their social origins and implications, the author indicates the body as the symbolic framework of society, a sign of its collective representations. An important contribution was also made by Pierre Bourdieu (1988), when he showed that body language is a mark of social distinction. Bodily communication holds a fundamental position in his argument and theoretical construction, which places food and cultural consumption and the way they are presented (including the consumption of clothes, beauty articles, hygiene and care and manipulation of the body, in general) as the three most important ways of distinguishing oneself, because they reveal the deepest determined and determinative structures of the habitus. In the words of the author:

> *The body is the most irrefutable objectification of class taste that manifests itself in various ways. In the first place, with regard to what is most natural in appearance, in other words, the dimensions (volume, stature, weight) and shapes (round or square, rigid or flexible, straight or curved, etc...) of its visible conformation, but that expresses in a thousand ways a whole relationship with the body, i.e. a whole way of treating the body, of looking after it, of feeding it and of maintaining it that reveals the most profound dispositions of the habitus.*
> (Bourdieu, 1988:188)

The reflections of Michel Foucault (1979) made a big impact (and still do) on studies on the body in the social sciences. In his work knowledge is pointed out as being an important power mechanism and a way of guaranteeing the development of industrial society. In coining the concept of bio-power (defined as a disciplinary technology aimed at achieving the utilitarian control of the body, thereby seeking to maximize its efforts and reduce it to the dimension of political economy) Foucault reveals that disciplinary technologies focused their power on the body, the main purpose being to train and make individuals docile in order to extract from them the forces necessary to serve the interests of capital. This is the body fully managed and put into service to ensure the great social driving forces function well.

Deriving from the reflections of Foucault are the theories about the man-machine, the idea that the body is a product of scientific manipulation (Rouanet, 2003), as well as about the post-human, or the possibility that humans have advanced in the evolutionary scale of the species and are now beings of another type, from the cognitive, cerebral and psychological points of view, beings that bear a new *sensorium* in the Benjaminian sense of the word, inhabiting the planet as if they are members of a last generation of humans (Kahn, 2003).

Leaving aside any elaborate speculation bordering on science fiction, what we are able to gather from the contributions made by the social sciences on the body, and here briefly related, is the fact that the body, its practices and representations can only be understood when they are connected to the social dynamic, in other words, understanding what a society does with its bodies and what is valued or devalued, is understanding a lot about the logic that constitutes the same.

Among authors who debate this theme there is agreement on the idea that concern with the body has been assuming an increasingly central place in contemporary social life. This centrality has even been inverting the body-fashion relationship to the extent that for the last thirty years or so of the 20th century it is no longer clothes that projects the body but it is the body that has started defining/molding fashion. First and foremost, it is necessary to *sculpture* an extremely slim body *in order to be in fashion.*

Body, fashion and identity

Concern with the appearance is linked to the split between being and appearing, which is typical of modernity. Fashion, which involves not only clothes but everything that has to do with the construction of appearance, is growing in importance as concern with the construction of identity becomes a crucial task for individuals.

Unlike at other moments in history, modern societies force individuals to define their own identities, which are no longer pre-determined by tradition and custom, as in traditional contexts. In a society that is constantly changing, the individual must rethink himself/herself and redefine their identity ever more frequently.

According to Anthony Giddens, within the context of modernity *self*—understood as self-identity—is the product of a reflexive project in which individuals take primary responsibility for themselves. In this perspective *reflexivity* has an important influence on the dynamic of modern life since *it has to do with the possibility of most aspects of social activity and material relationships with nature being radically reviewed in the light of new information or knowledge* (Giddens, 1997, 18).

We can clearly see an effort by the author to throw into relief the ephemeral nature and multiplicity of spaces and institutions that mark the modern condition, situating the individual as an active being in this process. Given the multiplicity and segmentation of scenarios that constitute social life, lifestyles are configured as spaces, or specific *action environments,* which the author calls *lifestyle sectors. In part due to the existence of multiple action environments, the choices and activities of lifestyle frequently tend to be segmented as far as the individual is concerned: the forms of action followed within a particular context may vary more or less substantially in relation to those adopted in other contexts.*

The constant process of redefining self implies the construction of a *lifestyle* in which the body assumes the central role: body regimes, such as diets and fitness, as well as the organization of sensuality in late modernity become open to continuous reflexive attention against a backdrop of a plurality of choice.

Therefore, the construction of *appearance*—involving adornment, postures and ways of dressing—begin to depend increasingly on body shapes and volume and become the central element in the reflexive project of *self;* this is the reason for the growing importance of fitness.To these practices we can add dressing, certainly one of the most important forms of development of the *auto-narrative of self,* or of the constant (re)definition of identity in the contemporary situation. For Giddens (1997), both the planning of life and the adoption of lifestyles have become integrated into bodily regimes, in other words, faced with contemporary complexity and fragmentation the body becomes the central element in the search for more stable meanings and points of references, perhaps because they are part of the only domain that is still controllable by individuals.

Body cult in numbers: the recent evolution of the beauty industry

In pointing out the material dimension of a phenomenon studied from the behavioral point of view we start with the assumption that cultural manifestations are constituted at the point where the two dimensions—symbolic and material—overlap. One must try, therefore, not to lose sight of the links between the field of symbolic elaborations and their base material, in which the social relationships between the players and economic institutions take place.

We can identify three diffusing and legitimizing instances of a slim and straight body standard in contemporary life: the culture industry, the fashion industry and the cosmetics industry. For purposes of this reflection we shall concentrate on the latter two.

Over the last decade of the 20th century and the first few years of this century, all sectors of the economy involved in the production and/or maintenance of beauty have been experiencing significant growth. Data presented by *ABIH-PEC* leave no doubt as to the impressive performance of the sector responsible for manufacturing personal hygiene, products, cosmetics and perfume, one of the main sources of profit for the beauty industry. In terms of revenue, in the period between 1991 and 1995 the sector grew by 126.6%, going form US$ 1.5 billion to US$ 3.4 billion. The strong growth trend continued into the second half of the 1990s and the first decade of the new century. Net revenues jumped from R$ 4.9 billion in 1996 to R$ 8.3 billion in 2001, and reached R$ 17.3 billion in 2005. In the first half of this decade (between 2001 and 2005) average revenue growth in the sector was 10.7% a year.

Until the middle of the 1990s the great increase in net revenue of companies in this sector can be attributed basically to three factors: (a) a reduction in the tax burden on cosmetic products[2], which motivated greater investments, a real price reduction and a broadening of the consumer base; (b) the impact of the *Real Plan*, which increased the purchasing power of a population that until then had been strangers to the cosmetics market; (c) the reduction in the tax payable on sales of cosmetic products, which expanded the consumer base even further and the per capita consumption of those who were already consumers.[3] To these we can add some more recent factors, such as the growing interest of men in beauty products,

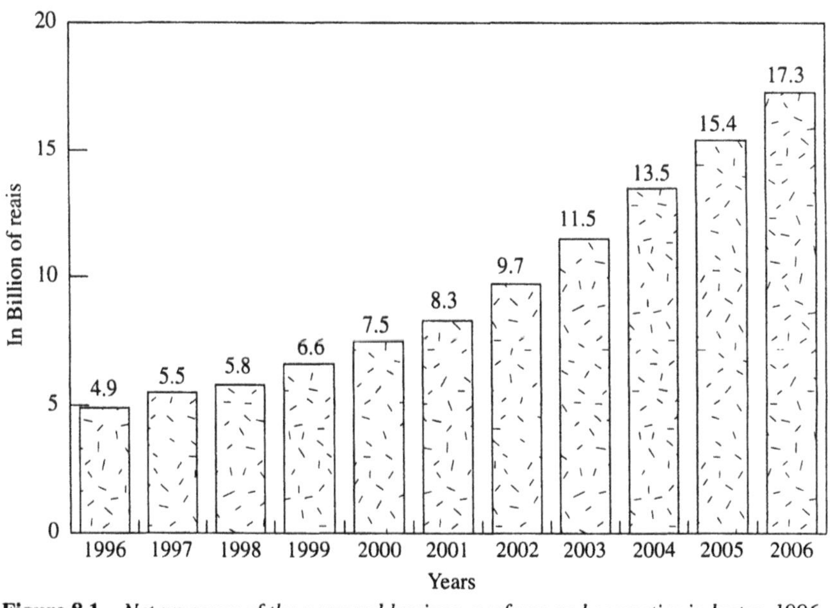

Figure 8.1 *Net revenues of the personal hygiene, perfume and cosmetics industry, 1996–2006.* Source: *ABIHPEC/SIPATESP, 2007.*

the increase in life expectancy, associated with the need to appear young, and the increase in productivity with the intensive use of technology, meaning that sector prices have been rising at a slower rate than inflation (Dweck, 1999).

To better understand the growth dimension of this sector it is necessary to put it in context relative to the global economy by comparing it with the Gross Domestic Product and with the performance of all industrial sectors. Figure 8.2, containing data from a study carried out during the 1990s, reveals the impressive performance of the sector, with growth levels higher than those of industry as a whole and of the Gross Domestic Product. The same trend has continued in the 2000s, with a growth rate greater than that of the GDP. Between 2001 and 2005, according to *ABIHPEC*, while average growth was 10.7% a year, the GDP grew by an annual rate of 2.2%.

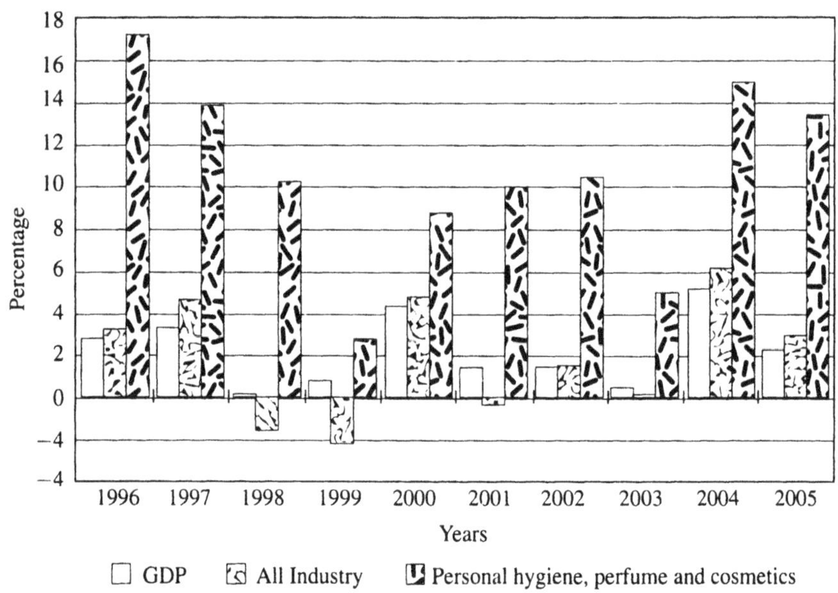

Figure 8.2 *GDP growth as a %, for all industries and for the personal hygiene, perfume and cosmetic industry (1995–2005).*

The level of employment is a great indicator for explaining the sector's performance and although it has its limitations it can be used as a point of reference. The fact that the level of employment falls in an industrial sector does not necessarily imply a fall in production and/or revenue, although growth in the level of employment implies an increase in production.

If the industry trend as a whole has been one of a systematic drop in the level of employment since 1985, the trend in the particular sector we are looking at has shown the opposite tendency, having grown constantly (albeit with a slight drop in 1992 and 1997, which can be attributed to the impact of the Collor Plan and the Asian crisis), and in 1999 was employing 52.41% more people than the average

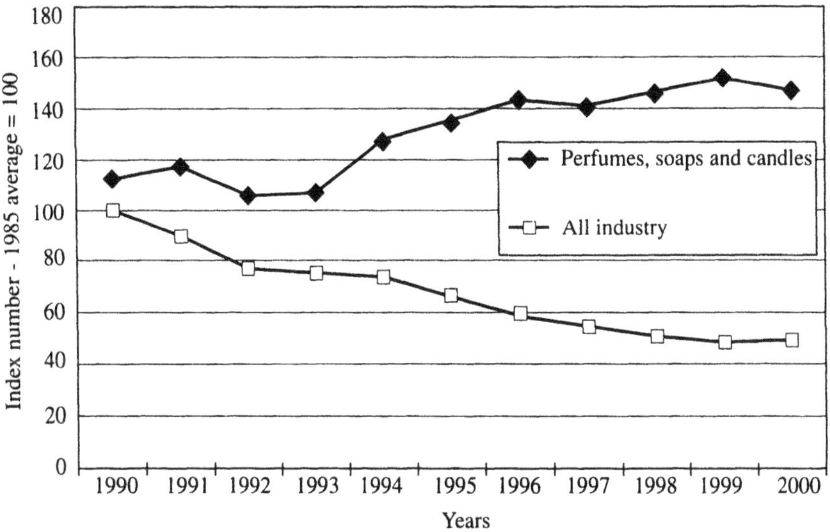

Figure 8.3 *Industrial employment level index: 1990–2000. Index number - 1985 average = 100.* Source*: DEPEA - Department of Documentation, Research and Evaluations. FIESP/CIESP.*

number employed in 1985, while in the same year there were almost 50% fewer people employed in industry as a whole than were employed in 1985, as shown in Figure 8.3.

The reduction in the number of jobs in industry, allied to growth in the third sector, is one of the aspects of *post-industrial* societies. Domenico De Masi (1999) points out that the fundamental factors in the transformations that have occurred in the world of labor in current societies are the growth in the third sector and the reduction in the employment capacity of industry:

As indicated in Figure 8.4 there has been a recomposition in the participation of each segment in the personal hygiene, perfume and cosmetics sectors. The items considered closest to the category, '*superfluous—perfumes and cosmetics—* increased their share in the sector's revenues, while *personal hygiene,* which is closest to the category of *goods of primary necessity,* reduced its share, which indicates a change in consumer behavior, which started valuing and incorporating these items into their consumption basket.

Comparing national numbers with the worldwide trend we can see the occurrence of a similar movement, with the participation of the *perfume* and *cosmetics* segments increasing and the *personal hygiene* segment reducing, as Table 3 indicates:

Brazil is the world's fifth largest market for cosmetics, the forth for shampoos and the third for conditioners, although according to the manufacturers consumption per capita is still low. Brazilian industry has concentrated its investments in the so-called global brands, in other words, products that are also sold abroad. The main companies that have been redirecting their investments in order to ex-

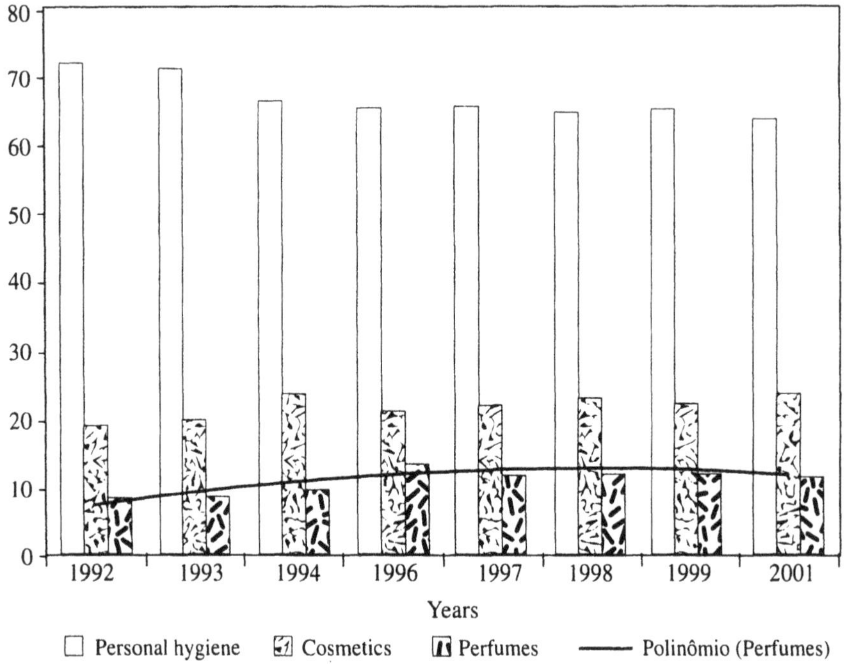

Figure 8.4 *Share by segment in personal hygiene, perfume and personal cosmetics sector revenues: 1992–1999.* Source: *Publicity handout from ABIHPEC/SIPATESP, Jan/Dec, 1997 and SIPATESP Report Dec/2000.*

tend their penetration abroad of products such as shampoos, conditioners and skin creams are Avon, Proctor & Gamble, Gessy Lever and Colgate-Palmolive.[4] In line with the trend in the sector as a whole the revenues of Avon, leader in the Brazilian cosmetics market, grew considerably in the 1990s, as shown in Figure 8.5. Brazil, which historically was 18th on Avon's world ranking, has held the second position in this brand's sales since 1996.

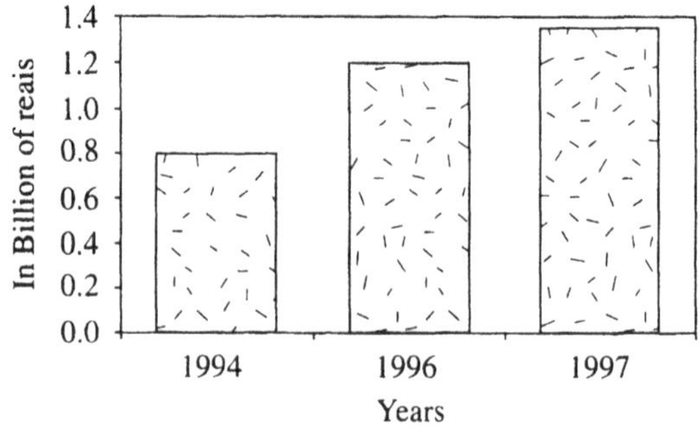

Figure 8.5 *Avon revenue.* Source: *Folha de São Paulo - Business Supplement - 8/11/1997.*

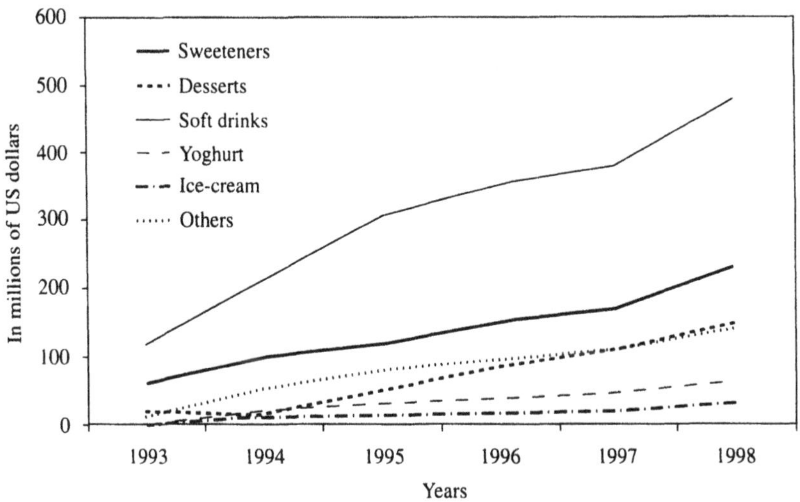

Figure 8.6 *Diet food sector revenues.* Source:*ABIAD, 2000.*

World revenues for the cosmetics industry grew by R$167 billion between 1990 and 1998. In Western Europe sector revenues were US$ 64 billion in 1998, growth of 15% relative to 1990. In the NAFTA region growth was 30% and in Japan, 21%. These three regions together were responsible for 75% of the sales volume in the sector in 1998.

The Brazilian market represents 4% of the world market, a share that is much greater than that characterized by any number of other products (1% to 2%). As a consumer market Brazil jumped from the sixth position it held in 2000 (behind the USA, Japan, Germany, France and the UK) to third in 2007, behind only the USA and Japan.[5]

The major international companies in the sector are basically organized in two ways: major conglomerates, which operate in different sectors of the economy, the main ones being Unilever, Procter & Gamble and Johnson and Johnson, and companies, the activities of which are concentrated in cosmetics and in some cases also in perfumes, like L'Oreal, Coty and Avon.

As to the performance of the diet foods sector, another important source of business in the fitness and beauty market, it is worth pointing out that, according to data from the Brazilian Association of Diet Food Industries (*ABIAD*, 1996), in December 1991 diet foods were consumed by 55% of the Brazilian population, most of them women (60%), in the 18 to 34 age band (69%) and belonging to Class A/B (63%).

Revenues in the diet food sector grew by 420.47% between 1993 and 1998, with the *desserts* segment having the most significant variation (640%). *Soft drinks* represent the segment with the biggest share in revenues in the sector: 43%. In the *services* sector, fitness centers and plastic surgery clinics are the most significant institutions in the beauty production market. According to the Union of

Water, Air and Land Sports Establishments of the State of São Paulo (*SEEATESP*), although there are no accurate figures it is estimated that the hard physical exercise market in Brazil turns over R$ 2 billion a year.[6] Brazil is already the largest fitness center market in Latin America and the second largest in the world by number of fitness centers. While in 1999 there were approximately 4,000 centers today there are more than 7,000, frequented by more than 2.1 million people. More than 50% of them are in the south-east, particularly along the Rio-São Paulo axis, 25% are to be found in the south and the other 25% are divided between the north, northeast and mid-west regions.

As far as plastic surgery is concerned it is worth remembering that Brazil is currently the world's largest market for this type of surgery, having overtaken the United States that used to lead the ranking of surgery for esthetic reasons. According to information supplied by the Brazilian Society of Plastic Surgery, in 2000, *350,000 people underwent at least one surgical procedure that had an esthetic purpose. If we do the sums this means that for every group of 100,000 inhabitants, 207 people were operated on in 2000. The United States, the traditional leaders in the ranking, recorded 185 people operated per 100,000 inhabitants in the same year.* Table 8.1 shows the evolution in the number of plastic surgeries carried out in Brazil.[7]

Table 8.1 *Number of surgical interventions for esthetic purposes carried out in Brazil.*Source: *Press Liaison Office - Brazilian Society of Plastic Surgery.*

Year	1994	1995	1996	2000
Number	100,000	120,000	150,000	350,000

There are currently 6,000 plastic surgeons in the country, but only 50% have the title of Specialist, in other words, a license to carry out surgery. Of the plastic surgical operations carried out in Brazil, 80% are for esthetic purposes, while only 20% are reparatory, i.e. they have a direct bearing on health.

Women, more so than men, have always sought out this type of surgical intervention. Recently the difference in the distribution of the number of surgeries carried out by gender has reduced. If in the early 1990s the proportion was 85% women and 15% men, today it is 70% women and 30% men. With regard to the most common types of surgery carried out by sex, in women, breast, abdomen and face operations predominate, while in men it is the stomach, eyelids and gynecomastia (removal of over-enlarged mammary glands).[8]

The growth in the fashion industry is another indicator of the contemporary obsession with corporal beauty care and presentation. The explosion in undergraduate courses in fashion in the country and the impact and repercussion of sector events, like the São Paulo Fashion Week, bear witness to the expansion of the concern with fashion within social life. In order to have an idea of the economic importance of SP Fashion Week, it is worth remembering that it drives business in the textile chain, the retail trade, service industries and exports. Just in space in the media SPFW is responsible for turning over more than R$ 600 million a year and every edition of the event has a budget of R$ 6 million. Unlike other fashion weeks, such as Paris or London, in which it is necessary to spend something around US$ 130,000, including renting space and hiring professionals to assemble the rooms where the fashion shows are held, fashion week in São Paulo is fully sponsored.[9]

The growth in the beauty industry, involving other activity sectors, cosmetics, plastic surgery and fashion, is a strong indication of the centrality that the body cult has been assuming in contemporary culture, which is becoming increasingly difficult to 'read' and interpret. Returning to the socio-anthropological principle that the body is the greatest sign of collective representations, we are trying to understand this real obsession with the construction of appearance as a strategy for constructing identity in a society in which fragmentation and ephemerity are increasingly taking the place of points of reference that were previously a little more fixed and stable.

It is worth pointing out that analysis of the economic sectors linked to the construction of appearance is taken as the starting point for a reflection about body cult, a worldwide phenomenon, regulated by symbolic representations about the body that is socially accepted as being beautiful and healthy, which is one of most significant cultural features of current society. Faced with the complexity of the links in this relationship between the economic and the symbolic, many questions still remain, awaiting responses that are a little more definitive. The only certainty we have is that revealing these links is one of the most enlightening keys there is when it comes to understanding contemporary culture.

It was with this in mind that we carried out field work of a socio-anthropological nature in 4 fitness centers in the city of São Paulo, Brazil. In these places we observed and interviewed people that frequent them and take part in five different types of physical activity: yoga, *capoeira* [*traditional Brazilian martial art/dance exercise*], body-building, gymnastics and gymnastics for the third age. It is worth highlighting that the body cult, as we understand it, is not limited only to the fitness center and involves the consumption of food, cosmetics and clothing. The fitness center was taken as the empirical working universe because it is a privileged place for understanding the phenomenon, since the people who frequent them religiously practice some type of physical activity.

The research field

In collecting the data we used three research instruments: observation and recording in a *field diary, a self-completed questionnaire*, mostly consisting of closed questions, thus seeking to build up a social-cultural profile of those who take part in each modality studied, and *interviews*, using a semi-structured script, with those who frequent the centers and with the instructors, with the aim of investigating in depth the issues detected during our observations and when the questionnaires were being applied.

Based on a map of the fitness centers that exist in São Paulo we selected those that have the following criteria:

- *Location*: fitness centers in the center and on the outskirts;
- *Modalities offered*: centers that, in addition to *physical exercise* (aerobics, local impact, step), offer courses that are of a *therapeutic* nature or are concerned with body awareness (yoga, hydrotherapy, tai-chi-chuan) and martial arts (practices that are oriental in origin, which combine self-defense, physical conditioning and a philosophy, the principles of which are equilibrium and physical and psychological well-being);
- *Size*: we sought to ensure we considered both large and small establishments, based on the number of students and the area in m^2.
- *Price*: we selected places charging different prices.
Therefore, the fitness centers visited were:
- *Formula*: located in an elegant shopping mall, situated in an elegant district in the city, this is the largest fitness center in Latin America (4,000 students, 60 instructors and covering $6000m^2$) and one of the most widely talked about in the city. It always has some innovation in terms of *fitness*: the most modern apparatus, the most complete training programs and exercise clothes that become fashionable; it is also the most expensive of all. The range of options is large: gymnastics, aerobics, low impact exercises, step aerobics, stretching, hydro-gymnastics, swimming, body building, judo, karate, jiu-jitsu, kung-fu, tai-chi-chuan, Olympic gymnastics, athletics, squash and jazz.
- *SESC- Pompéia*: located in a middle class district it is known for its alterative role in the city's cultural scene. The SESC sporting complex offers: gymnastics, sporting interaction, hydro-gymnastics, swimming (children, adult and third age), *capoeira*, judo, tae-kwon-do, kung-fu, tai-chi-chuan, yoga and sports' courses, such as basketball, volleyball, handball and dancing (belly, flamenco, afro, jazz, ballroom).
- *Dandy*: located on the outskirts of the city, it has 200 students and 7 instructors. It offers gymnastics, aerobics, low impact exercises, body-building, jazz, swimming, yoga, hydro-gymnastics, judo, karate, kung-fu, ballet, Olympic gymnastics and ballroom dancing.

Analysis of the interviews and the forms pointed to the need to consider a space frequented mainly by men, with the aim of exploring in more depth the discussion about male gender. Therefore, we subsequently included Bad Tiger in our sample; it is located on the outskirts in the north of the city and offers body-building, its main activity, as well as aerobics, karate and *axé [type of dance];* it has 120 students and 5 instructors.

We grouped the body cult activities into four modalities: *fighting/self-defense, dancing,* practices that mainly promote *body awareness* and practices/exercises that are exclusively *esthetic*. With the intention of considering the four groups we chose the following modalities in their respective locations:

- Fighting and self-defense: *Capoeira* - SESC, Formula and Dandy
- Body awareness: *Yoga* - SESC and Formula
- Esthetics: *Localized gymnastics*: Formula, Dandy and SESC
- Dancing: *Ballroom dancing* - SESC and Dandy

So, we had 10 groups, with practitioners from different modalities in four locations. We sought to identify the similarities, or regularities, and the differences or peculiarities of the body cult in each group we studied and, so, during the visits to the fitness centers we recorded our observations and impressions in a field diary and we interviewed those who frequent the centers and the instructors/teachers of Physical Education, who are understood to be *cultural intermediaries*, to the extent to which they *filter* their scientific knowledge about the body and its workings into language that is accessible to the students.

For the interview students who stood out as leaders (because of their extroversion, excellent attendance at classes and curiosity about my presence, which after became interest in the research) were chosen. The interviews were transcribed and organized by topic and cross-referenced with the information recorded in the field diary.

Motivation

Health, esthetics and the search for sociability networks were indicated as the strongest motivations for frequenting a fitness center, allied to a fourth reason for practicing physical exercise: *guilt,* resulting from the individual being attributed with the responsibility for the appearance of his/her body, with bodily defects and imperfections being understood as negligence and a lack of care with themselves.

> "For me this is a vice and it's almost possessive. I find it impossible not to do it; my body cries out for it. When I don't come here I feel guilty because of my laziness".[10]

> "I no longer know what it's like to live without doing gymnastics. Gymnastics leaves me feeling really great and when it ends I get that *feeling of having done my duty*. It's only then that I can go to work with a *clear conscience*, knowing that *I did something for myself on that particular day*".[11]

Sociability is another important motivation: in choosing to do some type of physical activity individuals are looking for people that have some affinity with them, it's a get-together with like-minded people, something that will help cement their social relationships, in addition to simply keeping fit or ridding themselves of guilt. Through physical activity they try to identify with others and distinguish themselves from other groups in the complex game that leads to the construction of their social identities.

To work with the hypothesis that body cult corresponds to a type of cultural consumption that expresses one of the dimensions of the lifestyle of the practitioners, we investigated their cultural tastes and preferences and we were able to perceive that each group we investigated has a peculiarity, to the extent that they share a common cultural universe that reveals itself when the space occupied in the fitness center, the type of clothes appropriate to the exercise being practiced, the music that plays in the background during the lesson and the conception of body as expressed by the group are observed. In the *body cult market* individuals define the type of physical activity to be practiced that is associated with the other *choices* made by the individual: style of musical, television program, reading preferences and types of leisure.

So those who practice yoga have the highest income and best educational levels and are characterized mainly by the *alternative bias* in their lives, so their choice of an unconventional type of physical activity is not by chance. They state that they read more books, newspapers and magazines and go to the cinema more regularly In terms of television program they mention interviews and documentaries as being the ones they most watch. Yoga is characterized, therefore, as being the activity most sought after by individuals who enjoy a higher socio-economic level, which is due to the fact that these are the individuals who have the conditions necessary for carrying out the main proposal of this particular practice: the search for self-knowledge through interiorization, which requires time and an absence of material concerns with survival.

As to this group's conception of body we noted a more apparent concern with body/mind integration, particularly more concern for the mind, or inner peace, which would presupposed as being necessary for having a beautiful body.

> "What yoga students here at Formula are looking for is to get rid *of stress*, they're looking for *relaxation and to educate their breathing*, to ensure they have time to observe themselves and calm down".[12]

> "You have that facility of dealing with people, like, sometimes you're in traffic, you see an argument, a fight, something like that and you also have something good to pass on (...) With yoga you're more *tolerant, and can stand day to day things better*".[13]

> "I was very agitated and very irritable and yoga helped me calm down, to be more at relaxed. I'm still irritable, but a lot less. I've improved a lot (...) Yoga works a lot with the *mental side, helps you concentrate more, to look inside yourself more. It's a type of self-knowledge*".[14]

> "It's *mentalization* that you're looking for and nowadays, in our agitated world, you *really need a strong mind, don't you?"*.[15]

Those who practice '*capoeira*' and who do body-building and those learning ballroom dancing are a very heterogeneous group from the age point of view, but they can be grouped together because of their desire to show off what they have learned/incorporated by practicing it. '*Capoeira*' practitioners is an adolescent group that is looking for a sociability space, an exchange of experiences and self-affirmation at a moment in their lives marked by the definition of their identities.

The ballroom dancers bring together different age bands in their search to dominate the dance steps and to gain the confidence they need to show themselves off. The group that does body-building, comprised fundamentally of men, is the one which best expresses exhibitionism as the main principal of the body cult. So, despite their age differences they are similar because they practice a type of physical activity that aims to show off in public what has been learned, as well as because they proved to be the groups most permeated by the gender question, because the first two are predominantly masculine, while for ballroom dancing the presence of both sexes is necessary. In dealing with cultural tastes/preferences these groups are in an intermediary position between *conventiona*l and *alternative*. They are used to going to the movies most of them do not like classical music and Brazilian country style music in this group enjoys less prestige than it does in the conventional gymnastics group.

But there are important differences: the young people that practice '*capoeira*' are "cool young things", who are not worried about esthetics—they have great bodies—and while they want to look after them they do not want to have to submit themselves to the mechanical movements and "contortions" of gymnastics:

> "I think it's very *square,* that thing of *everybody doing the same movements*, like *trained monkeys*. I see no fun in that at all. Here, even though we do it together, each one has his own particular body movements and when we're fighting then you can choose your blow and movement in accordance with the situation. You have to use your head, not just your body".[16]

The young *body-builders,* with their mechanical apparatus, are looking for an esthetic ideal that values the exuberance of muscle mass in order to show off their bodies in the street, at the club or by the swimming pool. This is a group that assumes the search for an esthetic ideal more clearly and with no qualms:

> "I came to do body-building in order to feel good when I look in a mirror, to go out into the street and notice that people are looking at me. That boosts my ego...".[17]

There is an assumption that the body should be exposed, like an object of contemplation and seduction.The *ballroom dancers* like traditional rhythms, do not read much and their main diversion is to go dancing. There is a more instrumental conception of the body, which is conceived of as an instrument of pleasure and expression through dance.The difference between gymnastics for the *third age* and *conventional gymnastics* is in the pace of the class: the exercises are the same, but the way they are developed is different. The purpose of the class for the third age is very different also, filling as it does the empty space left by the successive losses these people have suffered (the death of husbands/wives/friends, or children that marry and leave home). This means they need to increase their self-esteem, reorganize their lives and re-bind the ties of affection and friendship, while the conventional gymnastics group tries to work out hard and to mold their bodies, understood as being an object to be appreciated, although they conceal this expectation linked to the esthetic with talk of being in search of a healthy body.The third age and conventional gymnastics groups, although belonging to very different groups in terms of age, share a similar cultural universe: they like watching soap-operas and live TV studio shows, they do not normally read or go to the cinema and they like traditional style music, such as country and '*pagode*' [*a type of noisy and popular samba*]. The conventional gymnastics group has a peculiarity that is very revealing of this trend towards *average taste*: for all styles of music almost 50% of the replies fall into the intermediary category, '*like*'. The only exceptions are the two most modern rhythms, rap and funk, which reinforces the group's tendency towards traditionalism. With regard to their musical taste, for almost all styles the majority of the responses of the gymnastics group (35% to 50%) fall into the intermediary category '*like*'. Analysis of the information relating to the gymnastics for the third age group reveals that what the oldest really like are *traditional* rhythms: as far as the waltz, tango and bolero are concerned, on average 90% of the responses are concentrated in the categories '*like*' and '*like a lot*'. *Modern* rhythms, like rap and funk, on the other hand, are not appreciated.

Despite the specific aspects of each group we can identify common features: the conception that physical activity is a way of keeping the body looking beautiful and healthy and that discipline and dedication are indispensable when it comes to obtaining a good result runs right through all the modalities investigated. The fitness center perceived as a space for socializing is also a fact that is common to all groups, which seek to make friends, find partners and people to talk to, and finally the sensation of pleasure and well-being generated by physical activity that paradoxically is stated as an obligation that generates a feeling of guilt when not practiced.

Notes

1. Body cult is understood as being a type of relationship of individuals with their bodies, the basic concern of which is modeling it in order for it to be as approximate to an established standard of beauty as is possible. It therefore involves not only the practice of physical activity, but also diets, plastic surgery, the use of cosmetic products and everything else that expresses the concern with having an attractive and/or healthy body.
2. According to the President of SIPATESP, Edson Aleixo de Oliveira, as from 7/21/92, as a result of Decree Law 609, the rates of IPI on cosmetic products were reduced from 77% to 30% or 40%, except for sun-block and hair dyes, which were reduced to 20% and 10% (interview given in February, 1997).
3. SIPATESP report, December 2000.
4. *Folha de São Paulo* [newspaper], Business Supplement, 8/11/97.
5. According to a statement from João Carlos Basílio, President of *ABIHPEC*, made in February 2007.
6. *O Estado de São Paulo* [newspaper], Franchise Supplement, 7/14/98.
7. *Veja* magazine, 1/17/2001, Pg.84
8. Information supplied by the press liaison office of the Brazilian Society of Plastic Surgery.

9. Information about São Paulo Fashion Week was taken from FORBES Brasil magazine, year 7, edition 148, January/2007.
10. Statement by Paulo, a body builder at Bad Tiger.
11. Statement by Vera, who does gymnastics at Fórmula.
12. Statement by Mônica, a yoga teacher at Fórmula.
13. Statement by Odinéia, who practices yoga at SESC.
14. Statement by Milton, who practices yoga at Fórmula.
15. Statement by Odinéia, who practices yoga at SESC.
16. Statement by Luiza, who does *capoeira* at SESC.
17. Statement by Paulo, a body builder.

Bibliography

Barth F. (1998), *Grupos étnicos e suas fronteiras.* In: Philippe Poutignat; Jocelyne Streiff-Fenart: *Teorias da etnicidade.* São Paulo: Fundação Editora UNESP.

Castro A. L. (1998), 'Culto ao corpo, modernidade e mídia'. *Lecturas: educación fisica y desportes.* Año 3, n° 9, Buenos Aires, marzo.

Castro A. L. (2003), *Culto ao corpo e sociedade: mídia, estilo de vida e cultura de consumo.* São Paulo: Ed. Annablume.

Crane D. (2006), *A moda e seu papel social.* São Paulo, Ed. Senac.

Bourdieu P. (1982), *A economia das trocas simbólicas.* São Paulo. Ed. Perspectiva.

Bourdieu P. (1988), *La distinción: critérios y bases sociales del gusto.* Madrid, Taurus.

Bourdieu P. (2003), *Culto ao corpo e sociedade: mídia, cultura de consumo e estilos de vida.* São Paulo, AnnaBlume.

De Certeau M. (1994), *A invenção do cotidiano: artes de fazer.* Rio de Janeiro, Vozes.

Durkheim È. (1984), *Algumas formas primitivas de classificação.* In: Rodrigues J. (org). Émile Durkheim, São Paulo, Ática.

Douglas M. (1976), *Pureza e perigo.* São Paulo, Perspectiva.

Douglas M. and Ishwood B. (2004), *O mundo dos bens: para uma antropologia do consumo,* Rio de Janeiro, Ed. UFRJ.

Dweck R. H. (1999), *A beleza como variável econômica: reflexo nos mercados de trabalho e de bens e serviços.* Texto para discussão, n° 618, IPEA, Rio de Janeiro.

Entwistle J. (2000), *The fashioned body: fashion, dress and modern social theory.* Oxford, Polity Press.

Featherstone M. (1992), *The body: social process and cultural theory.* London, Sage.

Featherstone M. (1995), *Cultura de consumo e pós-modernismo.* São Paulo: Studio Nobel.

Foucault M. (1979), *Microfísica do poder.* Rio de Janeiro, Graal.

Foucault M. (1985), *O cuidado de si.* In: *História da sexualidade.* Rio de Janeiro, Graal, Vol. 3.

Garcia R. (2005), *Internacionaliação comercial e produtiva da indústria de cosméticos: desafios competitivos para empresas brasileiras.* São Paulo, *Revista Produção,* Vol. 15, n° 2.

Giddens A. (1997), *Modernidade e identidade pessoal.* Oeiras/Portugal, Celta Editora.

Geertz C. (1978), *A interpretação das culturas.* Rio de Janeiro, Zahar Editores.

George M. (1991), *Identidades passadas, presentes e emergentes: requisitos para etnografias sobre a modernidade no final do século XX ao nível mundial.* São Paulo, *Revista de Antropologia,* USP, 34, Pg. 197–221.

Hall S. (2000), *A identidade cultural na pós-modernidade.* Rio de Janeiro, Ed. DP&A.

Kahn A. (2003), *Morte do sexo?* In: Novaes, Adauto: *O homem-máquina: a ciência manipula o corpo.* São Paulo, Cia das Letras.
Lipovetsky G. (1999), *O império do efêmero: a moda e seu destino nas sociedades modernas.* São Paulo, Cia das Letras.
Mauss M. (1974), *As Técnicas corporais.* In: *Sociologia e Antropologia.* São Paulo, Edusp.
Martín-Barbero J. (1997), *Dos meios às mediações: comunicação, cultura e hegemonia.* Rio de Janeiro, Ed. UFRJ.
Ortiz R. (1994), *Mundialização e cultura.* São Paulo, Brasiliense.
Rouanet S. P. (2003), *O homem-máquina hoje.* In: Novaes Adauto, *O homem-máquina: a ciência manipula o corpo.* São Paulo, Cia das Letras.
Sant'anna D. Bernuzzi de (2001), *Corpos de passagem: estudos sobre a subjetividade contemporânea.* São Paulo, Eestação Liberdade.
Souza G. de Mello (1987), *O espírito das roupas.* São Paulo, Cia das Letras.
Simmel G. (1999), *La mode. Philosophie de la modernité. La femme, la ville, l'individualisme.* Paris : Payot.
Turner V. (1988), *The anthropology of performance.* New York, PAJ Publications.
Wilson E. (1989), *Enfeitada de sonhos.* Lisboa, Ed. 70.

Chapter 9
Because also I am worthy: the male body on the stage

Salvatore Polito

Abstract

This paper aims to propose a key of reading for the recent social tendencies about the masculine body compared to the experiences of life of every men, getting further from media cliché and stereotypes which influenced too some sociological reflections on the matter. Particularly the paper show that the male body goes more and more revealing, with the knowledge that it is impossible to avoid the physical attraction in the "seduction yoke". In practice, the daily life imposes to men the obligation to attempt on the same field of the opposite feminine, adopting the same weapons. But, to be able to do it, their bodies need to continuous remodelling in order to remain attractive and efficient in pursuing the success in all ambles of the daily life, included the sexual sphere or the one of couple intimacy, in which the incoherences of late modernity and its expressions are emphasized. What emerges from the paper, in a strict synthesis and through the examination of the data of some contemplated investigations, is that the masculine body became a great protagonist on social stage.

The body, which, from many points of view appears as the most concrete and the most available reality for anybody, looks more and more immaterial, and is looked as the main investment for the future (Duret e Roussel, 2006). In front of this, the present paper proposes some references about the last proceeds of transformation of the dominant socio-cultural patterns about the representation and social use of the male body. In particular, the aim of these pages is to offer a key for reading the actual consumption tendencies about the male body, far from Media's clichés and stereotypes which have also provoked some sociological reflections (Shilling, 2005; Connell, 2006; Welzer-Lang, 2006). Reading this paper, one will understand that the male body conquests more and more the absolute center of all the possible stages (media, institutions and—slowly—academics) or, more generally, present in the indefinite forms of diary interactions, and became one of the main actors in movement on the world stage.

The particular attention paid to the body of men in this paper, is the transversal sign of a kind of lack. In the push of the feminine movement in the sixties and seventies of the former century, a great importance was given to feminine identity and to the female body; but less deep were the research about masculinity[1] and male bodies: indeed these researches were looked as a space of privacy, in which cultural prohibitions and silent taboos are still present (Pieroni, 2002; Borgna, 2005). For sociologists, to emphasize this is very important in order to describe crossings between male and female identities (de Singly, 2006), as shows the fact that belonging to one gender class is often understood as a maturation journey which aim is to define, correct or achieve in a social way what was still undefined in the time of birth or childhood.[2]

The male body as significant area in the media

Ecce Homo (John 19, 5), literally meaning "Here is the Man", is the sentence told by Pontius Pilatus to the Jews when he shows them flagellated Jesus. Christian tradition understood in this sentence the unwanted confirmation by Pilatus that Jesus is the *Man* for excellence, the type of man who represents the highest realization of the mankind. Yet, abstractly defined by bio-medicine and ideologies which take inspiration in a rigid way from the differences of nature, the body of individuals seems an imperfect body in comparison with the Jesus' one, go through in limited spaces and times which are continuously moved and re-elaborated by the culture. However, as Jesus' one, the body of every individual gives to *read* the own social identity because it is defined by a specific kind of visibility, which allows to make itself partially visible (or invisible), according different ways, through the back, the hair, the (covered or not) shoulders and so on.

The body in fashion, for instance, is socially specificated because it is shown according specific stiles and fashions, through an exclusive or/and distinctive belonging declaration. By the contemporanean fashion, either in the street or in the fashion shows, we are accustomed to transparency or visibility, to "stage" of intimate's masculine undies, usually hidden. The masculine model of the 19th century *flaneur*, a kind of crossing between bohemian and tramp (like Baudelaire, for instance), who founds his urban insertion on anonymity and distance, is now obsolete; this is demonstrated by, for instance, the spot produced in 2004 by Sloggi in which Ben Cohen, looked as one of the most sexy among rugby players of England, gives himself to admire in undies inside the undressing room of a stadium. Indeed, if some years ago, undies stores were considered by men like "off limits", nowadays they are not few to visit them. Stimulated to look at themselves—and at other en—like objects of the consumerist desire (Argyle, 1992; Pezzella, 1996; Corrigan, 2004), men grant themselves pleasures which, formerly, were considered like taboo or only feminine. And possibilities of choice don't lack: from the jean boxer with elastic belt to those of most extravagant fabrics and colors, even to items which in the past rather belong to the word of feminine undies. In Italy, it is enough to visit one of the selling places of *Calzedonia* or *Intimissimi* to understand that! In the last years, because of provocations organized by catwalks which sustained the growing of masculinity consumption, the media have promoted a new model of the male, almost proportionate to each man: the *metrosexual*. This word designs men who have their roots in the metropolis (from which comes the word "metro"). Conservative or trouble-makers (Salzman, Matathia, O'Reilly, 2007), they are also "strong consumers of cosmetics, addicted to a sophisticated shopping and with a very refined sensitivity" (De Camillis, 2005, Pg.

24). This phenomenon is in relation with many products, not necessarily directed to men, even if in these products one may easily identify aspects connected with *male* essence. This occurs, for instance, with the Antaeus perfume, the main of masculine perfumes of Chanel, which is sold in a black flask with a deep red wrap, for a "strong and seductive character man", as says the spot. If the metrosexual shows an esthetic sense and quite a maniac care of body to the point to be looked as homosexual (Voight, 2003), the *retrosexual*[3] has the best grim, strength of mind and testosterone.[4] Then, in order to replace David Beckham, icon of the metrosexual movement, one thinks to another sex-symbol: Fabio Cannavaro, "true ice eyes" Neapolitan, who wring the scepter to Beckham and was recently loudly as "virile male who never needs to ask".At a more general definition level of masculinity models, through an investigation directed about the main Italian masculine reviews, (Men's Health, For Man Magazine and Fox Uomo) and integrating the results of other investigations, Boni (2004) show that, in these reviews, like more generally in the press of the last twenty years, the dominant more often proposed are the *new man* and the *new lad*. Both results of a very often produced and reproduced representation in the media which organize commercial operations in order to open new markets get more consistent benefits. More and more, the media show the bare man, "discovering" his bodyness. The *new man* is a model which was proposed like an answer to Feminism, in the sense that it integrates some critics of this Feminism. This man tries to recover a kind of this sensitivity which, in the past, his was compelled by traditions to contain and repress; or, more respectful with women, the new man takes a part in domestic activities and children education. His expenses are moved by a specific narcissism, consequence of rediscovering sensitivity, which invites him to a particular care of his own body, as testified by practices likes physical care or strictly following the fashion, which was in the past specific to women. The *new lad*, on the other hand, was born as a reaction against Feminism with the aim to recover traditional masculinity. The *new lad*, who makes a name for himself in the 90s, is the most represented in masculine magazines. In them, he appears like the supplier of tireless income, and par excellence, for his family of family, who devotes himself to sport, eventually sharing e bar with friends. In front of the first type, the *new lad* is conservative in fashion, and his shopping is done respectfully to rigid norms which define his "virility". This kind of man, opposite to the *new man*, is shown by James Bond performed by Daniel Craig in *007 Casino Royale*, which is different of the previous actors: blond, clear eyes with a hooligan face more than of a baronet, sure more "macho" (violent, bright and almost without soul) than Sean Connery, Timothy Dalton, Pierce Brosnan or George Lazenby.Masculine magazines contain advertisement for new practices, new behaviors' and new esthetical prescriptions for the body; they promote a kind of ironical disenchantment about sexual differences between men, and between men and women. The Pirelli spot who shows the runner Carl Lewis (8 golden medal and one silver in three Olympic challenges) on vertiginous heels in the pin, for instance, is inspired by a masculine identity that is debating between the imperative of its power and the verification of its brittleness, elements that at the same time they evoke a multiplication of the competences that strengthens instead of impoverishing the masculinity: "power without control is nothing", proclaims the spot. In this case, like in many others, the masculine body reveals a mutable experience, at least as the female one, eradicating itself from the masculine / female binary opposition. In other similar spots, it appears an endless collection of bodies disclosed that are a not a rigid declaration of traditional roles anymore (the violent, raider man), but a *meaning surface* of re-writing, or, sometimes of overturn of these roles (Alfano Miglietti, 2004). Nevertheless, it is a phenomenon recent what has aimed the reflectors of the media on the most intimate aspects of the masculinity and on the bodies of the men, revaluing the characteristics of it. An example of it is given by all the declinations of the television format of the reality that, however likely or projected custom, some men traditionally put in scene a hidden aspect: their emotional sensibility. Already, the men also cry. Difficult not to remember the weeping of Baresi, nicknamed "captain of ice", to the missed victory of Italy to the Championship 1994 World USA. How to forget, then, the dry weeping, noisy of Merckx, after the exclusion from the 1969 Italy Turn for a history of doping, to the microphone of Sergio Zavoli and in direct TV?Among the many media, the publicity is that which more than all work of the selections on the partner-cultural environment in which it acts, promoting and valorizing the terms proper of that linguistic vocabulary that it creates and it uses, putting so the alone accent on some themes, concepts and mental categories, strengthening the values that it promotes and emptying of meaning those that it ignores. (Codeluppi, 1989; 2007). For these motives it is not possible not to consider the publicity as agency able to put in evidence and to the meantime to redefine the new lines of the masculinities and the bodies of the men. The publicities are a lot of that have as protagonists men that suggest to other men as you/they must appear, in public and in private, as you/they must have care of itself and his/her own body, as you/they must live, that perfume you/they must use, as you/he/she must be their body and what the elements that define their sexuality are (Boni, 2004). The spot has become by itself an action of seduction of the spectator and the most important is that it is always more directed to the male. From their count, the advertising agencies assure that the naked male "sells" and not because he were more "macho", but because, abdicating a traditional image and accepting his own part of femininity, he shows that is sex appeal is not repressed. So solicited, the L.G., a known mark of appliances, placed three models in underpants in the Galerie Lafayette close to washers and refrigerators, followed by the alpha Romeo with a motorist without any dress driving his auto.

The aesthetical visibility of the masculine body from the maquillage to the surgery

There are practical of remodeling the body that acts more or less on its surface in depth, those that go from the makeup to the plastic surgery, passing in general for its care. In front of this, even if with critical tones not fully shared, Faludi (1999) sustains that the men have become by now victims of an ornamental culture, that have turned them into the time from "male producers" to "feminine consumers". In other words, the marketing of the masculine body is become a datum of fact.

About the particular relationship that the men have with the care and the beauty of their body, from a study conducted by Dano, Roux and Nyeck (2003) it has emerged that, despite considered for custom produced destined to the women, the cosmetic today are in full expansion among the men. The market of the masculine cosmetic has passed by a 30% growth in 2001 to over 50% in 2002 to world level, whereas the UNIPRO (Association Italian Cosmetic Industries) has said that the market of the masculine cosmetic in 2005 has confirmed the 2004 consumptions and has made a billing increase of 1,7%. The Packaged Fact (American society of searches of market) has esteemed for 8 billions and 800 millions dollars the world turn of business for masculine cosmetic in 2004, so much that the UNIPRO has said, for instance, that the consumption of the creams for the treatment reserved to the men has grown of 17, 5% (in 2004 growth had been of 30%) the 2005 bringing the total value of the masculine line to around 287 millions, with a growth of 2, 8%.[5] Even the Corriere della Sera has put in evidence this tendency, devoting many spaces of the magazine "Style" to publicity and services contemplated on lifting, peeling, anti-wrinkle creams for men.The success of the cosmetic destined to the men shows the slow-modern importance of the to appear and of the body "cosmeticated" and exploited of the men (Maffesoli, 1993; Lipovetsky e Roux, 2003). According Dano, Roux and Nyeck (2003, Pg. 506–509), more the use of the products by men is intensive (*passionate experience*) anymore, more it draws near us to the universe of the cosmetic, while the use is limited (*rational experience*) anymore, more the references of the universe are solicited "care and hygiene". In this dichotomy the relationship manifests itself between care of looking and masculinity, a relationship that is subscribed in the traditional masculine / female opposition. In fact, the binary distinction in this relationship is so marked to differentiate in a clear way the more emotional and female universe of the cosmetic (beauty) from the more functional and masculine one oriented towards hygiene and care (health).In the heterosexual communities and potentially sexist, nevertheless, the consumption of cosmetic from the men still suffers the effect of strong taboos. According Dano, Roux and Nyeck (2003), the heterosexual males are little that admits the use of the cosmetic, while for the gays the experiences of the cosmetic are more passionate and they involve more openly. From the analyze done by Dano, Roux and Nyeck, it emerges that the men find normal to care themselves in order to appear at the top or simply because they like it, but the experiences that they have with the cosmetic are of a rational type: "it is a matter of dose... if I put a normal quantity, it is not seen... I think that the man in whom it is seen that it has the fund shade... he will be judged"(Pg. 510) is the common thinking. In substance, it seems that the cosmetic and the worry for the beauty still stay, for big part of the heterosexual men, territory of sovereignty of the women or, even, of the gays. It is for this motive that the aesthetical matters that reenter in the care of the body are mainly looked by the men with a kind of "medical dimension". The linguistic physician register, sometimes mitigated by irony, allows them to treat the *female matters* (beauty and care of the body) inserting them in the masculine universe of the health.Let us enter the detail with some examples. The depilation first of all, that, as masculine practice, surely has a symbolic dimension, not so much because it is opposite to the social prejudice that typically considers it as a female ritual, but much more because shagginess has been for long time one of the highest symbols of the virility. To every way, the increasing appeal to the depilation from many men doesn't represent so much a discussion of the "vireo-control that replaces the patriarchy" (Badinter, 2004 ps. 32), but much more an adjustment to the new aesthetical canons, mainly directed to the hedonism: for some people, a masculine body depilated looks more beautiful and more pleasant to the sight and the touch. Another example is given by the cuts of the hair; this matter is a fundamental factor in the representation of the masculine bodily ideal. In truth, according to Synnott (1993), hair is loads of social and cultural meanings. First of all they are a clear indicator of kind and model of masculinity, distinguishing the main models from those who are inferiors or secondaries, whereas for every hair style the society prescribes different behaviors for the men and for the women. Obviously there are exceptions, from the moment that there is always somebody who rebels against the conventional norms. Exceptions that becomes constant in some western territorial contexts, underlining that the men are also released by the tradition. It is enough to think about the fact that if in the past the man went to the barber and the woman to the hairdresser, now and more and more, the hairdresser indiscriminately serves women and men.The aesthetical surgery, that once was almost exclusive prerogative of the superstars of the show, if not considered a sign of purely female vanity, has assumed notable importance for many men, for which the plastic form of the bodies and the taken care of aspect are remarkable more and more in the social relationships and of job particularly. It is shown by the fact that, for some interventions, the needs of the men has overcome those of women. According to the Censis, in fact, in 2003 they have been 12.000 the men—on 150.000 interventions—that in Italy asked a plastic surgery hoping to be facilitated in the job or under the sentimental profile. Today, men of all the ages and all the social bands look to the plastic surgeon to improve their own aspect. It is the case, for instance, of the interventions affected to the chin, particularly cared by men who suffer the presence of adipose accumulations of the skin. Then, more and more in the last years, there are those who search to improve the dimensions and the definition of the muscles and of the penis. Among these, the more performed are the plant of prosthesis to the calf and the on the chest. It deals with enough recent interventions, done after long discussions with an experienced surgeon.In front of a phenomenon in growth in comparison to the preceding years, the Harley Medical Group has noticed that in 2002 the males have represented 35% of the total of the patients which submitted to interventions of aesthetical surgery in the United Kingdom. The most meaningful number of men focuses in the band of age that has gone since 22 a 37 years. To today, the five in demand interventions have been: 1) non surgical enlargement of the penis; 2) rinoplastica; 3) liposuzione; 4) otoplastica; 5) treatment of the wrinkles. While the Laser Pallaoro Medical in Padua, comparing its own data with those published by the ASAPS, has noticed that the most demanded interventions of masculine aesthetical surgery in 2005 in Italy have been: 1) the liposuzione that occupies the first place among the interventions with well 25% of the masculine patients; 2) the rinoplastica that is positioned in second with over 18%; 3) the blefaroplastica that occupies the third place with few less than 17%; 4) the reduction of the masculine breast, to correct the ginecomastia, that occupies the fourth place with 14%; 5) the transplantation of hair that maintains unchanged its own importance with 9%.[6] These data notice that the loss of control of the own body, due to the natural aging, produces anxiety around the loss of that "fresh and young" identity which is desired to maintain; and this conducts to the unbridled search of surgical methods to deceive the time. From a purely sociological point of view, this notices that the aesthetical surgery is by now independent of the kind of affiliation from the age or from the social status.

Performance, bodies and masculinity

The identities, especially in terms of alterity, "historically vary and depend on their context of definition" (Dubar, 2001 ps. 3). In front of this, the own individuality and the body have become privileged objects of comparison with the others and the world (Lasch, 2001; Diani, 2003; Melotti, 2006). Of these, however, it is the body that must be more taken care of, modeling it "in a schizophrenia in which identity escapes to every definition and is lent therefore to all the manipulations and auto manipulations"(Weil, 1990, Pg. 121–122). Shortly, the individual is hocked to mould his own body to trace its contours of an ended product more and more, a performative body that becomes the objective among the most important of his own existence.

The body is the most visible part that we exhibit or we hide to the others the mean with which we communicate: the skinniness, the roundness, the musculature, the gestures, the power, the brittleness tell the others a lot of things of (and on) us. The good-looking physical, signals for instance, what type of sporting activity we practice and as and/or with how much appointment is developed. For some men understood even that, during the sporting activity, the body is visibly marked by traumas, that are made scars by to exhibit in some cases as I mark concrete of their will power during the sporting activity, that which "more males it makes". Carrier (2002), illustrating the specific qualities of the "mole-body", the body modified because of the personal wish, brings him/it that is really to the sporting context. Both to motive the intense sporting trainings programmed and the actions of the technical support and scientific of which is object, the mole-body is intrinsically the cause, the product and the tool of the sporting performance. For its results, this performance is the expression of the extreme point of cult for the actual body.The analysis of Sassatelli (1999; 2000) of the world of the fitness, interpreting the reflections of Foucault on the discipline of the bodies, it puts in evidence the ambivalence of the abilities of which the subjects are endowed in gym and the normative-institutional dimension of its care. The gym, to the peer of barrackers, factories, hospitals and schools, can be enumerated among those institutions that make all of this that is necessary for a utilitarian productive and economic transformation of the individuals, of their energies and of their bodies. These institutions, through specific and circumstantial disciplinary practices, increase the social value of the body. Among everybody, however, Sassatelli observes that the gym has ambivalent social results: from a side indeed it furnishes the possibility to actively appropriate of knowledge and ability to the participants, from the other one it defines the contours of such abilities, contributing to define the needs of whose satisfaction founds the right legitimacy.The gym promotes a well precise ideal of bodyness and is the chosen place in which a whole aspiration (related to the body) is picked up, as a tool to use in the world, as a physical support to act daily. In relationship to this, the fitness as sporting activity in itself is specified from component as the energy, the strength, the resistance, the agility, the elasticity, the tonicity of the muscles and also the same ability to sustain the exercises. To go to gym assumes a particular value that is moral and symbolic at the same time: who hocks mainly in a "natural" considered activity toward the rediscovery of the body and its expansion, can get a better, authentic and complete personality (De Camillis, 2005) becoming master of himself in the measure in which he can enjoy of a good body, which is able in front of every situation (Sassatelli, 2002). If it is necessary to this purpose, the individual will submit his own body to extreme, even risky works. In truth, as noticed by Breton (2002), "the western societies know a profusion of physical and sporting practices that aim at a risky appointment from the individual today. They are above all the men that undertake more these activities, even if the women are not always from less. [...] to finally feel the world beat inside of oneself, to touch it with one's own hands, with the whole proper body, become some internal necessities for a big number of westerners" (Pg. 408). Who enters the world of the fitness, in short it ends up contributing "to naturalize" and to reflexively confirm the image of a busy subject to build and to improve the really itself through the body and to entirely attribute the responsibility for the care of it to oneself. This is the dynamics that the value raises performance, subscribing it in the sphere of the individuality, where the desire of men for a perfectly graven musculature becomes an aesthetical objective, up to reach the hypertrophy in the case of the body-building. The aesthetical objective, embodying a symbolic value or rather that of a "body in form", puts in relationship the physical (and its aspiration) beauty to the universe of the health. To this double (aesthetics and health) dimension it needs, besides, to add a bystander of it: the social one. The ideal of the good-looking, granted to that intimate of the performance, is manifested, in fact, also in terms of professional success and of an ampler social realization, up to involve the most intimate circles. In such sense, for instance, the sexology has signaled that the *anxiety about performance* is one of the most frequent causes of the sexual (the impotence, the precocious ejaculation) diseases, with consequent repercussions on the relational life (Fernandez, 2004).The project-idea of a well taken care of and tonic body extends also to the consumption of particular foods and more generally to all the topic of alimentation, above all when it puts in prominence the ways according to which the choices can be made, or like a base to judge the best shape and size in relation to the health of the same body. So, the building of the ideal masculine body is achieved or tends to be achieved with a control and a kind of monitoring of the food and the alimentation: reducing the rations or diversifying among the different typologies, some of which must be reduced to the least one if not straight taken off. Interesting to know this when you remember that diet has always been considered a typically female matter: the intent "to maintain the line" in order to improve one's appearance and beauty. The inversion about this prejudice is in the idea that to submit oneself to a checked alimentary regime will get benefits as on the aesthetical plan on the health one. So the men, particularly those who spend a lot of time in gym and doing sport, revalued the diet in transversal way regarding to the category of the aesthetics and performance of their own body.

Conclusions

Through the analysis of contemplated searches, all along the work it emerged that the image and the social use of the masculine body knows the influence of processes of transformation that continually happen in the context of larger cultural models. Paraphrasing Maffesoli (2005), the desire to be able in front of every situation of life pushes every man to look for, to find again the sense of his own existence right in the acting of his experience and no more in a distant and ideal result. Equally, the increasing attention towards the care of the body, always more also for men, made necessary by what Giddens (1995; 1999) defines the *privatization of the body*, appears like a compensation to the carelessness for the social body. We are probably passing, as Maffesoli (2005) supposes, *from a logic of the identity to one of the identifications*, for which it becomes necessary to do right and to know how to manage some multiple identities or to take conscience of a complex identity that every of us interprets to make his own check up, with the others and with the world in which h is living (Morin e Pasqualini, 2005). It is the plurality of the cultural contexts in which we live, the crisis of the ideologies, the hobbies, the addiction to collection, the activities developed in the social one, the tension of the physical form which produces different identities that cohabit in the same individual.The polarity born from the ethic and from the politics is progressively and definitely replaced by a polarity that articulates around the hedonism (Maffesoli, 1993), fortified by an aesthetics that globally mobilizes the senses. In other words, a kind of epicurismo of mass is appearing, the tension to rejoice day for day, for which both men and women undertake to socially display how much they are able to be masters of present sense and to enjoy it. In this which is becoming more and more a personal disposition, the body—which is looked as the main tool which the individual can more or less freely use to get a confirmation of his own reality and to get a public recognition of his own social identity—becomes a receptor of feelings that assimilates experiences, so much that its ability to be stimulated from several sides transforms it into a tool of pleasure. If we look right, the meaningful transformations of the masculine experience with its own physical character point really towards a strong hedonism. Many men realize that they cannot escape to the necessity to be physically attractive in the *game of seduction*, also whereas they reveal their intimacy with the awareness that an attractive body is not, necessarily, what is identified with the evidence of muscles, that the "beautiful body" it is not always what expresses strength, wealth or power. Re-sizing the "masculine privilege" imposes the obligation to venture him on the same ground of the female counterpart, adopting its same weapons (care and beauty of the body for the pleasure to be sexy). In this sense for today's man occurs a particular narcissism which, in its non pathological expressions, appears as a re-appropriation of the self-love and the self-respect, but above all as a request of re-appropriation of this bodyness which for so long a time has been subtracted to him. This is evident from the fact that for men the tonic body, which continually requires to be modeled in order to remain attractive in all the circles of the daily life, included the sexual sphere or that of the couple intimacy, which also contains the contradictions of the late modernity and its expressions.

Notes

1. Cf. Piccone Stella, S. (2000), *Gli studi sulla mascolinità. Scoperte e problemi di un campo di ricerca*, in "Rassegna italiana di sociologia", XLI, n. 1, Il Muilino, Bologna, Pg. 81–107.
2. Cf. Piccone Stella, S. e Saraceno C. (1996), a cura di, *Genere: la costruzione sociale del maschile e del femminile*, Bologna, Il Mulino e Besozzi, E. (2003), a cura di, *Il genere come risorsa comunicativa. Maschile e femminile nei processi di crescita*, Milano, Franco Angeli.
3. The word was created by the American journalist M. Simpson, the same who, on the online tabloid on line *salon.com* had for first time applied the word metrosexual to men of the new millennium.
4. According to some sociologists of the consumptions, to both the models him coop replacing, even if slowly, that of the *uber-sexual*, or rather a man taken care of in the aspect, but not fanatical of the appearance, and more brought above all to the dialogue with the female universe, without for this to revert in the stereotype of the anti-macho (Cf. Salzman, Matathia, O'Reilly, 2007 pps. 69-100).
5. Data's which are available on the following websites: http://www.unipro.org e http://www.packagedfacts.com
6. For demographic close examinations on the brought data, consult the following websites: http://www.chirurgiaestetica.com/news/harley.htm e http://www.pallaoro.it/chirurgia_estetica/interventi_chirurgia_estetica.htm

Bibliography

Alfano Miglietti F. (2004), *Identità mutanti.* Bruno Mondadori, Milano.
Argyle M. (1992), *Il corpo e il suo linguaggio. Studio sulla comunicazione non verbale.* Zanichelli, Bologna.
Badinter E. (2004), *La strada degli errori. Il pensiero femminista al bivio.* Feltrinelli, Milano.
Boni F. (2004), *Men's Help.* Meltemi, Roma.
Borgna P. (2005), *Sociologia del corpo.* Laterza, Roma-Bari.
Carrier C. (2002), *Le champion, sa vie, sa mort. Psychanalyse de l'exploit.* Bayard Editions, Paris.
Codeluppi V. (1989), *Consumo e comunicazione.* Franco Angeli, Milano.
Codeluppi V. (2007), *La vetrinizzazione sociale.* Bollati Boringhieri, Torino.
Connell R. W. (2006), *Questioni di genere.* il Mulino, Bologna.
Corrigan P. (2004), *La sociologia dei consumi.* Franco Angeli, Milano.
Dano F., Roux E. and Nyeck S. (2003), *Les hommes, leur apparence et les cosmeçtiques: approche socio-seçmiotique.* In: *Décisions Marketing*, n. 29, Bruxelles, Pg. 7–19.
De Camillis S. (2005), *I maschi. La psicologia degli uomini.* Edup, Roma.
de Singly F. (2006), *Liberi insieme.* Armando, Roma.
Diani M. (2003), *La base relazionale delle identità di movimento.* In: Leonini L. (a cura di), *Identità e movimenti sociali in una società planetaria*, Guerini Studio, Milano, Pg. 121–135.
Dubar C. (2001), *La crise des identités: l'interprétation d'une mutation.* Presses Universitaires de France, Paris.
Duret P. and Roussel P. (2006), *Il corpo e le sue sociologie.* Armando, Roma.
Faludi S. (1999), *Stiffed: The Betrayal of the Modern Man.* Chatto and Windus, London.
Fernandez E. (2004), *Comunicamando.* Franco Angeli, Milano.
Galimberti U. (2003), *Il corpo.* Feltrinelli, Milano.
Giddens A. (1995), *La trasformazione dell'intimità.* Il Mulino, Bologna.
Giddens A. (1999), *Identità e società moderna.* Ipermedium, Napoli.
Grosz E. (1994), *Volatile Bodies: Toward a Corporeal Feminism.* Indiana University Press, Bloomington.
Lasch C. (2001), *La cultura del narcisismo.* Bompiani, Torino.
Le Breton D. (2002), *Il corpo in pericolo. Antropologia delle attività fisiche e sportive a rischio.* In: *Rassegna italiana di sociologia*, XLIII, n. 3, Il Mulino, Bologna, Pg. 407–427.
Lipovetsky G. and Roux E. (2003), *Le luxe éternel: de l'age du sacré au temps des marques.* Gallimard, Paris.
Maalouf A. (2005), *L'identità.* Bompiani, Milano.
Maffesoli M. (1993), *Nel vuoto delle apparenze. Per un'etica dell'estetica.* Garzanti, Milano.
Maffesoli M. (2005), *Note sulla postmodernità.* Lupetti, Milano.
Melotti M. (2006), *L'età della finzione. Arte e società tra realtà ed estasi.* Sossella, Roma.
Morin E. and Pasqualini C. (2005), *Riscoprirsi identità complesse.* In: *Studi di Sociologia*, n. 4, Vita e Pensiero, Milano, Pg. 411–419.
Pezzella M. (1996), *Narcisismo e società dello spettacolo.* Manifestolibri, Milano.
Pieroni O. (2002), *Pene d'amore. Alla ricerca del pene perduto. Maschi, ambiente e società.* Rubbettino, Soveria Mannelli.
Righetti P. (2003), *L'uomo, sesso "debole" della pubblicità. Iconografia dei generi e rappresentazione pubblicitaria.* In: *Tendenze. Consumatore & Consumi*, n. 27, Allaxia GPF, Milano, Pg. 3–4.

Salzman M., Matathia I. and O'Reilly A. (2007), *Il futuro del maschio*. Corbaccio, Milano.
Sassatelli R. (1999), *Interaction Order and Beyond. A Field Analysis of Body Culture within Fitness Gyms*. In: *Body and Society*, Sage, UK, n. 5, Pg. 227–248.
Sassatelli R. (2000), *The Commercialization of Discipline. Fitness and its Values*. In: *Journal of Italian Studies*, Special Issue on Sport, Stony Brook, NY, n. 9, Pg. 332–349.
Sassatelli R. (2002), *Corpi in pratica: "habitus", interazione e disciplina*. In: *Rassegna italiana di sociologia*, XLIII, n. 3, Bologna, Il Mulino, Pg. 429–457.
Shilling C. (2005), *The Body in Culture, Technology and Society*. Sage, London.
Synnott A. (1993), *The Body Social. Symbolism, Self and Society*. Routedge, London-New York.
Voight R. (2003), *Une femme comme les autres. Beaux Arts Magazine*. Hors série, La Société de la Fashion, n. 8, Octobre, Pg. 86.
Weil P. (1990), *Il nuovo Narciso*. Franco Angeli, Milano.
Welzer-Lang D. (2006), *Masch*.

Chapter 10
Spectacular boobs. A indepth look at sex-themed restaurants and their female waitresses

Natalie Haber-Barker

Abstract

Sexuality has become the desirable, the unknown, the forbidden and these identifying factors lead to its appeal. The market of selling sexual desire enables precisely this model. One aspect of the selling of sexual desire is sexually themed restaurants such as Hooters, the past Playboy Bunny clubs, Casinos, etc. The secret weapon to these restaurants drenched in desire is the waitress. From the Playboy Bunnies to the Hooters girls, the women are recruited to sell food, drinks and sexual desire. The most important attribute being the waitress and her costume. Horkheimer and Adorno (1972) explain this process as the spread of commodification and domination of money which transforms cultural forms into external and autonomous objects. For example, a women's body is transformed into a desirable, appealing and sexually gratifying commodity. The women, their attributes, in particular their large breasts and svelte body are the product within the culture industry, which in turn becomes its own advertisement." (Pg. 129–130) This paper will examine sexual desire in a capitalist economy and explore the discourse of sexual desire, the creation of the spectacle, and the consumption of the sex-themed leisure industry. More specifically, I will examine sexually themed restaurants such as Playboy Bunny Clubs, Hooters Restaurants while focusing on their all female wait staff.

Sexual desire as a commodity: "Sell me some sex today"

Foucault (1978) points out by creating a specific discourse around sexuality, it makes sexuality a public good and sexually themed restaurants used this thought process to their advantage. They used current ideas on sexual freedom and current economic situations for women waitresses to build their empire. They enticed us with the promise of fun, relaxation and release and promoted their establishments as an escape from our everyday reality to places fulfilling our inner needs and desires. Spokeperson Mike O'Neil (hooters.com) states, "Sex appeal is legal and it sells. Newspapers, magazines, daytime talk shows, etc. consistently emphasize a variety of sexual topics to boost sales. Hooters marketing emphasizes the Hooters girl and her sex appeal." This marketing strategy is nothing new, early literature discourses depict women as serving wenches.

French restaurants in the middle of Pleistocene featured elaborately painted walls with wildlife and boosted consumer attendance by featuring topless or nude serving girls, chosen for their robust figure which was popular during that era (Nicolaides, 2004).

As the sex-themed restaurant continues to sell us sexual desire, and we gladly consume it with promises of fun and stimulation, they create themselves as a bourgeois class. They are the owners of the means of social production and employers of wage-labor. They exploit women's bodies for the realization of profit. They entice their employees with the promises of good tips and added bonuses of boob jobs.

For instance, three out of the seven women I talked with while working at Hooters had gotten a Hooters store manager to "sponsor" their boob jobs. Hooters store managers sponsor boob jobs by signing credit loans and supplying down payments to plastic surgeons (Haber, 2000). The sex restaurant bourgeois continue to create more massive and colossal productive forces than their preceding generation.

For instance, Hooters has 375 locations in 46 States and Hugh Hefner is opening a Playboy Bunny Club casino in Las Vegas in 2006 (Hooters and Hefner, 2004 websites). As Marx (1867) discusses, the bourgeois creates enormous cities and has thus rescued a considerable part of the population from rural life. The popularity of Hooters throughout cities and small towns represents this phenomenon.

Additionally, Hugh Hefner has even created a virtual Playboy Club online, which allows customers to relive those wonderful days as keyholders in the comfort of their own homes. They have created an industry reaching not only rural individuals but people all over the world. For example, Hooters has thirteen international stores ranging from Austria to China. They recently opened a store in India and are scheduled to open one in Zagreb, Croatia.(Hooters and Hefner 2004, websites)

The sex restaurant industry (bourgeois) has torn away from the individuals their private sexual veil and reduced sexual desire to a money relation. Notably, like any bourgeois, they cannot survive without constantly revolutionizing the instruments of production, and thereby, the relations of production and with them the whole relations of society. For instance, Hooters not only sells food but T-shirts, jackets, baby bibs, golf bags, etc. They have an airline which flies to certain destinations in the United States. The Playboy Club is opening a site in China and holds elaborate reunions for all their ex-bunnies in Las Vegas and other international sites. (Hooters and Hefners 2004, website) Also, the bourgeois has given a cosmopolitan character to production and consumption. Hollywood actresses as Pamela Anderson and Lauren Hutton attribute there success to being a former Hooter girl or Playboy Bunny.

New avenues of business, and expansion of their markets which include dubbing Hooters a family place and making Playboy Bunny costumes in child sizes, create a need of constantly expanding market for its products chasing the bourgeois over the whole globe. Marx states, "It must nestle everywhere, settle everywhere, and establish connections everywhere."

Sexuality and ideology: "Why does it make us feel so good?"

In a market flooded with exploitation, not only of the women doing the job but also the individuals consuming it, I wonder how this actually constitutes sexual freedom? In other words, why do women become Bunnies or Hooters girls and for that matter why do we buy into it? As Marcuse (1955) points out, individuals identify themselves with the existence which is imposed upon them and have it in their own development and satisfaction. This identification is not an illusion but reality. Of course, the reality constitutes a more progressive stage of alienation. In other words, sexual desire becomes entirely objective, and the alienated subject or rather the women performing the work are swallowed up by its alienated existence. For example as Kathryn Leigh Scott (1998) writes in her memoirs, *The Bunny Years* , "In its own peculiar way, the Playboy Club in the mid-60's served as a magnet and crucible for women independently driven to achieve remarkable levels of success later in life. As the profiles demonstrate, the very impulses that led intelligent, ambitious women to defy convention and choose to work as Bunnies would later propel many of them into vastly rewarding, and often unconventional, personal careers." In other words the ideas of sexual freedom were propelled by the creation of specific sex symbols like the Playboy Bunny and later the Hooter girl. The women are dazzled by the prospect of freedom, wealth, and success. As one ex-Bunny states (ex-Playboy Bunny website), "I was making four times what I was making at Douglas (firm she worked for) by working for Playboy, plus I had time to spend with my daughter who was still under a year old. I was a single mom and determined to provide the best of everything for my little girl, Melissa." Marcuse (1964) describes this process as the false consciousness of their rationality becoming their true consciousness. More specifically, the women believe being a Bunny or Hooter girl represents sexual freedom and success. It's important to point out that the absorption of ideology into reality does not signify the end of ideology. On the contrary, advanced cultural societies are more ideological than their predecessor (Marcuse, 1964). The ideology the women believe is in the process of production. The productive apparatus and the goods and services which it produces "sell" and impose on the social system as a whole (Marcuse, 1964). This process is demonstrated by the extensive training sessions the women participate in before becoming Bunnies or Hooter girls. In addition, the management team relies heavily on the women's commitment to selling the product or rather accepting the ideology. For instance, Hooters managers' engage in contests in which the women try to sell a plain black pen for a large amount of money. The concept around the contest is simple: Hooter girls, with their sex appeal and charming personality, can sell anything to a customer (Haber, 2000).

The selling of sexual desire also creates a false consciousness for us as consumers. Individuals are subjected to one dimension and it's everywhere and in all forms. The sellers of sexual desire claim to bring us pleasure, fun, and escape, but in reality only continue to sell us their product. We buy into it because it invokes in us a certain fantasy not controlled by our reality. For example, Hooter girls are taught to be sexy, alluring, and seemingly appear available but never can be available. It's merely a fabrication of want and desire. Sexual desire is marketed as a commodity and this ideology not only exists in buying and selling of products, but it's in almost every aspect of our daily life, i.e. Monicagate and the Clarence Thomas trial. Selling sex as entertainment or scandal carries with it prescribed attitudes and habits, and certain emotional reactions which bind the consumers more or less pleasantly to the producers. The products (sexual desire) indoctrinate and manipulate and promote a false consciousness which is immune to falsehood (Marcuse, 1955). In other words, we buy into the philosophy because it makes us feel good. Sexually gratifying experiences tend to be a good way of life. However as a good way of life, it resists change and ignores controversial issues surrounding the goods. Hence, what emerges is a pattern of one-dimensional thought and behavior, in which ideas and aspirations are repelled or reduced to the term of the universe. It's simple; sex sells and makes us feel good in an otherwise disturbing reality of war, poverty and unemployment.

How are the thoughts of sexual freedom sustained? According to Gramsci (1971), dominant groups in our society maintain their dominance by securing "spontaneous consent" of subordinate groups through the negotiated construction of political and ideological consensus. As I previously noted, Marcuse (1964) explains this process as accepting a false consciousness as a true consciousness. More specifically, the dominant classes must get other classes to accept its own moral, political and cultural values (Gramsci, 1971). Sexually themed restaurant owners and their marketing teams sell us products which enhance our lives. They also convey to us a lifestyle defined by experiences. They are conveying an ideology of consumption rooted in sexual freedom by making us believe that consuming their products or working in their stores will in some way enhance our lives. In actuality, I suppose it is gratifying to watch tightly clothed women bring food and drinks, as well as, it is satisfying to earn substantial amounts of money by simply putting on a costume and serving people. For Gramsci (1934), it's not merely the enjoyment, but rather sexuality as a sport. He refers to the popular proverb, "the man is a hunter, woman a temptress". The gratification is in the sport and the catch, and sex themed restaurants are built on this particular ideology. This continued need for gratifying experiences drives us to want these experiences no matter the cost. This is an easy tool in relaying ideology and can extend to other aspects of societal norms and beliefs such as struggles between groups. For example, Hooter girls and Playboy Bunnies have been strongly criticized by women's groups claiming exploitation, as well as, perpetuating objectification of women's bodies.

On the other hand, the owners of these conglomerates portray it as successful employment for women and fun for consumers. In addition, women who have worked in these restaurants often claim empowerment, success and increased self-esteem. Ex Playboy Bunny Jeanne boasts (Ex-Playboy Bunny website), "The adventure of being Bunnies provided us with such wonderful opportunities and experiences; to this day it still gives me a tear of gratitude to Hef for creating the

Playboy Clubs." It is important to note that most of the women working in sexually themed restaurants view their work as satisfying, empowering and monetarily successful. Gramsci (1971) explains these thoughts as subordinated groups accepting the ideas, values and leadership of the dominant group not because they are physically or mentally induced to do so, but because they have reasons of their own. The thoughts of the Bunnies and Hooter Girls demonstrate Gramsci's (1971) belief that Bourgeois power is based on economic domination, as well as, intellectual leadership. Basically, women work at Hooters and Playboy Clubs, because it's good money and it carries with it a certain status and prestige. More importantly, these thoughts are conveyed by the sex themed restaurant owners and the women accept it as true.

Culture and spectacle: "The hunger for the bunny and all the others"

Hugh Hefner said it best when he dubbed the Playboy Bunny a cultural icon and sex symbol of the sixties. The creation of the Bunny revolutionized the restaurant business and lead to the explosion of sex-themed restaurant. The most important attribute being the waitress and her costume. Horkheimer and Adorno (1972) explain this process as the spread of commodification and domination of money which transforms cultural forms into external and autonomous objects. For example, a women's body is transformed into a desirable, appealing and sexually gratifying commodity. Adorno (1967) explains the phenomenon, "Cultural entities typical of the culture industry are no longer also commodities, they are commodities through and through...

Ultimately, the culture industry no longer even needs to directly pursue everywhere the profit interest from which it originated. These interests have been objectified in its ideology and have even made themselves independent of the compulsion to sell the commodity. Brought to bear is a general uncritical consensus, advertisements produced for the world so that each product of the culture industry becomes its own advertisement." (Pg. 129–130) The creation of various cultural sex icons such as the Playboy Bunny and the Hooter girl become not only an avenue for profit, but stand alone as independent advertisement for sexual freedom, monetary gain and desire. For example an article written on the Playboy Bunnies 20th anniversary states (Playboy magazine, 1980),"She's been called a sex symbol, a Girl Scout, an Occidental geisha, a "good girl dressed as a bad woman," "pure sin on sight," the embodiment of a perpetual male erotic fantasy. Her working uniform is arguably the world's most recognized. She was born 20 years ago. She's the Playboy Bunny, and ever since that memorable night in February 1960, she has been the subject of curiosity—then, and now, often manifested in goggle-eyed stares—and controversy. Back in 1960, purse-lipped Mrs. Grundys feared that the mere sight of these shapely young ladies would corrupt the morals of their sons. In 1980, militant feminists, some of them equally grim-visaged, complain that the Bunnies themselves are the victims of some sort of sexist corruption. Still, the Bunny has not only survived, she has multiplied. Triumphantly. Since 1960, upwards of 25,000 young women have worn the ears and tails of the Playboy Bunny" (pg 1) The Bunny has become an independent advertisement and an intrical part of popular culture. Playboy does not have to continue to promote it as a commodity; it is a commodity through and through.

The success of the Playboy Bunny Club was only the beginning of what I deem the sexually themed leisure industry, which delivers a standardization of sexual desire through gratifying experiences. The man with leisure has to accept what the culture manufactures offer to him (Adorno and Horkheimer,1972). Sexual desire becomes classified for us. This exists on many levels and in all forms such as pornography, clothing, internet sites, and sexually themed restaurants. Kant explains that formalism still expects a contribution from the individual, who is thought to relate the varied experiences of the senses to fundamental concepts, but the industry robs the individual of this function (Adorno and Horkheimer,1972). Sexually themed restaurants do it for us, they deliver the sexually appealing classification with the creation of the Bunny or the Hooter girl. Their main service to the customer is to do his schematizing for him. The prototype is the same for all the sex-themed restaurants; scantily clad women sell food and drinks.

The specific entertainment only appears to change but in reality is derived from the same prototype (Adorno and Horkhemier, 1972). Hooters is only a recreation of the Playboy Club, as much as, the Playboy Club was based on the ideas of Chicago's Gaslight Club and its girl servers. They market it to us under the idea of consumer choice, in which individuality itself becomes the ideology of the sexually themed leisure industry (Adorno and Horkheimer, 1972). For instance, other sex themed restaurants have opened such as "Global Cuisine", which features body sushi. Women lay down on tables, and their private parts are covered with leaves. The sushi is laid on the women's bodies and guests eat it off the women with chopsticks. It also features a chocolate bar with women painted in chocolate serving the guests (Los Angeles Times, 2003) Adorno (1972) explains Hooters and restaurants like it as a standardization of consciousness, and the mass production of a homogenized consumer identity. Men want sexy, attractive women to serve them food and drinks and the details of the specific restaurant are irrelevant only the conformity of the end product matters. The product is the sale of sexual desire. It is difficult to discuss the culture industry and the role of sexual leisure without mentioning Debord's ideas on the spectacle.

Debord (1994) explains the spectacle as not a collection of images, but rather a social relationship between people that is mediated by images. For example, when Hugh Hefner created the Playboy Club he wanted to bring his living room and life to the masses. As Debord (1994) states, "everything that was directly lived has moved away into a representation." The spectacle in the sense of sex-themed restaurants adheres to the principle of commodity fetishism, the domination of society by "intangible as well as tangible things," which reaches its absolute fulfillment in the spectacle, where the tangible world is replaced by a selection of images which exist above it, and which simultaneously impose themselves as the tangible par excellence (Debord , 1994, #36). In other words, the sex-themed restaurant replaces the real world of sexual gratification by selling an image or promise of sexual desire. The customers develop a relationship with the image. For instance, the women are seen as obtainable but separate from reality, existing for pleasure and fulfillment. Finally, as wage earners in the capitalist economy we are required to participate in the endless pursuit of attainment. The reality of this blackmail accounts for the general acceptance of the illusion at the heart of the consumption of modern commodities (Debord, 1994, #47). Sex-themed restaurants sell us an illusion, and we buy the product because it gives a feeling of pleasure. Debords (1994, #47) explains, "the real consumer becomes a consumer of illusions. The commodity is this factually real illusion, and the spectacle is its general manifestation."

Consumption and desire: the addiction to the image

It is impossible to discuss the impact of sexually themed restaurants and why we desire them without addressing the issues of sexual addiction. Freud (1961) devoted most of his life's work to this phenomenon, and the core of his theory begins with sexual desires. Capitalism has definitely made profits by inundating us with sexual images from playboy magazines to strippers, but why do we buy it? What makes us addicted to sexual images? According to Freud (1961), as individuals our id is the center of our impulses and desires, and of course uncontrollable sexual desires. Arousing sexual images, with no intimate contacts to an individuals i.e. girlfriend or lover, etc, allow us to escape into a world rooted in desire and fantasy. For instance, the Hooter girl/Playboy Bunny/cocktail waitress waits on us, dotes on us, and overall makes us feel wanted and desired. Their affection is not real, in fact they are trained to do so. Freud (1961) explains the fantasy as one mental activity which retains a high degree of freedom from the reality principle. The introduction of the reality principle enabled one mode of thought activity to split off. The fantasy is kept free from reality-testing and remains subordinated to the pleasure principle alone (Marcuse, 1955). The attention one receives from the women remains separate from reality and exists only for pleasure. It is highly unlikely Joe Smith, being waited on at Hooters, will date or become intimate with his waitress, but he will continue to frequent Hooters and rendezvous with his waitress in sexual fantasy. He is stimulated without any repercussions to his current reality. However, it is important to note that some customers are forbidden from Hooters or Playboy Clubs by their significant others. If they attended it would have some impact on their lives, but not in terms of acquiring the object of desire i.e. the waitress.Another important aspect in the addiction to the image is the use-value and exchange value of the aesthetically appeasing appearance of the waitress, and the ability of the restaurant to provide this to its consumers. The agents of capital are dependent on the factor to create or appear to create what consumers want. What do they want? As I stated previously, they want gratifying images of sexual desire. However, it is important to note the commodification of sex is still realized as direct or physical contact i.e. strip clubs or prostitution, etc. However, Haug (1986), points out there is a use-value in mere sexual illusion. It is sexually gratifying to the consumer and becomes illusory satisfaction. The satisfaction through sexual illusion reproduces a further demand and a compulsive fixation. This is demonstrated by numerous websites devoted to Hooters girls, and Playboy Bunnies, as well as, the various stalking crimes attached to customers' obsession with these women. Another example is the recent lawsuit filed by Chicago Hooter girls against employees watching them change into their uniforms. Some examples of devoted websites: Dale's website (www.skirtman.org/hooters.html) is devoted to pictures with Hooter girls, and boasts, "This page was taken offline for a while due to a threatened lawsuit by Hooters® of America. You are welcome to read their letter and my response. After hearing no further word from them, I have restored this page." The page continues with approximately 25 pictures of girls from various Hooter locations. J.D's website (www.geocities.com/TelevisionCity/Stage/7186/) chronologically shows pictures with Hooters girl from 1995 to present. He has special links to different kinds of pictures of the women such as the at beach and in the snow. A site devoted to the Playboy Bunny (www.forwardingsite.com/adults/playboy/playboy-bunny.html) advertises: What is a Playboy Bunny? Playboy bunnies are gorgeous women who once worked in Playboy Clubs.

Would you like to see Playboy Bunnies in their most private moments?So how does this compulsive fixation develop? Simple, a society rooted in sexual repression. As discussed earlier, Freud states that arousing sexual images with no intimate contacts to the individuals allows one to escape into a world rooted in desire and fantasy. Haug (1986) further explains that feelings of guilt and the angst they arouse block the way to the sexual object, and the commodity of sexual illusion acts as its replacement. In other words, I can't have the real thing, but I can have the image of it. It is not the sexual object which takes on the commodity form, but the tendency of all objects of use in commodity—form to assume a sexual form. It is important to point out the discrepancies within these thoughts.

It is not clear yet that the Playboy Bunny or the Hooter girl is an object, because to be an object implies some form of degradation. However, some studies have discussed the degradation these women feel in performing their job such as Hochschild (1983), Hall (1993), Haber (2000), LaPointe (1987), and Zagorrin (1997). Haug (1986) further illuminates the commodification of the waitress is that the sensual needs and means by which they satisfy consumers are non-specific. In a way they come to resemble money, which Freud compared to anxiety, where sensual needs are triggered by almost anything and they become freely convertible into stimulation from any source. Sexuality becomes the waitresses' assistant, and the exchange-value for the consumer transforms itself into sexuality (Haug, 1986). The background of sexual enjoyment becomes the commodity's most popular attire or rather the gilded backdrop drenched in sexual images against which the commodity appears, i.e. the sexually-themed restaurant.

Conclusion: sexual desire still sells; what happens now?

The concept of selling sexual desire seems simple enough and to most of us harmless. What's wrong with feeling stimulated, relaxed and having a fantasy experience? Sex themed restaurant conglomerates view it as an expression of consumer choice and individuality. In reality the sexually themed leisure industry perpetually cheats us of what it perpetually promises. (Adorno and Horkhemier, 1972) The promissory note, with its strategic marketing tools, draws on endlessly prolonged pleasure such as the promise of something more arousing and more real. However, the promise is actually all the spectacle consists of, which is the illusory (Adorno and Horkhiemer, 1972).

It just confirms that the real point will never be reached. As individuals we will never be ultimately satisfied because commodities and the culture will only convince us to want more. The creation of sex-themed restaurants will never arrive at the ultimate because the idea is to keep us wanting more. As Adorno and Horkhemier (1972) state, the diner must be satisfied with the menu, until a bigger or better one comes along. The sexual freedom of it all disappears because the freedom to choose an ideology, since ideology always reflects economic coercion, is only the freedom to choose more of the same i.e. Playboy club or Hooters. The most intimate reactions of human beings have been thoroughly reified that the idea of anything specific to themselves now persists as only an abstract notion. As consumers, as workers, we are sold the ideas of sexuality, freedom, and success.

We are told how to view these commodities and what feelings they will invoke in us. Our passion or hunger for these experiences is trickled down to us from the ruling class or the corporations selling it to us. Where's our agency in all of this? How do we begin to reject programmed thinking or for that matter the continued exploitation of all of us, consumers and workers. Gramsci (1978) advises, that we have to combine our interests with the interests of other groups and each group must make their contribution to a new socialist society. However, this doesn't solve the problem of advertisements, false consciousness and consumption. I suppose it is more than joining together against the sexual exploitation of all of us. We have to realize the tools they use in spreading hegemony. Sex-themed restaurants didn't dream up the selling of sexual images on their own, it can be seen in the media, advertisements, gender relationships, pornography, etc. Foucault (1978) comments that the way to defeat this apparatus of sex is not through 'liberation' of sexual desires because it does not change the need for a norm. More specifically, it merely replaces one standard for another such as sexual liberation or freedom. As I demonstrated throughout my paper, this movement continues to take what people believe to be private and make it extremely public.

The solution for Foucault (1978) is creating new ideas regarding bodies and pleasures therefore destroying the norm. He states, "The rallying point for the counterattack against the deployment of sexuality ought not to be sex-desire, but bodies and pleasures." Essentially, as individuals we are only able to see one type of sexuality, one discourse rooted in freedom or repression. As individuals we identify with the existence that is imposed upon us. The concept of sexuality is artificial, and the artificial nature of desire is only further perpetuated by the selling of sexual desire as a commodity, in other words, leading us deeper and deeper into discourse (Foucault, 1978).

Bibliography

Adorno T. (1967, 1989), *The Culture Industry Reconsidered.* Pg. 128–135, In: *Critical Theory and Society: A Reader.* Edited by Stephen Eric Bronner and Douglas MacKay Kellner. New York: Routledge.

Adorno T. and Horkheimer M. (1972), *Dialectic of Enlightenment.* New York.

Herder and Herder Debord Guy (1994), *The Society of the Spectacle.* Translated by Malocom Imrie. London and New York, VersFoucault, Michaela. (1978) The History of Sexuality, Vol. 1. New York, Pantheon.

Freud S. (1961), *Civilization and It's Discontents.* New York and London, W. W. Norton & Company Gramsci Antonio (1971), Selections from the Prision Notebook, edited and translated by Quintin Hoare & Goffery Nowell Smith. London, Lawrence and Wishart.

Gramsci A. (1934), *Some aspects of the Sexual Question.* In: *The Gramsci Reader:IX Americanism and Fordism.*
(Article taken from www.marxists.org/archive/gramsci/editions/reader).

Haber N. (2000), *Personal Observations as a "Hooters Girl"* .

Hall E. (1993), *Waitering/Waitressing: Engendering the Work of Table Servers.* Gender and Society. 7(3)(September). 329–346.

Haug W. F. (1986), *Critique of the Commodity Aesthetics.* Southeast Minneapolis, MN: University of Minnesota Press.

Hochschild A. R. (1983), *The Managed Heart.* Berkeley and Los Angeles, CA: University of California Press.

LaPointe E. (1987), *Waitressing: The politics of the labor process.* Paper presented at the annual meeting of the American Sociological Association, August, Chicago, Illinois.

Marcuse H. (1955), *Eros and Civilization.* Boston: The Beacon Press.

Marcuse H. (1964), *One Dimensional Man.* Boston: The Beacon Press.

Marx K. (1867), *Capital.* In: Robert, Tuckers (1978). *The Marx and Engels Reader.* New York and London: W. W. Norton and Company.

Scott K. L. (1998), *The Bunny Years.* Los Angeles: Pomenerate Press.

Zagorrin/Washington A. (1997), *Sexism will be served.* Business. 150 (15) (Oct. 5).

Internet sites used in this paper

www.explayboybunnies.com
www.hooters.com
www.dearauntnettie.com (article by Ditty Nicolaides on French restaurants)
www.Hughhefner.com
www.forwardingsite.com/adults/playboy/playboy-bunny.html
www.geocities.com/TelevisionCity/Stage/7186/
www.skirtman.org/hooters.html
www.kathrynleighscott.com
www.playboy.com (article posted on website in playboy magazine, 1)

Chapter 11
Southern bodies: bodies, gender and sexuality in world society

Raewyn Connell

Abstract
Sociological theories of embodiment, like theories of globalization, are mainly constructed by writers from the global metropole (i.e. western Europe and the USA), and then circulated to the rest of the world. When the making of world society is seen from the periphery, very different emphases appear. This essay suggests how embodiment may be seen in southern perspective, focussing on questions of gender and sexuality. The destruction of indigenous gender regimes, the organization of bodies into racial hierarchies and new labour forces, the proscription and reorganization of sexual identities, and the changing technology of military dominance, are among the issues that become central to the discussion of embodiment.

The sociology of the body

In the last two or three decades, the body and its social control and representation have become important topics in sociology. No sociological congress is complete without discussions of body images, sexuality, fashion, body modification and the like. The sociology of embodiment, as proposed by Turner (1984), has become part of sociology's disciplinary culture.

The greatest influence on this intellectual movement, by far, is the work of the French cultural historian Foucault. In two immensely influential books, *Discipline and Punish* and *The History of Sexuality*, Vol. 1, plus a range of other writing, Foucault taught a generation of social scientists to study the disciplining of the body as a fundamental technique of social control in modernity; and to understand identities, including sexual identities, as constructed discursively. From these starting-points have flowed the project of understanding bodies as socially constructed, that continues today.

Though not many practitioners of the sociology of the body acknowledge it, this is not the first time that bodies and their differences have been important to sociology. In the evolutionary sociology of the second half of the nineteenth century, when the intellectuals of Europe and the European settlements in north America were constructing a new science of society, questions about bodies and society were of the first importance. They were, however, shaped differently.

Sociology, in its first two generations from Comte to Hobhouse, was an intellectual enterprise, and from the 1890s a university discipline, that centred on the understanding and explanation of social "progress". It was inherent in this enterprise, as I have argued in "Why is classical theory classical?" (Connell 1997), to picture a contrast of primitive and advanced societies and institutions. For instance, the bulk of Spencer's great *Principles of Sociology* (1874-77) is taken up by narrating the progress from primitive to advanced forms of every kind of social institution that Spencer could think of: family, state, economy, etc.

It is no accident that such an intellectual project should take shape among the intellectuals of Europe and north America, because the states they lived in were, at that time, the centres of the greatest imperial expansion the world has ever known. A flood of information, stories and images flowed back to the metropole from the frontiers of overland and overseas conquest: the "westward expansion" of the United States, the overland conquests of Russia, the overseas conquests of Spain, France and Britain, the Netherlands, Portugal and later Italy. The texts of Spencer, Durkheim, Ward, Weber, Tönnies, Hobhouse, Sumner and a host of others are packed with examples from the colonized or semi-colonized world. The very fact of colonial conquest delivered the contrast between "primitive" and "advanced", in the minds of the colonizers. It is not too much to say that imperialism was a defining influence on the making of sociology.

A great deal of what interested sociologists in the tales from the frontier was what we would now classify under the sociology of the body. For instance Sumner's famous *Folkways*, dating from 1906, has many passages dealing with fashion and body modification, abortion and infanticide, cannibalism (a whole chapter here!), sexuality, incest, ritual uncleanness, prostitution, sports, diet, and so on. Perhaps half the book concerns bodies, their control and representation. Ward's *Pure Sociology*, published in 1903, is much more abstract, but more explicit about the importance of biological knowledge for sociology. Its longest chapter deals with "the phylogenetic forces", i.e. sexuality and gender.

It is not surprising that a social formation that was built from conquest and population movements should generate concerns with sexuality, reproduction and gender relations. The very concept of "phylogenetic forces" signals this concern. But the body issue that imperialism most insistently forced on the attention of intellectuals was the issue of race. Colonial societies were structurally divided on racial lines, and the divisions hardened, rather than softening, during the nineteenth century. Racial difference was, in popular ideology in the late nineteenth century metropole and in settler colonies, both the explanation and the justification of imperial power. The "white race" was assumed to be more advanced, and was the bearer of progress in the world.

It is not surprising, then, that the science of progress should also be concerned with race. Some sociologists regarded race as the main framework of sociological knowledge. This was true, for instance, of the pioneering text *Sociology, Based upon Ethnography* (1881) by Letourneau, who was to hold the first chair of sociology in the world. In this book, the hierarchy of black, yellow and white races provides the framework for a vast assemblage of data from the colonized world, collected by travellers, scholars, colonial administrators, and so on. Evolution, to Letourneau, is both bodily and social, with the Aryan race showing the highest development of the faculty of reason. One shudders a little at the language, but Letourneau was no proto-Nazi; his book ended with a little Utopia picturing the growth of liberty and care.

Sociology from the 1870s to c. 1910 therefore dealt with many of the issues that have come into focus again in the "sociology of the body" since Foucault. The most striking difference is that sociologists in the earlier period drew most of their facts from the colonized world, and arranged them according to a theory of progress, while sociologists in the later period draw almost all of their facts from the metropole, and arrange them according to an ironic doctrine of social control. Foucault's two great books, notably, confine their idea of modernity to Europe (with a little acknowledgement of north America), and ignore the colonized world completely.

Is such a framework adequate for contemporary sociology? I consider it is radically inadequate to construct social theory by ignoring the lives and experience of the large majority of the world's people. In the light of recent work on the construction of social theory from the global periphery (Connell 2007, Alatas 2006) I offer some thoughts about decolonizing the sociology of the body; or, less polemically, exploring what is involved in making the sociology of the body relevant to world society.

Modernity, globalization and imperialism

Our first task is to re-configure the idea of modernity as a framing concept for the social analysis of embodiment. The currently popular idea of multiple modernities is a step in this direction; but a simple pluralization is not enough. We have to understand any regional experience of modernity in a relational way, as Domingues (2008) does in *Latin America and Contemporary Modernity*, taking the relationship with the global metropole constantly into account.

Most social-scientific discussions of "globalization" and its social consequences are produced in the metropole, the rich, capital-exporting countries of the global North. Naturally enough, they build on the experience of people and institutions in the metropole. The very term "globalization" came into common use in the 1980s to describe the field of business opportunities and strategies as seen from the boardrooms of multinational corporations—as the world-spanning institutions of modern capitalism were then called.

As the word was popularized in media, filtered into political ideology, and came into use in the social sciences, it reflected this origin. It stood for the idea of a growing interconnection and homogenization of the world—the shrinking of distance and time, the blending of cultures, the lowering of barriers, the convergence of political systems and economies.

Theorists of globalization such as Giddens, Bauman and Beck constructed their analyses by projecting pre-existing accounts of modernity or postmodernity onto a global screen.

The term "globalization" is now so compromised that many people have looked for a better language to describe the undoubted changes of the last three decades. I find it helpful to think of an emerging, and still only partly integrated, world society. World society has a history—the history of interaction between local cultures and economies, which goes back into pre-history and in which current "globalization" is only the latest episode. The modern configuration of world society was produced, over the last five hundred years, by European imperialism which in several stages conquered most of the world's people, and in two cases produced peripheral states (Russia and the USA) which rose to world power.

Seen from outside the metropole, contemporary world society is not marked by homogeneity, it is marked by inequality. Massive inequalities of income, nutrition, health and education are abundantly documented by UN development agency reports (not favoured reading for metropolitan social scientists). Massive inequalities of power have recently been demonstrated to us by the US attacks on Afghanistan and Iraq, pulverising the political and social order in those countries.

In contrast to the metropolitan image of media-tized globalization, homogenized by its virtual existence (to put it more bluntly, dominated by media representations and electronic communication), world society seen from outside the metropole is embodied, material, and institutionalized. For instance global capital markets, seen from the point of view of debtor countries, are not the vehicles of a happy integration of formerly distinct national economies to gain economies of scale. They are above all the vehicles of domination, by which economic development in poor countries is mortgaged to lenders from rich countries, while supervisory institutions like the IMF use indebtedness to control local policy, especially to limit social investment and redistribution.

I do not think there is a single "southern" viewpoint. The world outside the metropole is immensely diverse, ranging from rich colonies of settlement like Australia to areas of mass poverty like Bangladesh; from long-established postcolonial states like Colombia to areas which have still not quite shaken off colonial rule, like parts of the South Pacific. Yet it is valuable to think about world society from any southern standpoint, to shake the dominance of metropolitan thought and start new agendas.

This intellectual shift concerns, above all, our understanding of modern history. Metropolitan theories of social change are built on an assumption of historical logic, of an unbroken sequence of states of society (primitive to modern, modern to postmodern; feudal, capitalist, socialist; etc.) in which one stage emerges by an intelligible process from the one before.

By contrast, the colonized world is marked by a fundamental discontinuity, the advent of the overwhelming power of the colonizers. Here, change did not emerge by an intelligible logic from what was happening before. The postcolonial world is marked in intimate depth and detail (to a degree metropolitan theorists cannot be expected to understand, except by a mighty effort of imagination) by the historical legacy of this encounter, this unbelievable, unintelligible, disruption.

Imperialism and embodiment

We are accustomed to thinking of imperialism only in class terms, as a stage in the history of capitalism. This is, more or less, the approach of the "World System" school launched by Wallerstein. But each of the stages in the history of imperialism was gendered and embodied.

The sailors, soldiers, traders and governors of early colonialism were with very few exceptions men, and the imperialist states were patriarchal institutions which created modified patriarchies in the colonies. The creation and management of empires involved the movement and control of bodies, from the military expeditions of conquest to the transport of labour forces, such as the African/American slave trade. The emergence of racial hierarchies and modern racism, a characteristic product of empire, provided schemes for classifying bodies and policing boundaries between them.

De-colonization is also an embodied process, as we see in the violence of wars of independence. If we follow the argument of Fanon (1968), this is as much a psychological as a military struggle, in which submission to the colonizer's hierarchy of bodies must be broken. The demolition of apartheid in South Africa is the most dramatic recent example of this wrenching process.

It is not a once-for-all change. On the contrary, as the history of de-colonization in Latin America shows, racial and class hierarchies can certainly be re-established in post-colonial regimes. Connections with metropolitan power, including new imperial centres, reinforce this over time. Indeed, we may now be seeing an intensification of this process through electronic media, as images of idealized white bodies circulate globally. Instances are the popularity of "Xuxa" as a media star in Brazil, cosmetic surgery among the affluent in Asia, and the predominance of "white" models in advertising, which can be seen from Japan to Chile.

Post-colonial development is, to a significant extent, a matter of the political economy of bodies. Nutrition is one dimension. The "green revolution" was one moment in the growing control of transnational corporations over the world's food supplies. The current promotion of genetically-modified crops is another, using farmers' struggle for a better livelihood to produce a technical dependency which also has grave environmental consequences. The size of population is another. Attempts are being made in many parts of the world to control population growth through a mixture of incentives, propaganda and authoritarian interventions.

But populations are hard to control, partly because they keep wandering about. Populations are shifting, both in massive rural/urban migrations creating the new mega-cities that are such a striking feature of the contemporary world, and in long-distance migrations which have now become a key political issue in the rich countries. It is striking that a new regime of tight controls over labour migration is being erected, at the same time as the free global circulation of bourgeois bodies is being promoted. Carceral systems impinge heavily on colonized populations. Just one illustration: in Australia's Northern Territory, where about 30% of the population are Aboriginal, 83% of prisoners in the overcrowded gaols are Aboriginal (Murdoch 2008).

Colonizing power, and the disruption of local communities and economies, had a formidable effect in disabling colonized bodies. This issue has been brought into focus in an important recent paper by Meekosha (2008), who shows how the sociology of disability needs to be re-thought in Southern terms. A large majority of disabled people live in the global periphery, and their bodily situation has been created by a cascade of damaging impacts, from initial conquest to the modern arms trade, and ranging across land dispossession, famines, nuclear testing and contemporary sweatshops.

In global arenas such as mass communication and the travel industry, the diversity of human bodies is presented as a spectacle of difference—not of inequality or common need. This is strikingly illustrated by the construction of regional and national identities in tourism. In tourism, affluent travellers consume embodied difference as a form of entertainment. Local identities are packaged for them, using photogenic versions of traditional dress, art, religion and cuisine, and ignoring the daily realities of life for the local population. Thus tourists to Istanbul visit the Blue Mosque and the Covered Market, watch Dervish dancing and drink Turkish coffee. But they are steered well clear of the angry and overcrowded suburbs where millions of migrants from the countryside have clustered, looking for work.

In other global arenas, such as the international military and security system, bodies are addressed as targets of control. The neo-liberal world order has evolved beyond the old systems of colonial control, but still needs from time to time punitive operations to smash movements or regimes that the dominant powers find objectionable. This is done by dealing out wounds and death to the bodies of the local population, especially the supporters of the local regime. Estimates of the deaths resulting from the recent punitive treatment of Iraq are controversial, but it is reasonable to estimate half a million deaths from boycott, military attack, destruction of infrastructure, destruction of social order and trade—with many and perhaps most being deaths of children.

Letourneau thought that the white race showed the highest development of the faculty of reason, and reason in its most elaborated form is currently materialized in the aerospace technology. Air attack has become the key military technique of punitive expeditions, with clear implications in terms of the social evaluation of bodies. Since the time the British pioneered "air control" of colonial populations in the 1920s (a decade before the German attack on Guernica), it has been characteristic of air attack that it involves a high proportion of "civilian" deaths. Despite claims about "precision bombing" (World War II) and "smart bombs" (Desert Storm), air attack has always involved an assault on a population. On the other hand, in the conditions of these punitive interventions, against poor countries with little in the way of anti-aircraft weapons, casualties among the attackers are very low. Thus the form of the intervention maximizes the damage to third-world bodies and minimizes the damage to first-world bodies.

Gender and sexuality on a world scale

Colonial and neocolonial violence has mostly been dealt out by men, in masculinized institutions such as armies and colonial states. The targets have often been the bodies of women, not only through specific acts of violence including area bombing and mass rape, but also through the destruction of subsistence livelihoods and the creation of gender-segregated labour forces.

We have come to understand gender as a structure of social relations (Connell 2009). It is now necessary to view that structure as having a global dimension; that is to say, we must see gender as a structure of world society. We do not need to assume that gender is a homogeneous or tightly-organized structure at the world level, as early theories of patriarchy did. Essentialist myths about women persist, for instance in ecofeminist thinking and in gender in development work; but as Cornwall (2008) shows, these are insupportable.

Yet we cannot fall back on a "mosaic" model of distinct gender cultures around the world. As Bulbeck (1998) shows, feminisms from different parts of the world have been in complex dialogues with each other. All we need assume is that significant linkages between gender orders do exist.

In fact such links have existed throughout the history of modernity. Imperialism and globalization change the conditions of existence of local gender orders, as they affect every other structure of local society. For instance, the linking of previously-separate production systems changes the flow of goods and services in the gender division of labour, as seen in the impact of industrially-produced foods and textiles on household economies. Colonialism itself often confronted local patriarchies with colonizing patriarchies. The turbulent aftermath has seen both challenges to patriarchy, and contestation between rival patriarchies, as seen in southern Africa (Morrell 2001).

The links that constitute a global gender order seem to be of two basic types: the interaction of gender orders, and the creation of new arenas of gender interaction. Between them, these create specific gendered histories of embodiment (Connell 1998). Imperial conquest, neo-colonialism, and the current world systems of power, investment, trade and communication, brought very diverse societies in contact with each other. This has often been a violent and disruptive process. Missionaries and colonial governments attacked local gender arrangements, such as the third-gender "berdache" tradition in North America, which did not fit the colonizers' templates.

The destruction of indigenous systems of social control, interacting with the colonizers' models of masculinity and industrially-produced alcohol, has led to high levels of domestic violence and alcohol dependence in many indigenous communities. The gender division of labour has been re-shaped by slavery, indentured labour, migration and resettlement. In the contemporary world, the institutions of masculine violence are linked by an arms trade that amounted to $US 45.6 billion in 2006 (Stockholm International Peace Research Institute 2008)—nearly double what it had been a decade earlier.

The gender practices re-constituted by such means form new patterns which are, so to speak, the first level of a global gender order. They can, in themselves, be the focus of significant social struggle—such as the struggles in many Islamic countries around the influence of Western models of femininity. Yet if the argument of Mies (1986) and others is correct, the broad tendency of this process is to produce gender orders sharing structural features of the gender order of the metropole, notably women's economic dependence on a male wage earner as the basis of the gender division of labour.

The second type of link involves the creation of institutions that operate on a world scale: transnational corporations, international state agencies, international communications media, and global markets. Each of these institutions has its internal gender regime, and each has its gender dynamic—constitution of interests, gender politics, and processes of change. World-spanning institutions thus create new arenas for gender formation and gender dynamics.

The largest business organizations, corporations operating in global markets, deploy very large resources and employ very large numbers of people. They typically have a strong gender division of labour, and a masculinized management culture. United Nations agencies, the European Union, and a multiplicity of other international agencies and agreements both regulate gender issues globally (e.g. through development aid, education, human rights and labour conventions), and themselves possess multicultural gender regimes. They are perhaps more complex than transnational corporations because of their more diverse staffing (Gierycz, 1999).

It is familiar that there is a gender dimension to multinational media corporations, which circulate film, video, music and news on a very large scale. This is also true of the more decentralized media (post, telegraph, telephone, fax, the Internet, the Web) and their supporting industries. All contain gender arrangements and circulate gender meanings. Similarly international markets—capital, commodity, service and labour markets—have an increasing "reach" into local economies. They are often strongly gender-structured and are very weakly regulated.

Just as with gender, we have been taught, by the last generation of sociologists and historians (e.g. Weeks, 1986) to see sexuality as social process and as history—not as unchanging nature. Weeks has persuasively argued, for the metropole, that sexualities have undergone a notable "speciation", with a diversification of identities and sexual communities. Is this true of the rest of the world? How are we to understand sexual identities and politics in the moment of history marked by the global triumph of neo-liberal politics, the free-market agenda, the economic dominance of transnational corporations and international capital markets?

In an important survey of sexuality and globalization, Altman (2001) argues that it is a mistake to see globalization as a force homogenizing sexual regimes around the world. This is a view that has been taken by those who see the demographic transition producing Western-style heterosexual romance and couple relationships everywhere, and by those who see a US-style gay identity as simply displacing indigenous forms of same-gender sexuality, with San Francisco style "gay communities" developing around the world.

Altman does not deny the cultural and economic power of the global North. But he convincingly argues that its impact on local sexual regimes is to diversify, rather than homogenize, sexualities. A range of compromises, intermediate patterns, and even completely new identities, spring up in the interplay between local society and international bourgeois culture. Even the "gay communities" of cities such as Kuala Lumpur are configured very differently from those of the USA.

I would add to Altman's argument a line of thought about the dynamics of neo-liberalism. Neo-liberalism self-consciously promotes human rights. Its emphasis on individual freedom, protection against arbitrary state power, anti-discrimination laws, privacy laws and equal opportunity measures, has been heavily qualified by the "war on terror" and religious neoconservatism, but remains influential in global arenas. There has been significant advancement, through education and professions, for women and for ethnic minorities. There has been more tolerance for sexual minorities, especially gay communities. There has been a relaxation of laws controlling sexuality. Though some religions maintain a highly restrictive sexual code, the actual sexual repertoires of metropolitan populations continue to diversify.

But neo-liberalism also undermines human rights, in ways that have implications for sexuality. To a striking degree, the "free market" agenda of neo-liberalism lacks popular support. There has never been a popular movement demanding more labour market insecurity, more economic inequality, fewer public services—core items on the neo-liberal agenda. Neo-liberal politics has a chronic problem of legitimacy.

Given the need by neo-liberal leaderships to draw boundaries around a normative community, all "minorities" capable of being targeted are under permanent threat. A consequence of the speciation of sexualities is that increasingly visible sexual communities are available for targeting. In Australia this takes the form of periodic panics about "pedophiles"—actually a rare psychological disorder, which has become an excuse for ignoring the far more common fact of sexual abuse of girls within the family.

What is emerging worldwide, I hypothesise, is a two-tier sexual regime. Among the professional and managerial classes, the main beneficiaries of neo-liberal globalization policies, sexuality is increasingly individualized and libertarian. (There is statistical evidence from metropolitan countries, e.g. Laumann *et al.* 1994, of libertarian sexual attitudes being more common among the more highly educated and socially privileged.) The participation of these groups in an international secular culture, with high rates of personal travel as a result of global business and tourism, has broken down many of the social controls over sexuality formerly exerted by religion and local cultures and kinship systems.

In its place has arisen a sexuality based on a market in attractiveness and an individualized consumption of pleasures. Hence the rising attention in this class to personal appearance and desirability (e.g. the spectacular growth of cosmetic surgery as an industry). Men as well as women are subjected to the market pressures. Hence the growth of body-culture among privileged men, ranging from high-tech gyms to Armani suits to Viagra pills and penis enhancement surgery. Though heterosexuality remains normative, other sexualities, especially gay sexualities, are allowed as acceptable personal "choices". This regime of sexuality is part of the wider bourgeois culture to which the emerging elites of developing countries, from the Pacific Islands (Hau'ofa, 2007) to China, are recruited.

In the second tier, among the bulk of the population living in poverty, sexual regimes develop which are based on local traditions as modified by contradictory pressures from global media, economics and politics. At the same time as international media circulate images of glamour, desirability and personal freedom on an unprecedented scale, the imperatives of the neo-liberal neo-colonial world order create a threat-driven sexual politics in which denial and repression, fear and guilt are prominent. Evangelical versions of Christianity and Wahhabi versions of Islam share these traits, and have converged politically in some forums such as the United Nations conferences on population and on women.

Conclusion

In the immensely complex arena of world society, systematic inequalities define conflicting interests, but do not determine whether social groups will mobilize around them, nor what strategies will be pursued in response to difference and inequalities. There are often multiple paths open, as Moghadam (1999) has argued in relation to gender and national identity in the Islamic middle east and north Africa. Some countries in this region have followed an emancipatory model, where gender equality is part of national development; others have followed a patriarchal model where national independence has meant the empowerment of men and the domestication of women.

Several global trends suggest a democratisation in the politics of gender, sexuality and bodies. They include the gradual closing, on a worldwide scale, of some of the educational and economic gaps between men and women; the movement and mingling of populations, producing many patterns of ethnic diversity; the diversification of sexualities and unravelling of some of the old systems of social control, at least among the globally privileged.

Yet we cannot be optimistic about these trends, in a world that also sees a growth of economic inequality; a masculine predominance, often amounting to a monopoly, in major institutional power; a shift towards violence and away from consent and negotiation in international affairs. The world of markets is a world in which new forms of power politics seem to be developing, which do not look like the old fascism or military authoritarianism but which may be, in the long run, just as dangerous.

Knowledge of society has a role to play in this world, though it is not, as our sociological forebears thought, the role of mapping the triumphal march of progress. We have to settle for a more complex and ambiguous agenda, in education, in policy work, and in the development of social knowledge itself. In *Southern Theory* (2007) I have argued that social science can play a significant role in global democracy, but to do so must democratise itself on a global scale, ending the intellectual hegemony of the global metropole. In this paper I have tried to show how this project might affect our understanding of social embodiment. Developing an understanding of embodiment on a world scale is a significant part of the task of constructing a genuinely global social science.

Bibliography

Alatas S. F. (2006), *Alternative Discourses in Asian Social Science: Responses to Eurocentrism*. New Delhi, Sage Publications.

Altman D. (2001), *Global Sex*. Chicago, University of Chicago Press.

Bulbeck C. (1998), *Re-Orienting Western Feminisms: Women's Diversity in a Postcolonial World*. Cambridge, Cambridge University Press.

Connell R. (1997), *Why is classical theory classical?* American Journal of Sociology, Vol. 102 no. 6, Pg. 1511–1557.

Connell R. (1998), *El imperialismo y el cuerpo de los hombres* [Imperialism and men's bodies]. In: Teresa Valdés and José Olavarría, ed., *Masculinidades y equidad de género en América Latina*, Santiago, FLACSO-Chile, Pg. 76–89.

Connell R. (2007), *Southern Theory: The global dynamics of knowledge in social science*. Cambridge, Polity Press.

Connell R. (2009), *Gender: in World Perspective*. Second edition. Cambridge, Polity Press.

Cornwall A. (2008), *Myths to live by? Female solidarity and female autonomy reconsidered*. Pg. 145–63, In: Andrea Cornwall, Elizabeth Harrison and Ann Whitehead, ed., *Gender Myths and Feminist Fables: The Struggle for Interpretive Power in Gender and Development*. Oxford, Blackwell.

Domingues J. M. (2008), *Latin America and Contemporary Modernity: A Sociological Interpretation*. New York, Routledge.

Fanon F. (1968), *The Wretched of the Earth*. New York, Grove Press.

Foucault M. (1977), *Discipline and Punish: The Birth of the Prison*. New York, Pantheon.

Foucault M. (1980), *The History of Sexuality*, Vol. I: An Introduction. New York, Vintage.

Gierycz D. (1999), *Women in Decision-making: Can We Change the Status Quo?* Pg. 19–32, In: Ingeborg Breines, Dorota Gierycz and Betty Reardon, ed., *Towards a Women's Agenda for a Culture of Peace*. Paris, UNESCO Publishing.

Hau'ofa E. (2008), *We Are the Ocean*. Honolulu, University of Hawai'i Press.

Laumann E. *et al.* (1994), *The Social Organization of Sexuality*. Chicago, University of Chicago Press.

Letourneau C. (1881), *Sociology, based upon ethnography*. London, Chapman and Hall.

Meekosha H. (2008), *Contextualizing disability: developing southern/global theory*. Keynote address, 4th Biennial Disability Studies Conference, Lancaster University, September 2008.

Mies M. (1986), *Patriarchy and Accumulation on a World Scale: Women in the International Division of Labour*. London, Zed.

Moghadam V. M. (1999), *Gender, national identity and citizenship: reflections on the Middle East and North Africa*. Comparative Studies of South Asia, Africa & the Middle East, Vol. 19, no. 1, Pg. 137–157.

Morrell R. ed. (2001), *Changing Men in Southern Africa*. Pietermaritzburg, University of Natal Press.
Murdoch L. (2008), *'Sausage factory' NT courts cited in crime squeeze*. Sydney Morning Herald, Pg. 27–28 December 2008, Pg. 6.
Spencer H. [1874-77] (1893-96), *The Principles of Sociology*. 3 Vols. New York, Appleton.
Stockholm International Peace Research Institute (2008), *SIPRI Yearbook 2008*. Oxford: Oxford University Press.
Sumner W. G. [1906] (1934), *Folkways: A Study of the Sociological Importance of Usages, Manners, Customs, Mores, and Morals*. Boston, Ginn.
Turner B. S. (1984), *The Body and Society*. Oxford, Blackwell.
Ward Lester F. (1903), *Pure Sociology: A Treatise on the Origins and Spontaneous Development of Society*. New York, Macmillan.
Weeks J. (1986), *Sexuality*. London, Horwood & Tavistock.

Chapter 12
How do I change his nappy? Disabled mothers and cyber-community

Claudia Malacrida

Abstract

Literature on people with disabilities' use of the internet typically falls along several lines. The first argues that people with disabilities, who often face physical challenges in accessing the public sphere, and emotional challenges in terms of stigma in 'regular' interactions, will find an accessible and non-stigmatizing forum in the internet. In this literature, it is argued that people with disabilities can overcome or elide their bodily differences by presenting themselves however they wish to, thus transcending the body and engaging in a wide range of possibilities for self-presentation and connection. A second argument, typically drawing on psychological and support group literature, is that the internet offers people with disabilities 'communities of practice' that act as public forums where people can share experiences, contexts and histories of joint interest that will engender relationships on-line, and improve social support and psychosocial well-being. A third argument is that the internet provides a space for people with chronic illness and disability to share their personal narratives and, through producing their own stories and incorporating the stories of similar others, they can construct a new, empowered identity. In all of these arguments, there is an explicit or inferred erasure of the corporeal. Instead, the body is conceptualized as elided, transcended, or reconstituted through online talk. In this study of an on-line forum for parents (almost exclusively mothers) with disabilities, I find that the body is instead foregrounded and made a central fact of on-line communications. Bodies are described in detail with particular attention to the functional aspects of mothering in hopes of gaining information relevant to specific disabilities. Bodies are also foregrounded as a way to formulate and present an identity as a person who struggles against and overcomes difference in order to be a good mother and a competent adult. Thus, the mothers' web-based interactions do lend support to theories about the potential of on-line sites as spaces where resistance to stigma, advocacy against systems, and constructing positive identities can occur, but they indicate that the body remains an active presence on-line for these disabled parents.

Bodies, disability and cyberspace

Ironically, at a time when sociology is finally begin to address its elision of the body, technological developments in communications have made it possible for the body to disappear again. In cyberspace, the argument goes, the body has become either an avatar to be used in performing and producing new identities that transcend traditional bodily categories, or it is made completely invisible, freeing humans to interact outside of traditional regimes of bodily privilege. Thus, the internet offers a democratic and potentially freeing space where the body can be left behind and social change can be made possible virtually, if not necessarily in actuality. These ideas have been taken up outside of Sociology in the field of Disability Studies; here, the position has been that the differently-abled body is a body that will benefit most from these freeing virtual interactions and possibilities. The internet and web-based interactions can putatively provide a space to people with disabilities where the body will not matter, where its capacities will not be limited as in real space, and where stigma can be avoided by strategic self-presentation.

These concepts are examined through a qualitative study of on-line interactions on a web-based forum for parents with disabilities, using qualitative data analysis to examine the qualities of parents' interactions on-line. I choose to examine the intersections of parenting and disability because both of these constructs are profoundly embodied. Parenting, and particularly mothering, is a social role that traditionally has been privileged to women who are able-bodied, so an on-line forum that focuses on parenting with a disability will provide an illustrative case study into the ways that bodily 'problems' are elided or overcome in on-line interaction for people with disabilities who are or seek to become parents.

Before discussing the implications of disability, parenting and disembodied cyber-possibilities as illustrated by interactions on the on-line forum for parents with disabilities, it is important to understand the theoretical claims that inform this study. In its early days, the internet was heralded as a technology that could provide a utopian arena in which the body would be a malleable entity to be presented in ways that suit fantasy and transcend earthly bonds (Sullins, 2000). At its ultimate, the internet has been conceptualized as a means to achieve a complete erasure of the body, overriding bodily stigma such as race, gender, age and, by extension, disability (Haraway, 1991; Stone, 1991). Donna Haraway's early writing on women's bodies can be seen as a foundational argument in this theoretical perspective. She argues that the potential of post-humanism and the cyberworld can be exemplified in the possibilities of the cyborg, "a cybernetic organism, a hybrid of machine and organism" that offers a way to transcend the bounds of history and materiality, providing us with a way out of problematic gender and possibly class relations (1985[1991]: 149). Haraway argued that, with cyborg technologies, traditional distinctions between private/public, woman/man, nature/machine, self/other, culture/nature, God/Human could be overarched. When reading Haraway's work, it is clear that she believes that cybernetics and cyberspace offer a politics of the body that is both revolutionary and liberatory; in her manifesto, the cyborg offers us a way out of problematic gender, race, sexuality and class relations. The cyborg, while frequently an actual machine-human interface such as that between a computer and its user, can also be a virtual entity that the user deploys in cyberspace (Stone, 1991). In either version, the anchor of the body is released so that the individual who writes and acts in cyberspace becomes a floating signifier, unbounded by traditional gendered, classed, racialized or sexualized meanings and oppressions. Building on this idea, the physical body in much of this theorizing becomes not only surmountable, but is actually conceptualized as cumbersome 'wetware' or 'meat' that serves little purpose except to complicate the purity of 'incorporeal interactions' that are unfettered, playful, creative and evolved beyond those that are possible with current human forms (Sullins, 2000).

These ideas have been expanded upon by disability scholars, who see the possibilities of cyberspace as a means of averting the stigmatization embedded in everyday interactions faced by people with disabilities. Cyberspace, in these formulations, also offers a means of presenting and developing new identities for people with disabilities. In societies that value 'normalcy' and 'beauty', people with disabilities who are 'different' and 'unusual' are almost certain to encounter stigmatization and experience negative and personally undermining interactions with others – indeed, stigmatization is a social phenomenon that is often argued to be one of the core burdens of living with disability (Davis, 1997; Hall, 2002; Shakespeare and Watson, 2002; Wendell, 1996). Erving Goffman has argued that the body is a sign vehicle that people actively manipulate so as to convey desirable social information about themselves both through intentional 'expressions given' and unintentional 'expressions given off' (Goffman, 1959). For people with disabilities, intentional and unintentional bodily signs are vulnerable to being stigmatized, devaluing the person in the eyes of others and subjecting them to negative and often punitive reactions by others. This is likely in the case of bodies with visible disabilities that carry 'discrediting stigma' that the individual is not able to hide from others, but it is also the case for bodies with invisible disabilities that carry 'discreditable stigma' not immediately evident to others (Goffman, 1963). Thus, people with visible disabilities face social burdens in trying to elide their discrediting stigma in everyday interactions, while people with invisible disabilities bear the burden of working to avoid exposing their discreditable stigma in interactions with others. Either way, it has been argued that managing stigma comprises a burden for those who live with bodily difference in societies that value youth, beauty, symmetry, strength, and capability. Thus, strong normative pressures mthat position attractiveness and wholeness as synonymous with 'humanness' (Shildrick, 2002), online media can provide people with disabilities with alternative possibilities for positioning themselves socially, eliding the stigma of everyday interactions and opening up possibilities for new and more positive identities and life chances (Bowker and Tuffin, 2007).

Gendered aspects of stigma and cyberspace

In a culture where women are specifically targeted with heightened normative bodily expectations, and where women in many ways 'are' their bodies, women with disabilities suffer additional stigmatization because of the gap between ideal feminine bodies and disabled women's different bodies (Taub, Fanflik, and McLorg

2003; Taub, McLorg, and Fanflik 2004). Empirical evidence to support the additional stigmatization of women with disabilities can be found in disabled women's gendered outcomes concerning employment, income, and vulnerability to abuse. Women with disabilities are likely to achieve social inclusion and acceptance than are non-disabled people or men with disabilities, as they earn less, have lower educational attainments, and are more likely than either of these groups to experience abandonment and victimization in their intimate relationships (Asch, 1992; Asch, Rousso, and Jefferies, 2001; Government of Canada, 2008; Schur, 2004). Further, women with disabilities are more likely than men with disabilities to be parents, and they are highly likely to do so while facing systematic barriers to acceptance in their communities for a number of reasons. First, these women, like all people with disabilities, are likely to experience stigmatization simply because they have bodily differences, but they are also likely to face social sanctions because of their presumed inappropriateness as mothers and their heightened likelihood of parenting alone, living in poverty and being socially isolated (Asch, 2001; Asch, 1992; Malacrida, 2009a; Schur, 2004).

It could thus be argued that women with disabilities may benefit more than men with disabilities in accessing the disembodied and purportedly stigma-free world of cyberspace. In the forum described in this research, almost all of the participants were mothers, reflecting the continuing reality that in western societies, women are more likely than men to engage in parenting practices and to count parenthood as a core aspect of their adult identity (McMahon, 1995).

Access to democracy and the public sphere

Stigma management and self-presentation are not the only barriers that cyberspace can reduce and perhaps override for people with disabilities. In the material world, public transport, buildings, homes, road and path systems, and weather pose often insurmountable barriers to people with disabilities who seek to participate in their communities and societies (Oliver, 1990; Shakespeare and Watson, 2002). This lack of access to the public sphere is not limited to shops, streets, workplaces and homes, but it also impedes admission to playgrounds, schools, medical facilities, and offices that provide information and services to citizens and parents with disabilities, hampering parents' abilities to provide full and appropriate care to their children. Poor access to information about services that would support disabilities and families has severe implications for the health and well-being of all family members; when parents are not supported in achieving personal health and well-being or in accomplishing their parenting roles, this negatively affects everyone in the home (Blackford, 1993; Malacrida, 2007; Prilleltensky, 2004). Poor access to playgrounds, schools, playgroups and health facilities can mean that parents with disabilities are barred from providing safety to their children and from offering their children full inclusion in their neighbourhoods and social networks (Malacrida, 2007). A final aspect of the inaccessible public sphere lies in the increasing privatization of services and information concerning disability and family; this privatization has led to significant fragmentation, where service delivery is offered by a vast array of providers whose details, policies and programmes constantly shift, turning the obtaining of information, services and accommodation into something of a moving target (Malacrida 2009b). Because of this fragmentation, people with disabilities are actually increasingly challenged to access services and information that might assist them in their parenting roles.

Virtual space offers access to social interaction and the public sphere in ways that may not be possible with actual space; it has been argued that the internet can act as a virtual public forum where democratic practice, inclusion and full citizenship can be made available to all people who are able to access the technology (Froomkin, 2003).[1] For people with disabilities specifically, on-line possibilities include access to information about policies, programmes and services, opportunities to network with other people with disabilities or with non-disabled people who share other commonalities, and the ability to do things like work for pay, attend school, or engage in hobbies that would be difficult to access in actual space.

On the other hand, it has been argued that interaction in virtual space often fails to lead to democratic or meaningful social interaction. Because of the fleeting and impersonal nature of the internet, the creation of true community is difficult, and the possibilities for social change are limited (Gold, 2003). In this critique, the absence of a body as a material platform for social interaction, is not seen as a solution to social inequality, but is instead seen as a problem in achieving positive social outcomes. Nevertheless, regardless of whether theorists are arguing that cyberspace offers a range of new possibilities for connection, self-presentation and advocacy, or that cyberspace fails to deliver on these fronts, in both arguments the absence of the body is seen as a central aspect of cyber-communications.

Social networks and access in virtual space

A further argument relating to cyberspace and disability draws on psychological and support group literature that focuses less on concerns about self-presentation and citizenship, and instead examines the kinds of interactions that can offer positive social opportunities for people with disabilities. Here, the claim is that the internet offers people with disabilities 'communities of practice' (CoPs), or disability-specific public forums where people can describe their experiences (challenges and triumphs), explain the contexts of their lives, and discuss their shared histories in ways that will engender relationships on-line and improve social support and psychosocial well-being (Anderberg, 2007; Anderberg and Jonsson, 2005). CoPs offer assistance in coping with difficult situations and linking people who share similar challenges, and are characterized by several factors.

Shared interests, knowledge and group expertise are brought together on-line to help build relationships with other members and to develop joint learning. CoPs also tend to accrue knowledge and relationships through continuous and shifting interactions relating both to these shared knowledge stocks and to consequent social actions.

Thus, support is only one of the anticipated outcomes of a true CoP while emancipatory practice is a second key element; CoPs are more than just sites where people get together to share stories in virtual space, but the knowledge shared and developed on them can help people to understand how to engender change in actual space (Wenger, 2000). For people with disabilities, some of the practices that optimally will ensue from building and participating in a CoP could be learning from other people with disabilities about the systems that oppress, and finding ways to overcome these systems at both the immediate and the policy and programme level (Anderberg, 2007). Thus, the argument goes, the internet offers possibilities for democratic process, access to a new public sphere, and possibilities for emancipatory change. In the study discussed here, I consider the on-line forum for parents with disabilities to be a positive example of a disability-related CoP.

The necessity of a democratic process that includes people with disabilities is particularly pressing. Since the mid-1970's, when the UK-based Union of the Physically Impaired Against Segregation (UPIAS) argued for a social model of disability[2], it has become increasingly clear to disability scholars and activists that that while physical, sensory and intellectual impairments might result in different ways of moving, communicating and living, social factors such as inaccessibility, lack of accommodation and prejudice are the truly disabling aspects of people's lives (Oliver, 1990). Only in recent years have developed countries begun to establish anti-discrimination legislation, with the United Kingdom approving the Disability Discrimination Act as recently as 1990 (Drake, 1999) and further refining its disability rights legislation in 2003. However, despite legislated human rights for people with disabilities, delivery on these promises has been spotty in terms of services, workplace accommodation, accessibility, freedom from poverty, and protection from discrimination.

This is particularly the case when it comes to pregnancy and parenting. Parents with disabilities encounter significant prejudice from their family, friends, neighbours, and professional service providers. A US study of parents with disabilities found that approximately 15% of the parents surveyed reported efforts to remove their children, while another US study found that 7% of 300 undergraduate students agreed that people with disabilities should not become parents (Kirshbaum and Olkin, 2002). Concerns about parental competence are particularly highlighted when the parents have severe mobility challenges, or when parents have intellectual disabilities, brain injuries, or mental health diagnoses (Booth and Booth, 1994; Malacrida, 2007; Malacrida, 2009b; Tringo, 1970). Thus, studying the practices of a CoP centering specifically on parenting and disability is a highly salient choice for exploring the social and democratic possibilities of cyberspace for people with disabilities.

The primary caveats about the emancipatory potential of cyberspace for people with disabilities center on issues of access; here, the democratic possibilities offered for disabled people on-line are seen to be compromised by unequal access to computer technologies for a number of reasons. First, the costs of internet access are prohibitive for many people with disabilities, particularly those who live in developing countries, those who rely on state financial support regardless of their location, and those who have intellectual or cognitive disabilities (Friedman, 2005; Guo, 2005). As well, there has been a significant literature examining accessibility to the internet in terms of bodily impairments. For example, despite legislation in the United States and Australia, many websites remain inaccessible because they use graphics that blind readers cannot read (Goggin, 2004; Heim 2000; Waddell 2000) or they use language that is not accessible to people with intellectual disabilities or brain injuries (Friedman, 2005). Inaccessibility also occurs because the information provided is presented in ways that are too complicated, or in sites and links that are fragmented and difficult to trace, and this has been particularly problematic in terms of government and service-provider websites that provide information about health, policy, services and programmes that are of critical importance to people living with disabilities (Davis, 2002; Vernon and Lynch, 2003). This is perhaps why CoPs like the disabled parenting forum are so important; they have the potential to overcome some of these access issues by providing an information clearing-house function that enables people with disabilities to circumvent some of the barriers that plague service providers' sites.

Narratives, identity and the internet

A final perspective on the utility and emancipatory potential of the internet is that it can provide a space for people with chronic illness and disability to share their personal narratives so that, through producing their own stories and incorporating the stories of similar others, they can construct a new and more empowered identity. Narrative theorists have argued that the power of shared narratives stand as a corrective to dominant 'medical model' understandings of the body and its experiences. The medical model conceives patients or clients as little more than their conditions, and their subjective knowledge is regarded as irrelevant to the understanding or treatment of illness and disability (Frank, 2002; Sakalys, 2000). Narratives have the power to reclaim and refashion knowledge by making one's stories public and thus they offer individuals an opportunity to bear witness to their experience, to affirm personal perspectives, and to challenge dominant helping professional ideologies (Balshem, 1993; Charmaz, 1990; Frank, 2002; Kleinman, 1988; Sakalys, 2000). Narratives further provide a politicized reading of relations of power, offering the individual an opportunity to bear witness to harms suffered, and drawing on the perspectives of subordinated individuals to expose the workings of power and domination within the medical/human services encounter

(Balshem, 1993; Sakalys, 2000). Thus, narratives that are made public and co-constructed have the potential to offer new, powerful identities to the story-tellers, and they are also capable of leading to social change and refashioning knowledge from the perspectives of individuals rather than the systems that govern them. For people with disabilities, where knowledge and experiences of the body have been dominated by medical hegemony, narratives hold a powerful potential in creating new truths and non-pathological identities.

On the contrary, it has been argued that expectations for meaningful social change stemming from shared on-line narratives are more limited off-line, both in terms of producing intimate social relations and in terms of individuals' relationships with broader social systems (Seymour and Lupton, 2004). The argument here is that while sharing stories and information with understanding others is indeed facilitated through internet sites, the fact that such sites are typically contained to issues of disability can reproduce the marginalization and separateness of disabled people as 'special and apart' while simultaneously reinforcing members' identities as disembodied and disabled (Seymour and Lupton, 2004). In this view, the transcendence of the body that is promised by virtual interactions is not actually delivered; while the body is indeed invisible on-line, it remains present as an invisible imperative that defines on-line interactions. Further, the margins that exist in actual space for people with disabilities are replicated in cyberspace – CoPs can acts as cyber-ghettos where disability acts as an overarching identity category (Seymour, 2004).

Another concern is that in cyberspace there are heightened likelihoods of problems of self-misrepresentation and misunderstanding precisely because of the disembodied qualities of cyber-communication. Here, it is worried that the lack of non-verbal communication and bodily cues in cyberspace can lead to miscommunications on the one hand and the potential for dehumanized and impersonal interactions on the other; without face-to-face interaction there is less social responsibility towards others while at the same time there is heightened vulnerability because people's interactions are characterized by self-disclosure and intensity (Morahan-Martin and Schumacher, 1999; Turkle, 2007). Thus, while it is possible to make quick connections on-line and engage in uninhibited exchanges earlier in cyber-interactions than might occur in actual interactions, these connections may also be shallow precisely because they elide the body (Kang, 2007).

In all of the above arguments, there is either an explicit or inferred understanding that cyberspace is synonymous with an erasure of the corporeal, with the body conceptualized as elided, transcended, or reconstituted through online talk. Bodies and their representations in virtual space are seen as central to formulating identities, making connections, transcending limitations, and creating democratic spaces for the exchange of knowledge and the movement toward real social inclusion. In analyzing the data on mothers' interactions in the parenting support network here, it is thus important to examine how consistently and how deeply bodies are elided, to speculate on the impact of bodily presentation on the depth and meaningfulness of the parents' on-line communities of practice, and to explore how the body and identity are portrayed and deployed in terms of social change.

The study
An on-line parenting support group

Data were gathered over a nine-month period from a British-based on-line forum for parents with disabilities. As discussed earlier, I consider the on-line forum to be a good example of a Community of Practice (CoP). This CoP was established and is run by people with disabilities and is used primarily by parents or people with disabilities wishing to become parents who are from the United Kingdom. In addition, parents from other countries, family and friends of parents with disabilities, support workers, professionals and researchers occasionally participate in the forum. It is publicly accessible, but registration is necessary in order to post or to send private correspondence to forum posters, and although it is intended for people with disabilities who are or who wish to become parents, it is open to anyone wishing to join. I joined as a researcher seeking background information and potential research participants for a face-to-face study concerning mothers with disabilities in Canada and the United Kingdom, however, I am not a parent with disabilities and do not contribute to the discussions. As such, my involvement on the group was situated firmly as an observer, rather than as a participant or co-constructor of knowledge and identity within the forum. In turn, although this may mean that I understand less from a personal perspective about participation in this particular on-line forum (Anderberg, 2007), it may also mean that my presence did not disrupt or inhibit the discursive flow in participant exchanges.

Upon registering for the group, one is sent a welcoming email with a list of guidelines indicating that the purpose of the forum is to provide a safe space to share concerns and information and also to teach each other about innovative ways to parent.

Members are assured that they are free to share opinions, but that personal attacks and negativity do not belong on the forum, particularly since so much negativity concerning disability and parenting exists outside the group. The Forum is quite active; archives on the forum indicate an average of slightly less than two postings per day over the past two-and-a-half years, but the forum is more accurately characterized by periods of inactivity with occasional spikes of up to a dozen daily posts when a topic is particularly interesting. The group is not moderated, and even so, there were virtually no instances of personal attacks, defamatory language, or judgmental responses that took place during the nine-month observation period. It is possible that the positive and supporting tone of the forum stems, at least in part, from the tone set by the non-profit group that hosts the forum.

Access to the forum is made through this group's web pages, which offer information relating to government resources and policies, speakers' bureaus, conferences and workshops, programmes and products specific to parents and to disability, articles and personal stories about and by parents with disabilities, and information about political advocacy for parents with disabilities. The tone on the group's webpage is very much one that promotes members' self-advocacy, dignity and sense of entitlement; there are numerous statements to the effect that parenting is a fundamental human right, and that parents who are different should not think of themselves – or allow others to think of them – as 'less than' any other parents.

The analytic process for this project has been both inductive and deductive. The data presented here are embedded in a more extensive study into motherhood, policy, and social control relating to mothers with disabilities in Canada and the United Kingdom (Malacrida 2007; Malacrida 2009a). In the early stages of the research project, I employed an inductive/crystallization approach (Miller and Crabtree 1998).

Data were collected, coded, and analyzed reflectively so that, following the principles of theoretical sampling in grounded theory, new questions and foci for enquiry and analysis were informed by emergent themes in the data (Charmaz 1990; Glaser and Strauss 1967; Strauss and Corbin 1994). This first round of data coding was accomplished by me and a research assistant using *Atlas-ti,* a qualitative data software package. As the research project developed, I considered the data more deductively. Huberman and Miles (Huberman and Miles 1998) argue that "tighter designs" using a more theoretically informed analysis and offering a more explanatory approach are appropriate once the researcher has a grounded familiarity with the topic. Thus, the data for this article have been further analyzed in consideration of extant theory about identity, the body and stigma in cyberspace, access to democracy and the public sphere in virtual space, and the uses of narrative and identity in on-line interactions, as outlined in earlier sections of this chapter.

Overview of the support group data

Over the nine-month observation period, 38 topic threads were presented, totaling over 250 posts. The topics discussed in these threads belonged to categories in the forum entitled *Parenting Stages*, *Specific Disabilities/Situations*, *Independent Living*, *Benefits, Equipment & Care*, *General Discussion* and a separate section for *Professionals, Media & Research.* Of these, by far the most active were the three categories relating to *Parenting Stages*, *Specific Disabilities* and *Benefits, Equipment & Care.*

Discussions relating to *Parenting Stages* were particularly active on fertility, pregnancy and adoption questions, and concerning challenges in providing infant and early childhood care. Here, parents and parents-to-be (in actuality, almost exclusively these were mothers and mothers-to-be) sought advice on what kinds of bodily changes specific to disability they should expect while pregnant, what kinds of supports they would be well-advised to put into place in anticipation of the baby's arrival, what kinds of relationships they might expect to engage in with service provides (including medical personnel, midwives and family and social service workers), and how to manage things like taking children out on public transport and accessing schools and clinics. They also sought information about how to deflect stigmatizing reactions to their pregnancy and motherhood. These posts were often anecdotes about reactions to pregnancy announcements by family and friends, or descriptions of interactions in the public sphere where pregnant women or mothers of children had felt themselves to be judged or scorned because of their dual status as mothers and disabled women. The postings were also made preemptively, in anticipation of avoiding or defusing such interactions. This focus undoubtedly reflects the reality that mothers with disabilities continue to face stigmatizing public perceptions of them as inadequate or inappropriate in the role of mothering. For example, helping professionals and family members often discourage women with disabilities from becoming pregnant, expressing concerns that they will not be competent mothers or that their disabilities will be passed on to their children (Killoran, 1994; O'Toole and Doe, 2002; Patterson and Satz, 2002; Gerodetti, 2003). Once disabled women become mothers, they face challenges in assuring the public of their appropriateness as parents and their capacity to mother adequately (Reinelt and Fried, 1993; Fitzmaurice, 2002; Grue and Laerum, 2002). Mothers and mothers-to-be on the list thus expressed an awareness that their bodies in real space were discredited signifiers, and sought the advice and support of others on-list to find ways to work around these stigmatizing encounters.

It is perhaps unsurprising that of those threads relating to *Specific Disabilities*, the most active discussions were those relating to invisible or marginal disabilities such as Chronic Fatigue or Fibromyalgia, since these are bodily issues that do not seem to be as well recognized or served through traditional policy and service provision in the United Kingdom as may be the case with more visible and established diagnoses or disabilities. Obtaining information about services and supports about Chronic Fatigue or Fibromyalgia may be more difficult in the UK for these diagnoses than for more 'traditional' and recognized disabilities because in many professional circles, these are not yet accepted categories. These 'newer' and invisible disabilities are highly stigmatized because of their lack of legitimacy and assumptions that there really is nothing 'wrong' with people who have invisible and indeterminate diagnoses (Blackford, 1999; Blum, 2007; Leicester and Cooke, 2002). Therefore, informal supports and the construction of positive identities through on-line interactions may be more salient to these parents than for parents with visible and more socially 'accepted' disabilities. In the *Specific Disabilities* threads, discussions focused on how to access services, how to dealing with unhelpful or gatekeeping professionals who were reluctant to refer parents to disability services, and how to overcome specific disability-related bodily challenges in achieving parenting.

In the discussions about *Benefits, Equipment & Care*, complaints and suggestions about social service interactions, discussions on how to obtain funding, transportation challenges, and how to access services, equipment and supports dominate. Challenges and triumphs were shared in stories about dealing with problematic professionals, finding one's way within the system, and locating resources outside of the system.

In all of these discussions, the primary posters and responders were female. As was noted earlier, gender is an important consideration in terms of its salience to parenting (where mothers are more likely to retain custody or be engaged in full-time parenting than fathers) and disability (where bodily norms are more stringently applied to women's appearances and activities). Thus, it is perhaps not surprising that almost all of the correspondence during the nine-month observation period, regardless of the thread, was conducted by mothers rather than fathers.

In the following sections, I will discuss these discussion threads in terms of the analytic categories laid out in theories about cybercommunication. and the public sphere in virtual space, and the uses of narrative and identity formation in on-line CoPs. These are; questions concerning the body and stigma in cyberspace, access to democracy and the public sphere in virtual space, and the uses of narrative and identity formation in on-line CoPs.

Bodies, stigma and cyberspace

In light of the claims by disability and cyberspace theorists that one of the more positive aspects of cyberspace rests in its ability to elide or override bodies that, in actual space, are routinely stigmatized and excluded, it is striking how much the body is foregrounded on the forum. Almost every thread in the data begins with a description of the body. These descriptions are offered in significant detail, similar to the following excerpts:

- I'm about 8 weeks pregnant and have CP (Cerebral Palsy) which affects my balance and speech but I am able to walk unaided...
- I have been an amputee for years, however...I have developed...chronic pain and on 'good' days use a stick to help me get around...
- I have hip dysplasia and I'm in constant pain. I also suffer with depression and restless leg syndrome...

These introductory descriptions of the body are often followed by a question about how to provide care, an account of a situation that needs resolution, or an offer of advice or information. In this light, foregrounding the body is not only reasonable, but actually necessary to inform others about what bodily attributes are posing challenges, causing worries or being overcome in achieving parenting. Thus, a mother-to-be who tells the group her diagnosis or explains the physical manifestations of her disability is also clarifying exactly why and what she faces in terms of challenges. Likewise, a mother who prefaces her advice with information about her body is laying claim to her own embodied legitimacy in terms of knowing the implications of certain kinds of disability on parenting practice. Therefore, particularly in those bodily descriptions followed by information or advice, the disabled body is actually positioned as a platform for expertise and knowledge; *this particular body*, in other words, is the body that permits one to legitimately speak and know about parenting and disability. The differently-abled body in the forum operates as the body that *legitimates* the individual in this virtual space, rather than delegitimating or stigmatizing her as often occurs in actual space.

That the differently-abled body is *delegitimated* in actual space is also made abundantly clear in forum contributions. Descriptions of stigmatizing situations and encounters are prominent in the on-line discussions. In five separate threads during the observation period, women who were pregnant described stigmatizing responses from their family, friends and support workers concerning their pregnancies, indicating that the pregnant and disabled body remains a cultural taboo. For example, one mother wrote:

> I'm worried about people's reactions and thinking we're doing the wrong thing. I just told my support worker as it will affect the care we get and her reaction was "oh my god how are you going to cope." I assured her we'd thought it through thoroughly and she said "Well you obviously didn't plan it but are you happy?" I didn't dare tell her it was planned.
> Some people think I'm being selfish as we are not 100% healthy but they do not know the full story. My mum works for social services with families who have problems and after answering all her questions and explaining things she said she was very happy for us.
> So my question is, how do I deal with all the negativity I'm getting? We're not going to be the first disabled people in the world to have a child but people seem to be acting like it! (And this support worker actually works with disabled people... so her reaction has worried me.)

In addition to describing stigmatizing interactions, anticipated situations are also described, discussed, and hopefully avoided through on-line processing. Several people wrote in discussing their plans to become pregnant and asked for input from already-parenting members about how to handle the inevitable negativity. For example, three women who were simply trying to decide whether or not to become pregnant, wrote asking for advice on how to manage unhappy parents, parents-in-law and social workers. One of these women's posts is offered below as a typical example:

> Both myself and my partner are wheelchair users - I have a rare kind of Dystonia and my other half has Limb Girdle Muscular Dystrophy - and we have spoken about the possibility of having a baby together... I am going to make an appointment with my neurologist to talk it over with him first, as I'm on medication and need to know if I can still take the tablets whilst pregnant. I also have to speak to another consultant about the arthritis in my hip (as I have pain in it) to see what he thinks. I want to know everything I can, first, before we make a decision....I know it will be difficult and tiring but this is something that we both want...I do have reservations about my family...I really do know that if I told them about this desire to have a child, they'd not be happy with it. Can anyone relate to this? Or give me some advice on how to bring it up without stirring their suspicions?

In this post, the woman does not deny but instead foregrounds her differently-abled body and her knowledge about it as way to assure others that she knows herself and is prepared to overcome bodily limits. In many ways, this reads almost like a practice run for the confrontation this woman expects will occur when her family learns of her pregnancy plans, and in it she conveys a long list of bodily accomplishments that she has either made or plans to make in order to legitimate her choice.

In response to this and the other similar requests for information and advice, the on-line offers of advice follow a similar pattern to the requests; women preface their advice or their illustrative stories with their own descriptions of the body. Women thus found their capacity for knowledge on the basis of their bodies; the disability described is typically similar to, or is even slightly more challenging than, the questioner's, and this provides the claim to expertise on which the advice rests. Further, the 'expert' stories are begun with statements outlining that "this is the body that I live with," and then giving an example of how that body has been mastered or worked around. The following is a typical response, given in this case to a question from a woman with Fibromyalgia who wanted advice about adopting:

> I have severe rheumatoid arthritis and fibromyalgia and we adopted a little girl at 20 months. I use crutches to get around and I found it difficult at the very start but I had lots of support from my husband and I quickly learned to adapt. My daughter is now 6 and is a happy, polite, well behaved child and doing very well at school despite having her own hidden disability.
> I spend lots of time with her, playing and we go on various days out, to the zoo, beach, etc.
> Fatigue is an issue for me but I have learned to pace myself and if we go out for the day I know I'll need to take it very easy for the next day or two after. Everything can be worked around.

In this response, the body acts as the legitimating ground on which the rest of the thread rests, and it is acknowledged as a source of both strength and weakness. There is no denial that this is a body that poses certain challenges to parenting, but those challenges are not insurmountable, and indeed, the body is more than capable of good parenting. Thus, in stories like this, the body, far from acting as a stigmatizing presence or a personal attribute to be elided or obfuscated, is positioned as the basic structure on which information, expertise and identity is based.

Access to democracy and the public sphere

A second aspect of on-line interactions on the parents forum relates to claims that the internet can produce CoPs where information can be shared and social opportunities can be accessed that would be otherwise unavailable for people with disabilities. However, CoPs are more than simple meeting places that provide social opportunities, since ideally they will have the capacity to act as democratizing spaces where important and emancipatory knowledge can be constructed, and where social change can begin. In light of these internet possibilities, it is thus not surprising that a significant portion of on-line discussion in the parents' forum focuses on rights talk. Here, parents are encouraged to understand their rights and entitlements, their children's rights to services, accessibility in the public sphere, access to transportation, the right to adopt, and that they need not accept the challenges of dealing with schools, social services, and other systems that provide relevant services. It is important to note that, even in discussions about rights and democratic inclusion, the on-line postings are not abstract but instead focus on bodily matters.

A core aspect of the on-line parenting group's discourse concerning rights focuses on the body in the private sphere, touching on topics such as one's right to become pregnant, the right to care for child, and the right in both childbearing and child-rearing to have the body accommodated/compensated for by social programmes. For example, a mother with a mobility impairment writes, "I just found out I'm pregnant and I'm so scared social services will say we can't cope. We both use wheelchairs but are able to walk a little. We can do a lot of things we just get tired easily." In response, she receives numerous congratulatory messages telling her not to be afraid, that this is indeed her right and a normal, adult expectation, and that others have felt similar fears but have found the parenting experience to be rewarding, very positive, and that their children have been well cared for without compromise. In short, her request is met with a discourse that normalizes disability pregnancy as a natural and happy human experience.

In addition, her request is met with more than just personal assurances and private encouragements, but is instead given responses that outline examples of how other parents have managed to accomplish parenting, and how they persisted and obtained the support necessary to accomplish parenting despite barriers. Below is a good example of such a response:

> Adult services at the "ss" [Social Services] are there to redress the balance, i.e. level the playing field to empower you as a Disabled person to become the best parent you can be; providing that there is no issue regarding neglect and they do appropriate assessments to fulfill their legal duty of care to you and your family, then there is no reason in the world why you can't be a fantastic parent. To call into question your abilities as a parent due purely to your disabilities is to Discriminate against you on the basis of your Disability, and that is unlawful... I have just written to the secretary to the minister for the disabled...Like you I have been coming up against brick walls locally... so I phoned [provides the telephone number] and spoke to the office of the minister for the disabled [provides a name of a contact]. I simply explained my needs and that I couldn't get them met locally. I explained I understood the law said that I was entitled to certain things that I couldn't get. Social inclusion etc...The lady has said she will look into it and get back to me if I write an e-mail. I did... [provides email address]...

In this response we can read a politicized narrative about rights and entitlements, accompanied by concrete examples of ways to get around a system that, although ostensibly designed to serve, can often be difficult to navigate and terrifying to negotiate for parents who are vulnerable to judgment and even apprehension of their children. This response also provides a sophisticated means of breaking down able/disabled and public/private dualisms, in its assertions that the social system is responsible for supporting differently-abled bodies in achieving their parenting rights.

The intersections between the public and private sphere is echoed in numerous threads concerning access to public spaces, to public transportation and to centres such as schools, parks and playschools that their children attend. These threads also explore such policy-related issues as obtaining a fair and adequate assessment for funding and service provision, how to lodge a complaint concerning services received, how to access information on specific services, how to obtain a new social worker when one is unhappy with the current one, what one's rights are to funding, housing, personal assistance, accommodation in the workplace/school/public sphere, the use of workers for care provision to children, and a host of practical issues that impinge on parenting with a disability. In short, the forum acts a clearinghouse for knowledge that is clearly not readily available or accessible through traditional service providers' sites.

The following exchange concerning an inaccessible washroom exemplifies a typical exchange concerning access and the public sphere:

Mother #1: I went in [provides the name of the restaurant and its location] and they have taken the baby changing unit out of the disabled toilet, the other toilets are up stairs and only in the female toilets. I am a wheelchair user with a 3 year old and 5 month old baby. How do I change his nappy? I try to go out on my own with the children. The buses are great and always make me welcome. Most shops are good but I can't go for food. It was for my daughter's birthday and it spoilt it!

Mother #2: I hope you complained to the management. Do they not realize that disabled people can have children too? I only have the use of my right arm, due to a stroke, and a very lively and quite heavy 22 month old, makes nappy changing a bit difficult in public places. I dread going to somewhere where the baby changing is in the ladies, rather than the disabled loo, cos my husband can't come in to give me a hand.

Mother #3: ...Seeing this post today made me realize that I shouldn't have to feel this way when accessing services so I am going to officially complain.

In the above, in addition to a simple complaint about lack of access, there is also an element of politicization relating to the lack of accommodation that differently-abled bodies experience in the public sphere. The first mother actually posts information about the name and location of the restaurant as a way of redressing its non-compliance to accessibility laws (and these outings of non-compliers are fairly common on the forum). In light of the literature concerning accessibility to on-line information and its fragmented and shifting locations and content, providing such information is a strong form of resistance to structural barriers for people with disabilities, and poses a grass-roots construction of a democratic sphere.

The second mother invokes a discourse about rights and accessibility, framing this in terms of the shared burdens of differently-bodied mothers, and the rights of parents to specific services that will level the playing field for parents with disabilities. The final post indicates the political potential of such discussions in mobilizing others. In this thread, a complaint is made public, a problem is legitimated and connected to broader politics, and resistance is taken up through the changed consciousness of the third mother. Thus, the parents' forum does have the potential to overcome some of the social barriers disabled parents experience to full participation in their societies and the democratic process.

While this may be a forum for micro-politics, it is clear from the threads concerning not only restaurant access, but programme provision, social service agencies and other public institutions that the forum is a mobilizing arena for emancipatory knowledge production and considerable social change. Thus, despite some critic's fears that disability sites can act as virtual ghettos that simply reinforce disability as a marginal category, these forums can also facilitate political action, and can produce meaningful solidarity.

Narratives, identity and the internet

A final perspective on the utility and emancipatory potential of the internet is that it can provide a space for people with chronic illness and disability to share their personal narratives and, through producing their own stories and incorporating the stories of similar others, construct a new and more empowered identity that stands in distinction from dominant understandings of the body and bodily experiences. One of the primary dominant understandings of disabled bodies is that they are not entitled to, capable of, or safe in parenting. Recall that disabled mothers face challenges in parenting due to an increased likelihood of parenting in poverty or without a partner, heightened susceptibility to abuse, inadequate or inaccessible housing, and public perceptions of them as inadequate or inappropriate to parenting. In addition, mothers face misperceptions from professionals and bureaucrats about their capability to parent; this is particularly reflected in the myth of the 'upside-down family', where outsiders presume that mothers with disabilities depend upon their children for care rather than being the adults in the family (Booth and Booth, 1998). It is thus not surprising that mothers with disabilities face challenges in obtaining approval and support for their parenting decisions, and heightened probabilities of negative family interventions and child apprehensions (McConnell and Llewellyn, 2000; Swain and Cameron, 2003).

Concerns about these dominant misperceptions and disabled parents' own worries about their capabilities are frequent postings on the parenting forum. Parents often request information about how to avert negative perceptions about their parenting capacity in the public arena, and the responses to these queries typically follow a redemptive narrative that resists dominant perceptions of disability as synonymous with a deficit model of parenting. Instead, the narratives work to reclaim a positive identity as a disabled parent. For example, in response to a young mother-to-be's query about how to approach her worker for parenting support without raising suspicions that she may not be competent, another mother wrote:

> I totally understand your fear, I have been through it all myself...After 5 years of no support, partially through fear of asking for it and partially from having the social worker from hell, I demanded in-home support [she then goes on to describe how much, from whom, and how she did it]...
> My kids are 6 and 8 now, and being a parent is the hardest but most rewarding job I ever did. Don't let your fear of dealing with these agencies rob you of the joy of being pregnant or make you fearful of asking for help. I have... [over the years]remained the main care provider, because the fact of the matter was, with the right level of support physically, I am just as good a parent as an able bodied person- well better, but I am prejudiced! Plus I get to spend all my spare time and energy on my kids without the distractions of work. I know it is easy to say, but a lot of parents- to-be feel as you do now without facing the obstacles we face.

This mother's narrative accomplishes a number of positive outcomes. First, she provides a concrete example of positively dealing with a powerful opponent and overcoming both bodily and systemic obstacles. She also provides a detailed story about how she did this, and in the telling, informs her listeners of how they, too, might approach such a task. In addition, she describes her mothering not in terms of deficit or dependency, but instead in terms of surpassing 'typical' able-bodied parenting because of her dedication, privileged at-home status, and her resourcefulness in getting her and her children's needs met precisely because she *does* have disabilities. Thus, in her description of overcoming bodily and systemic limitations, she lays claim to the identity of a 'good *disabled* mother', which is an identity not readily available to women with disabilities in the current culture of disablism[3].

We can also presume that these new, empowered identities help other parents in their negotiations for improved services and supports in the world outside of the on-line forum, as many parents attest to this in their reply posts in various threads.

This construction of a positive disability and parenting identity is a central goal for many members of the forum, as is expressed eloquently by another mother, who wrote:

> I am involved in the forums partly for company for myself and to have someone else who understands me, but also to try to see disabled parents don't go thru that same ordeal of assessment, feeling useless, like a bad parent, feeling like you are on your own and a failure as a parent and as a human being. It is the system that needs to change and not us.

Thus, for this woman and others, the on-line forum offers the promise of a democratic space, in which mothers' own stories from their own perspectives provide the opportunity for connection, the exchange of vital and otherwise difficult to obtain information and the basis for meaningful social change. The forum offers its members the capacity to rewrite the conditions of their lives and to see themselves not as women with bodies that pose problems, but as women whose bodies are deserving of more than is currently on offer from the systems of 'support' available to them outside the on-line forum. Additionally, these narratives provide the means of constructing on-line solidarity amongst members, who can celebrate their shared differences and triumphs and learn from them.

The shared narratives, the non-stigmatizing ways that bodies are used on-line to normalize difference and naturalize disabled parenting, and the ways that knowledge and consensus are constructed in the forum provide these parents not only with a real and meaningful platform for advocacy and social change outside of the virtual arena of the parenting forum, but they also open up the possibility of a new, positive, and otherwise rarely acknowledged identity as a good, disabled mother. Finally, in light of the forum's discussion about off-line self-advocacy, it is clear that these identities and possibilities do not remain contained in an on-line ghetto, but they actually make a difference for parents and their families in the 'real' world.

Notes

1. In speaking about computer-mediated interaction, Jodi Dean draws on Habermasian theory to argue that the Net is not a 'proper' public sphere, where people negotiate and contest matters of common concern to reach consensus. Instead, she argues that it is actually damaging to democratic processes. In its current guise, the net is characterized on the one hand by lack wherein people cannot all equally access the Net or evaluate information and truth mediated on the web. On the other hand, the net is characterized by excess, where there is, if you will 'too much' democracy, since anyone can participate in knowledge and truth production regardless of their qualifications to do so. Dean instead poses an ideal-type of on-line neodemocracy that would focus on issue, not actors, on credibility over rationality/self-presentation, and on outcomes rather than process.
   Dean, Jodi. 2003. "Why the Net is Not a Public Sphere." *Constellations* 10:95-112.
2. The social model of disability argues that it is not physical 'impairments' or 'conditions' that comprise the disabling aspects of people's lives, but rather it is social institutions, norms and structures that fail to accommodate difference that comprise the truly disabling conditions (Oliver, 1996).
3. Disablism is a term first used in the UK within the disability rights movement to describe the discriminatory, oppressive or abusive systems and interactions that arise from the belief that disabled people are inferior to others (http://www.scope.org.uk/disablism/). I am arguing that in this case, these systems operate to delimit women's capacity as women and as mothers.

Bibliography

Anderberg P. (2007), *Peer Assistance for Personal Assistance: Analysis of Online Discussions about Personal Assistance from a Swedish Web Forum for Disabled People.* Disability & Society, 22, Pg. 251–265.
Anderberg P. and Bodil J. (2005), *Being There.* Disability & Society, 20, Pg. 719–733.
Asch A. and Fine M. (1992), *Beyond Pedestals: Revisiting the Lives of Women with Dis-*

abilities. In: *Disruptive Voices: The Possibilities of Feminist Research*, edited by M. Fine. Ann Arbour: The University of Michigan Press.

Asch A., Rousso H. and Jefferies T. (2001), *Beyond Pedestals: The Lives of Girls and Women with Disabilities.* Pg. 13–48. In: *Double Jeopardy: Addressing Gender Equity in Special Education*, edited by H. Rousso and M. L. Wehmeyer. Albany: State University of New York Press.

Balshem M. (1993), *Cancer in the community: Class and medical authority.* Washington and London: The Smithsonian Institution Press.

Blackford K. A. (1993), *Erasing Mothers with Disabilities through Canadian Family-related Policy.* Disability, Handicap & Society, 8, Pg. 281–294.

Blackford K. A. (1999), *A Child's Growing Up with a Parent Who Has Multiple Sclerosis: theories and experiences.* Disability & Society, 14, Pg. 673–685.

Blum L. M. (2007), *Mother-blame in the prozac nation: Raising kids with invisible disabilities.* Gender & Society, 21, Pg. 1–25.

Booth T. and Booth W. (1998), *The Myth Of The Upside Down Family.* Pg. 146–168, In: *Growing Up With Parents Who Have Learning Difficulties*, edited by T. Booth and W. Booth. London, Britain: Routledge.

Booth T. and Booth W. (1994), *Parenting under pressure: Mothers and fathers with learning difficulties.* Buckingham: Open University Press.

Bowker N. I. and Tuffin K. (2007), *Understanding Positive Subjectivities Made Possible Online for Disabled People.* New Zealand Journal of Psychology, 36, Pg. 63–71.

Charmaz K. (1990), *Discovering Chronic Illness: Using Grounded Theory.* Social Science & Medicine, 30, Pg. 1161–1172.

Davis J. J. (2002), *Disenfranchising the Disabled: The Inaccessibility of Internet-Based Health Information.* Journal of Health Communication, 7, Pg. 355–367.

Davis L. J., (1997), *Constructing Normalcy.* Pg. 9–28, In: *The Disability Studies Reader*, edited by L. J. Davis. London: Routledge.

Dean J. (2003), *Why the Net is Not a Public Sphere.* Constellations, 10, Pg. 95–112.

Drake R. F. (1999), Chapter 6: *Contemporary British policy in the international context.* Pg. 93–108, In: *Understanding Disability Policies.* London: MacMillan Press.

Frank A. W. (2002), *Why Study Peoples' Stories? The Dialogical Ethics of Narrative Analysis.* International Journal of Qualitative Methods, 1, Pg. 1–9.

Froomkin A. M. (2003), *Habermas@discourse.net: Towards a Critical theory of Cyberspace.* Harvard Law Review, 116, Pg. 749–871.

Glaser B. and Strauss A. (1967), *The Discovery of Grounded Theory: Strategies for qualitative research.* New York: Aldine de Gruyter.

Goffman E. (1959), *The Presentation of Self in Everyday Life.* Garden City, New York: Doubleday Anchor Books.

Goffman E. (1963), *Stigma: Notes on the Management of Spoiled Identity.* New Jersey: Prentice-Hall, Inc.

Gold G. (2003), *Rediscovering Place: Experiences of a Quadriplegic Anthropologist.* the Canadian Geographer/Le Geographie canadien, 47, Pg. 467–479.

Government of Canada (2008), *Participation and Activity Limitation Survey 2006: Labour Force Experience of People with Disabilities in Canada.* Pg. 1–22, edited by Social and Aboriginal Statistics Division: Statistics Canada.

Hall K. Q. (2002), *Feminism, Disability, and Embodiment.* NWSA Journal, 14.

Haraway D. (1991), *The Biopolitics of Postmodern Bodies: Constitution of Self in Immune System Discourse.* Pg. 203–229, In: *Simians, Cyborgs, and Women: The Reinvention of Nature*, edited by D. Haraway. New York, New York: Routledge.

Heim J. (2000), *Locking Out the Disabled.* PC World, 18, Pg. 1–5.

Huberman A. M. and Miles M. B. (1998), *Data Management and Analysis Methods.* Pg. 179–210, In: *Collecting and Interpreting Qualitative Materials*, edited by N. K. Denzin and Y. S. Lincoln. Thousand Oaks, California: Sage.

Kang S. (2007), *Disembodiment in Online Social Interaction: Impact of Online Chat on Social Support and Psychosocial Well-Being.* CyberPsychology & Behavior, 10, Pg. 475–477.

Kirshbaum M. and Olkin R. (2002), *Parents with Physical, Systemic, or Visual Disabilities.* Sexuality and Disability, 20, Pg. 65–80.

Kleinman A. (1988), *The illness narratives: Suffering, healing and the human condition.* New York: Basic Books.

Leicester M. and Cooke P. (2002), *Rights Not Restrictions for Learning Disabled Adults: A Response to Spiecker and Steutel.* Journal of Moral Education, 31, Pg. 181–187.

Malacrida C. (2007), *Negotiating the Dependency/Nurturance Tightrope: Dilemmas of Disabled Motherhood.* Canadian Review of Sociology, 144, Pg. 469–493.

Malacrida C. (2009a), *Performing motherhood in a disablist world: dilemmas of motherhood, femininity and disability.* International Journal of Qualitative Studies in Education, 23, Pg. 1–17.

Malacrida C. (2009b), *Parents with Disabilities.* In: *Canadian Families Today*, edited by D. Cheal. Don Mills, Ontario: Oxford University Press Canada.

McConnell D. and Gwynnyth L. (2000), *Disability and Discrimination in Statutory Child Protection Proceedings.* Disability & Society, 15, Pg. 883–895.

McMahon M. (1995), *Engendering Motherhood: Identity and Self-Transformation in Women's Lives.* New York: The Guilford Press.

Miller W, L. and Crabtree F. B. (1998), *Clinical Research.* Pg. 292–314, In: *Strategies of Qualitative Inquiry*, edited by N. K. Denzin and Y. S. Lincoln. Thousand Oaks, California: Sage Publications.

Morahan-Martin J. and Schumacher P. (1999), *Incidence and correlates of pathological Internet use among college students.* Computers and Human Behavior, 16, Pg. 1–17.

Oliver M. (1990), *The Individual and Social Models of Disability.* Paper presented at Joint Workshop of the Living Options Group and the Research Unit of the Royal College of Physicians, Retrieved January 25, 2007. (http://www.disability-archive.leeds.ac.uk/authors_list.asp?AuthorID=133&author_name=Oliver%2C+Mike).

Oliver M. (1996), *Defining Impairment and Disability: Issues at Stake.* Pg. 29–54, In: *Exploring the Divide*, edited by C. Barnes and G. Mercer. Leeds: The Disability Press.

Prilleltensky O. (2004), *My child is not my carer: Mothers with physical disabilities and the well-being of children.* Disability & Society, 19, Pg. 209–223.

Sakalys J. A. (2000), *The political role of illness narratives.* Journal of Advanced Nursing, 31, 1469–1466.

Schur L. (2004), *Is There Still a "Double Handicap"?* Pg. 253–271, In: *Gendering Disability*, edited by B. G. Smith and B. Hutchison. New Brunswick: Rutgers University Press.

Seymour W. and Lupton D. (2004), *Holding the Line Online: Exploring Wired Relationships for People with Disabilities.* Disability & Society, 19, Pg. 291–305.

Shakespeare Tom and Watson N. (2002), *The Social Model of Disability: An Outdated Ideology?* Research in Social Science and Disability, 2, Pg. 9–28.

Shildrick M. (2002), *Embodying the Monster: Encounters with the Vulnerable Self.* London: Sage Publications Ltd.

Stone Allucquere R. (1991), *Will the Real Body Please Stand Up.* Pg. 81–118, In: *Cyberspace: First Steps*, edited by M. Benedikt. Cambridge, Massachusetts: MIT Press.

Strauss A. and Corbin J. (1994), *Grounded theory methodology: An overview.* Pg. 273–285, In: *Handbook of qualitative research*, edited by N. K. Denzin and Y. S. Lincoln. Thousand Oaks, California: Sage Publications.

Sullins J. (2000), *Transcending the meat: Immersive technologies and computer mediated bodies.* Journal of Experimental & Theoretical Artificial Intelligence, 12, Pg. 13–22.

Swain P. A. and Cameron N. (2003), *'Good enough parenting': Parental disability and child protection.* Disability & Society, 18, Pg. 165–177.

Taub D. E., Fanflik P. L. and A McLorg P. (2003), *Body Image Among Women with Physical Disabilities: Internalization of Norms and Reactions to Nonconformity.* Sociological Focus, 36, Pg. 159–176.

Taub D. E., McLorg P. A. and Fanflik P. L. (2004), *Stigma management strategies among women with physical disabilities: contrasting approaches of downplaying or claiming a disability status.* Deviant Behavior, 25, Pg. 169–190.

Tringo J. L. (1970), *The Hierarchy of Preference Toward Disability Groups.* The Journal of Special Education, 4, Pg. 295–306.

Turkle S. (2007), *Authenticity in the Age of Digital Companions.* Interaction Studies, 8, Pg. 501–517.

Vernon R. and Lynch D. (2003), *Consumer Access to Agency Websites: Our Best Foot Forward?* Journal of Technology in Human Services, 21, Pg. 37–51.

Waddell C. D. (2000), *'Electronic Curbcuts': The ADA in Cyberspace.* Human Rights Winter, Pg. 22–24.

Wendell S. (1996), *The Rejected Body: Feminist Philosophical Reflections on Disability.* New York and London: Routledge.

Wenger E. (2000), *Communities of Practice and Social Learning Systems.* Organization, 7, Pg. 225–247.

Volume Two

Mapping Bodies in Networked Space

First Part

The Body and its Prosthetics: Laboratories

Chapter 1
...*a body under the bushes*: ghosts in the machine

Alan Dunning, Paul Woodrow

The body and its selves

The Einstein's Brain Project is a collaborative group of artists and scientists who have been working together for the past twelve years. The aim of the group is the visualization of the biological state of the body through the fabrication of environments and simulations. The Project develops and presents systems and installations using analog or digital interfaces to direct the output of the human body to virtual environments that are constantly being altered through feedback from a participant's biological body. The core of the Einstein's Brain Project is a discursive space that engages with ideas about the resituation of the body in the world and its digital cybernetic and post-human forms.

The Project's work is focused on how representations[1] of the biological body might be manifest in the world through mediatized spaces. These representations conflate the virtual, symbolic and imaginary worlds through the use of interactive virtual environments and installations that promote a high degree of disorientation and the moment to moment construction of a self.

This is:

> ... the self which enters into social relations with other selves. ...The growth of the self arises out of a partial disintegration,—the appearance of the different interests in the forum of reflection, the reconstruction of the social world, and the consequent appearance of the new self that answers to the new object (Mead, 1913).
>
> The cognitive questions asked by most artists in the 20th Century are; How can I interpret this world of which I am a part? And what am I in it? The post-cognitive questions since then are: Which world is this? What is to be done in it? Which of my selves is to do it (Higgins, 1999).

The Project suggests that these selves are not limited by the boundaries of the body, but are engaged in a complex interplay of states at the limits of and beyond the ego boundary, shifting from subject to object, and from self to other in an intricate dance that never fully establishes the relationship between body, machine and world, as it remains uncertain as to who is made and who does the making. It suggests that the body/mind is a function of the technological environment, and the mind and the self extend out into the world.[2]

> ...life in a mediated age sets in train a complex series of opportunities and constraits which carry important implications for contemporary forms of emotionally embodied experience. Here the tensions between re-presentation and reality, emotional involvement and detachment, together with the associated high-tech problems of dualism in cyberspace, suggest a number of dilemmas for the individual at the turn of the century; dilemmas which defy easy answers or simple (re)solutions. These developments in turn, suggest the possibility or having critically to rethink existing sociological concepts: from the 'framing' of 'reality' to the dilemmas of the emotionally embodicd self, and from the problems of 'emotion work' to the transformation of 'feeling rules' and the vicissitudes of 'surface' and deep acting in cyberspace (Braidotti, 2003).

Recent advancements in technology, computing science and medicine have opened up new possibilities of representation for the artist, who can now use novel and surprising ways of imaging the body in ways that question our traditional notions of world, mind and body. The interdisciplinary nature of this investigation has forced artists to recognize the necessity of deep collaboration with practitioners of other disciplines, such as medicine and computer science, in order to create, develop and use the new languages and technologies of bodily representation in sophisticated ways. The Project proposes new interdisciplinary and collaborative methods, using art as one just one mode of inquiry, to create more comprehensive and total versions of the body, to explore and develop novel forms arising out of interdisciplinary collaboration.

The body is recursively and cybernetically constructed. It affects its environment even as it is affected by its environment. Bruce Sterling calls attention to our chemical relationship with the world:

> A human body can be understood as a sponge of warm saltwater within a shell of skin so everything we emit ends up partially within ourselves. Some artificial substances are bioaccumulative: our metabolisms preferentially suck them out of the biosphere and try to make structure out of them, These processes are involuntary and tale place beneath our awareness (Sterling, 2005).

Arthur Kroker describes an even more permeable and diffuse body:

> No longer the Cartesian thinking subject, however, but a fractal subjectivity in an ultramodern culture where panic science is the language of power: no longer ratiocination to excess. but parallel processing as the epistemological form of postmodern consciousness...; no longer the geometrically-focused and self-regulating body, but technologies of the body immune as key features of a libidinal economy that produces toxic bodies and designer aesthetics as its necessary conditions of operation: and no longer univocal (grounded) perspective, but the fatal implosion of perspective into the cyberspace of virtual technology (Kroker, 1987).
>
> ...the body simultaneously implodes into its own senses, and then explodes as its central nervous system is splayed across the sensorium of the technoscape. No longer a material entity, the postmodern body becomes an infinitely permeable and spatialized field (Kroker, 1989).

The limit of the body is a complex interplay of perception, imagination and chemistry. Commonly held beliefs characterize the body as separating us from the world—our skin being the limit of ourselves. In spite of this belief, our body is pierced with a myriad of openings. Each opening admits the world: stardust gathers in our lungs, gases exchange, rainwater moves through our blood vessels and external magnetic fields flow over and through our skin, muscle and bone. We are continually linked to the world and other bodies by strings of matter and invisible forces. We also project our bodies into the world: we speak, we breathe, we write we leave footprints and trails of cells and we absorb the trails of others. Thought of in this way, the body enfolds the world and the world enfolds the body; and the notion of the skin as the outer boundary of the body dissipates. The body, as here, but not there, and its defining sense of the other is thus a mental construction.

Given the complex relationship between the construction of the self and the construction of space, the body in a world has moved towards a hybrid state, composed of biological organism and machine, in which it is not always precisely clear who makes and who is made. The boundary between organism and non-organism, actor and non-actor, self and non-self has been abandoned, and our postmodern bodies are artificial constructions of technologies and technological discourses that are continually decaying and being renewed. The body is so inextricably enmeshed with its surroundings and the technologies that support it, that representations of the body become indistinguishable from the mechanisms of its representation and erasure.

This interplay of phenomena and energies leads to the non-body, the body without skin, without ego boundary, as the evolutionary consequence of the super-distribution of identity over increasingly broad and responsive real-time networks. The body enfolds the world and the world enfolds the body—the notion of the skin as the boundary to the body falls apart. The body, as here not there, and its defining sense of the other are a mental construction—every perception of the other is a creation and every invocation a re-creation.

Given this indeterminate state, the body is struck by considerable ontological anxiety. Faced with issues concerning the potential loss of identity, being and becoming, the Project conceptualized a series of theoretical body/mind models that were realised through a physical interface to computer code, allowing a participant in an installation to interact with the work's content and form, as they attend to bodies of greater or lesser physical dimension.

The interface—ALIBI (Anatomically Lifelike Interactive Biological Interface)—is a life-size human body filled with sensors that allows participants to interact with computer systems by touching, speaking to or breathing on the body. The body is covered with thermochromic paint that changes colour when touched. Used in conjunction with a brain wave recording device built into a Heads Up Display (HUD) the body interface not only provides access to computer code, but also conflates this with the participant's bio-signature, as it monitors the biological and physical output of a participant's own body including EEG data, skin temperature, electrical resistance, speech, gesture and motion.

The body appears male, but on closer inspection has varying degrees of maleness, femaleness and androgyny. In fact, the body was cast from several people, each contributing an arm, leg, genital or other body part, and assembled into its final Frankensteinian form. Its final gender is uncertain—its decisive genital area distorted beyond recognition.

This interface is at one and the same time beautiful and uncanny. The blue is reminiscent of Yves Klein's International Klein Blue, a distinctive ultramarine which Klein registered as a trademark colour in 1957 - a color he described as giving "rise to a sense of immersion in a space greater than infinity. The blue is the invisible becoming visible (Weitemeier, 2001) and the figure itself is cool and perfect, recalling the most beautiful of Greek sculptures. Yet, on close inspection the body begins to fall apart. Underlying this beauty is an increasing sense of misalignment, akin to the feelings we experience when looking at a mutation or mutilation. The body for all that it is brought together as a seamless representation, in which diversity and distinction are concealed, finally breaks apart, revealing itself as fatally marked by division and difference.

> Though the term 'monster' must be reserved for animate or organic entities. I would argue for a structural analogy between the organic monster and the technological or anthropomorphic other. Body-doubles, robots or automates have in fact the same metamorphic effect on the human observer as the freaks .... They are objects of wonder and terror, loathing and desire. As body-doubles. they represent a re-assembly of organic parts, often arranged in a new order. This arrangement follows the rule of organizing organs on monstrous bodies by excess, lack or displacements. Quite often, for reasons of convenience as well as playfulness, the rearrangement of organic parts in the mechanical body-double expresses a fantastic array of alternative body-shapes, bodily functions, morphologies and sexualities. ... As such it embodies the paradox of an irreducible singularity that serves as the model of expression for the innermost human faculties, while it renders them as external functions autonomous from the subject (Braidotti, 2003).

In the cycle of installations The Madhouse, The Mnemonic Body and Pandaemonium, ALIBI was placed in the centre of the room and participants touched, stroked and breathed upon particular areas to produce corresponding images, at times related to the body function and at other times related to metaphor and simile, generated in a virtual world. This world and its images were simultaneously seen by the participant through a Heads Up Display and by an audience as the images were projected on a screen in the gallery.

When viewed by other observers the participant resembles a pathologist examining a corpse. The image seen in the glasses are degraded black and white images. Distorted views of people, buildings, objects in motion, lights, appear as semiotic ghosts, faded memories of the city or low resolution recordings dredged up from some deep and long-forgotten time well. Attached to the HUD are electrodes that monitor the participant's brain waves, and feed these in real time into a three-dimensional virtual reality environment that manifests them as visual and aural equivalents that are projected onto the second screen and amplified into the space. This virtual reality simulation is in constant flux—a living phenomenon that evolves in infinite space and time and responds in real time to a participant's mental activity. What a participant sees is a real-time manifestation of his or her brain and biological activity. When the participant is at ease the environment and sounds are fluid, slow paced and smooth, as he or she responds to the images the environments and audio become jagged, fast paced and increasingly raucous. When a participant is again able to reach a tranquil state the work returns to a languid condition. As the participant touches the body sounds and images are synthesized in real time, modified by the biological output and fill the gallery space.

As biodata is streamed into the system, the local visualized data is radically modified, and its characteristics—its *life*—more and more distinctly independent of the participant. This *life* is a mix of machine code and biological process that is constructed and reconstructed so rapidly that it becomes increasingly difficult to be sure of what is being created, and quite where this creation resides. What is produced is a representation of a technologized bodymind—a representation which moves away from the traditional notion of the physical body towards the conception of a dynamic body; that is subject to the fluctuations of time and space and an uneasy co-existance of machine and biological body, that is unstable, contingent and dynamic.

In a world of objects, the subject is characterized and limited by boundaries and frames, perceived as unvarying and outside of natural morphic transformations. Now it is possible to view ourselves as dynamic entities continually engaged in perpetual iconoclastic, biological and social renovation and construction. We are able to recognize the transformations operating on our bodies and can acknowledge the hypermorphic and transorganic nature of our image of ourselves. In "The Saturated Self", Kenneth Gergen describes these selves existing in

> . . . a state of continuous construction and reconstruction. It is a world where anything goes that can be negotiated. Each reality of self gives way to reflexive questioning, irony and ultimately the playful probing of yet another reality (Kenneth, 2000).

ALIBI is a container for a series of body/mind models These models are situated between the imaginary and the symbolic, and together with the interface provide a tangible connection to worlds that are quintessentially incorporeal.

The Project has conceptualized ten body/mind models at play in the works of the Project: The Conscious Body, the Absent Body, the Active Body, the Mnemonic Body, the Amplified Body, the Data Body, the Coded Body, the Fractal Body, the Permeable Body and the Monstrous Body. For the purposes of this paper we will look at five of these models: The Conscious Body, the Absent Body, the Mnemonic Body, the Amplified Body, and the Auratic Body.

The Conscious Body—the body minded brain—is a body linked to consciousness, the awareness of both the external and internal functioning that can be felt, sensed, and expressed through non-verbal feelings—all those conditions which originate an apparent continuity in the brain. This is Edelman's and Damasio's conception of the movie-in-the-brain—the sense of self in the act of cognition. According to David Bohm

> ...consciousness involves continuous processes of enfoldment and unfoldment. For example, in the case of matter, entities such as atoms may represent continuous enfoldment and unfoldment, which manifests as a relatively stable and autonomous entity that can be observed to follow a relatively well-defined path in space-time (Bohm, 1980).

In this body, it is the brain that is the focus of subjective indeterminacy as perception transforms behaviors, utterances and contexts into new dimensions of intensity, and as the brain regulates, selects and discards data.

The Absent Body is a virtual body that must be consciously and repeatedly reanimated to maintain its purchase in the culture and fabric of a virtual world. While bodies are to varying degrees absent in daily life (forgotten, veiled or masked, or even no longer in existence) in virtual environments the surrogate body—a replicant body with all the difficulties of establishing one to one correspondence with a real world—becomes the only means for agency in cyberspace. Always on the threshold of emergence this absent body is the epitome of indeterminacy, characterized by Brian Massumi as abstract, incipient, recursive and atemporal (Massumi, 2002). The absent body is a living paradox where opposites intermingle, associate and collide.

The Mnemonic Body is a repository for memory and events. The body functions like a moving object rather than the static subject of important information when considered as a recording medium—as a meat tape. In Holonomic Brain Theory, Karl Pribram (Pribram, 1997) postulates that memories are enfolded within every region of the brain rather than being confined to specific atoms or cells. Pribram's collaborator, the physicist David Bohn, described the process in the following way. One may indeed say that our memory is a special case of the process described above, for all that is recorded is held enfolded within the brain cells and these are part of matter in general. The recurrence and stability of our own memory as a relatively independent sub-totality is thus brought about as part of the very same process that sustains the recurrence and stability in the manifest order of matter in general (Bohm, 1980).

Brian Massumi expands these notions:

> The body doesn't just absorb pulses or discrete stimulations; it infolds contexts, it infolds volitions and cognitions that are nothing if not situated, intensity is asocial, but not presocial—it includes social elements but mixes them with elements belonging to other levels of functioning and combines them according to a different logic (Massumi, 1996).

Adding that this can only happen if the trace of past actions, including a trace of contexts, were conserved both in the brain and in the flesh.

The Amplified Body is the body projected into the world. As the processes of the body are made visible and audible, the body is transformed to take on altered and enhanced personalities and shapes and is able to move in new, experimental (although often problematic) ways. While this shares some technical similarities to Stelarc's performances around the amplified body, the Project's use of this term is expanded to recognise a body that is able to move through the world by virtue of its largely unacknowledged and unnoticed relation to the code that is the interface between humans and machines.

> As the technological nonconscious expands, the sedimented routines and habits joining human behavior to the technological infrastructure continue to operate mostly outside the realm of human awareness, coming into focus as objects of conscious attention only at moments of rupture, breakdown, and modifications and extensions of the system No longer natural, human-only language increasingly finds itself in a position analogous to the conscious mind that, faced with disturbing dreams, is forced to acknowledge it is not the whole of mind. Code, performing as the interface between humans and programmable media, functions in the contemporary cultural Imaginary as the shadowy double of the human-only language inflected and infected by its hidden presence (Hayles, 2006).

The Auratic Body is a body that is expressed by electronic and statistical means. Demographic databases construct a surrogate body. This is a pareidolic body an imagined and imagining body, that searches for signs and symbols and meaning in randomness and noise. This is the self replicating body, self mutating, whose raison d'être is the making of meaning in the face of arbitrary and misleading signs.

The contemporary auratic body is a function of the technical image—a ghostly, spectral body, with no mass or matter save the photons that comprise light. It is an ineffable emanation produced as a function of the improbable arrangement of moving electronic particles on a screen. At any one moment moving pixels on a screen are producing pure noise, it is only over time, and after the fact recognition, that images emerge.

Given this it seems that all projections and light emanating screens can be considered as containing an auratic component, a true ghost within the machine. As an extension, what these manifestations suggest is that we ourselves must be, in part hallucinations, phantoms, our body image only an auratic projection of the brain. What this problematizes is the making of meaning in a world that is self-reflexively constructed through technical images.

> ...traditional images are mirrors. They catch the vectors of meaning approaching us from the world, codify them and reflect them, codified, on a surface. Therefore it is correct to question them on their meaning. The technical images, on the other hand, are projections. They catch meaningless signs that approach them from the world (photons, electrons), and they codify them to give them meaning. Therefore, it is wrong to ask what they mean (except for the meaningless answer: they mean photons). The technical images need to be asked why they mean what they show, as what they show is only a function of what they mean (Flusser, 2000).

The novel and transformative movement of these bodies through the Project's works is hypermorphic and transorganic.

Hypermoprphism is the tendency of world components to change from one form to another so rapidly and continuously that although they never fix on a form they appear stable. This is akin to Bergson's notion of "duration" as the continuum of becoming itself. The appearance of stability is at issue here. While components may appear stable, even static the deep structures are in constant flux and upheaval. Such stability is a surface characteristic—attached to conventional space and time—and is largely responsible for our tendency to categorize and identify, but it is never able to represent the true dynamic nature of matter and perception. It suggests that there is an alternative to the notion of apprehending discrete moments of a passing reality in a process that comprises independent moments rather than an interpenetration of images, instead suggesting a continuous remaking of the lived world without frames or direction.

Bergson reveals that:

> Instead of attaching ourselves to the inner becoming of things, we place ourselves outside them in order to recompose their becoming artificially. We take snapshots, as it were, of the passing reality.... the mechanism of our ordinary knowledge is of a cinematographical kind (Bergson, 1998).

Transorganic is the nature of forms oscillating hypermorphically between the extremes of organic and non-organic. This suggest the rapid approach of a postorganic society in which communicating bodies undergo a transformation as they interface with the postorganic, the classical (hardware-interfaced) cyborg and the postclassical (software-interfaced), transorganic data-based cyborg or personality construct (Tomas, 1991).

> The new convergence between body and architecture through electronic technology is described as postorganic. Our horizon is characterised by a paradigmatic reversal of perspective making it essential to overcome the logic of opposition between the organic universe of the body and the mechanical universe of technology, in a new logic of complexity in which the life of the body and forms meet through the machine. The word machine is used here to denote a symbolic value that, at this time of revolution, underlines the importance of a diachronic point of view to understand the role played by technique in the relationship between the body and architecture. A continuous thread binds (and opposes) the perspective machine and the electronic machine as strategies of vision, in other words, intermediate instruments between the body and space in understanding and depicting reality (Palumbo, 2000).

These hyper-bodies can exist only in categorical, deeply nested, multiplicitous environments. Spectator and spectacle are entwined, entangled in coexistent space. Perception enfolds us in matter and synthesizes the perceived object and us.

> The modes of representation and being-in-the-world are intimately intertwined in practice, for example in the way their relation can be superimposed on the relation between subject and object: if the body is conceived as an object, representations of the body are the site of subjectivity; if the body is conceived as subject, representations are objectifications of the body (Csordas, 1997).

If the body moves hypermorphically and transorganically between these two conceptions we move towards a version of the cyborg, "a hybrid creature, composed of organism and machine". We enter a world in which it is no longer certain what are the limits of our bodies. Caught, like the deep-sea diver, in the rapture of the deep, we are aware only of the painful and unbearable separation from ourselves, surrounded as we are by amputated limbs and severed psyches. In our desire to reunite and recuperate our unnaturally divided bodies we are reconfigured as body-world. The body is always on the verge of coming into being but never fully formed. It exists only in a retrospective past and unforeseeable future, with no place in the present. This is Body Degree Zero.

Body Degree Zero comes to represent a functional alternative, to the Cartesian or Lacanian constructions of self: no longer minds fed back onto the self-reflection of psychology or philosophy; rather "non-bodies" fed back upon themselves, experiencing the data derived from an immanent physical presence that both is and is not theirs; heartbeats that form not only the internal rhythm of living but an external soundscape; brainwaves that, stripped of the illusion of privacy, begin to gyrate on electronic screens; and McLuhan's prophecy fulfilled, as the data-skeleton of autonomic bodily processes becomes clothing worn, subject to all the same rules of designer fashion and aesthetic self-fashioning.

In a series of performance works entitled "Body Degree Zero" the traces of real-time and past bodily activity generate spaces and forms to realise visualizations of an immanent, past and future mindbody, that might never have existed at all.

> How could this be so? Only if the tract of past actions including a trace of their contexts were conserved in the brain and in the flesh, but out of mind and out of body understood as qualifiable interiorities, active and passive respectively, directive spirit and dumb matter. Only if past actions and contexts were conserved and repeated, autonomically reactivated, but not accomplished; begun, but not completed. ... pastnesses opening onto a future, but with no present to speak of. For the present is lost with the missing half-second, passing too quickly to be perceived, too quickly, actually, to have happened (Massumi, 1996).

Two participants are wired to various bio-electrical sensors, ranging from electrodes that track brain waves to sensors attached to the skin that record galvanic skin responses. The sensors track the participants' activity and bio-electrical output and visualize this in real time in the form of synthesized and manipulated images that are back projected onto two large screens in front of the participants. Accompanying the images are sounds generated from the activity of the participants. Local environmental conditions are monitored and streamed into the system, where they modify the participants' data. As the resultant mix becomes more and more complex and its' characteristics more and more an amalgam of the three data entities, it becomes increasingly uncertain what is being represented and exactly what space this new recombinant form demands, as the participants' identities are lost in the ensuing environmental data storm.

Random numbers create noise in the system, creating errant and misleading signs. Some visualizations are easily identified: the throbbing of a shape is a beating heart; the increase in particle production is a rise in GSR. Others are suggestive rather than indicative, taking the form of erratic, distorted, cinematic or historical images that are streamed into the space or mapped onto primitives or of shapes morphing from one form to another too rapidly to be comprehended. Similarly, sounds are generated in real time using a software synthesizer and are an index of peak voltages, frequency and percentage differences. Data are mapped onto harmoniousness—values close to whole numbers are harmonious, while fractional values produce enharmonic timbres. Data sampled over longer periods are mapped onto the spatial properties of the sounds and images, the degree of movement and the orientation of objects, images and sounds varying proportionally with respect to the general dynamism of the participants.

In this particular installation, the participants act out David Cronenberg's Shivers—a film dealing with body horror and social paranoia. The screenplay and movie of Shivers are streamed as text and image into the projected visualization. Both the movie and the screenplay are indexed in terms of their emotional and dramatic peaks and troughs, and, during the performance, the moving images and texts are reordered in response to incoming bio data from the participants. As data from the participants indicates a rise or fall in bioactivity, so texts and images relating to corresponding highs and lows in the film are sent to the screens the inverse.

This is

> ... a data-constipation in which the virtual lodges itself unyieldingly into the material body, or perhaps again a relation—now, not between bodies and world, but between perception and information with body-degree-zero as the unwritten script that might perhaps only be defined in terms of the variables allowed to inform the shapes it takes. It seems that perhaps what is crucial to the formulation of zero-degree content is precisely the same zero-degree that finds its way into Barthes' zero-degree writing—not in the sense of a reliance on interpretive presence for the formation of content, but rather in the immanent state of information without audience (Hiebert, 2004).

More recently the work of the Einstein's Brain Project, has developed generative systems in order to reference the ideas inherent in EVP (Electronic Voice Phenomenon) to examine ways in which we construct worlds, and bodies in worlds, through pareidolia, (a psychological phenomenon involving a vague and random stimulus—often an image or sound—being perceived as significant), apophenia (the seeing of connections where there are none) and the gestalt effect (the recognition of pattern and form).

William J. Mitchell speculates that the emergence of digital imaging has forced us

> ... to adopt a far more wary and more vigilant interpretive stance—much as recent philosophy and literary theory have shaken our faith in the ultimate grounding of written texts on external reference, alerted us to the endless self-referentiality of symbolic constructions, and confronted us with the inherent instabilities and indeterminacies of verbal meaning (Mitchell, 1994).

Mitchell is cautioning us about the uncertainty of the values we can assign to the photographic, and the complexities of image production and visual truth in the digital era, or as he remarks, the idea of photographic documents as truthful reports about things in the real world.

> ...as we enter the post-photographic era, we must face once again the ineradicable fragility of our ontological distinctions between the imaginary and the real, and the tragic elusiveness of the Cartesian dream.
> ... the photographic falsifier holds up not a mirror to the world but a looking glass through which the observing subject is slyly invited to step, like Alice, into a place where things are different—where facts seem indistinguishable from falsehoods and fictions and where immanent paradox continually threatens to undermine established certainties. To grasp precisely how this can be so, we must consider not only how photographs and pseudo-photographs are made, but also how they are used—how their potential uses are established, how they are appropriated and exchanged, how they are combined with words and other pictures and made to play roles in narratives, and how they may have the effect of creating beliefs and desires (Mitchell, 1994).

Jonathan Crary suggests this is as new model of vision:

> The rapid development in little more than a decade of a vast array of computer graphics techniques is part of a sweeping reconfiguration of relations between an observing subject and modes of representation that effectively nullifies most of the culturally established meanings of the terms observer and representation. The formalization and diffusion of computer-generated imagery heralds the ubiquitous implantation of fabricated visual 'spaces' radically different from the mimetic capacities of film, photography, and television (Crary, 1990).

Virilio thinks of these as the products of vision machines and what he calls sightless vision, and considers digital forms of representation leading to the division of representation and perception:

> Now that they are preparing the way for the automation of perception, for the innovation of artificial vision, delegating the analysis of objective reality to a machine, it might be appropriate to have another look at the nature of the virtual image. This is the formation of optical imagery with no apparent base, no permanency beyond that of mental or instrumental visual memory. Today it is impossible to talk about the development of the audiovisual without also talking about the development of virtual imagery and its influence on human behaviour, or without pointing to the new industrialization of vision, to the growth of a veritable market in synthetic perception and all the ethical questions this entails. This should be considered not only in relation to control of surveillance, and the attendant persecution mania, but also primarily in relation to the philosophical question of the splitting of viewpoint, the sharing of perception of the environment between the animate (the living subject) and the inanimate (the object, the seeing machine) (Virilio, 1994).

The recent work of the Einstein's Brain Project uses these divisions between subject and object, between animate and inanimate, the notion of fabricated space, and the conflation of fact and fiction, to raise questions about what it is to recognize a pattern and how and what meaning can accrue.

In the Sound of Silence, two projectors project images onto the walls of a room. Audio fills the room. One projection shows video static overlaid with text;

the other shows black and white images of what appear to be blurry and indistinct images of human faces. Ambient noise fills the space. Just at the threshold of recognition can be heard what appear to be human speech in different languages.

A CCD camera is turned on, but enclosed in a light tight box. It records no image and receives no light. Its input is adjusted with maximum gain and brightness to reveal the video noise inherent in the system. This noise forms the optical equivalent of audio noise and is used in a similar way to provide a medium that can be modified by external forces to produce images and sounds. The video noise is mapped to audio by sampling pixels in a QuickTime matrix and using the values to manipulate a stream of white noise. Voice recognition software parses the modulated noise and translates any sufficiently voice-like sounds into its nearest vocal equivalent, which is sent to the screen as text and rendered into audio by speech synthesizers into Italian, German, English or Spanish.
Face tracking algorithms using a cascade of Haar classifiers scan each video frame and look for any combination of pixels that form the basic characteristics of a human face. These are areas that can be loosely characterized as eyes, nose and mouth with a sufficient degree of symmetry. When the software finds such a combination of pixels and symmetry the area is zoomed to full screen, its contrast and brightness adjusted, blurred and desaturated to clarify the found images.

The images produced are only occasionally reminiscent of human faces. More often than not the images produced are recognized as indeterminate organic forms with a suggestion of volume and space, but fail to resolve themselves into anything recognizable. But occasionally, images are produced that are strikingly like a face, although in actuality containing only the barest possibility of being so.

This work uses the ideas found in Electronic Voice Phenomenon (EVP) to explore ideas about the construction of a world delineated by presence and absence, and pattern and randomness.

EVP is the recording of errant noises or voices that have no explainable or physical source of origin. These recordings are made when the recorder is alone, or under controlled circumstances. Most often white or pink noise is used as a medium that is acted upon by other electromagnetic forces. This electromagnetic medium produces wave forms that are, occasionally, like human speech. It has been argued that the voices are simply subjective interpretations—that we tend to hear voices in random patterns of sound, in the way we recognize forms in random visual patterns. For others, the voices are genuinely mysterious, opening up the possibility of communication with the dead. Spirit communication, through electronic recording, dates back to the 1940s. Originally called *Raudive Voices*, after parapsychologist Konstantin Raudive, these kinds of recordings were renamed "electronic voice phenomena".

In this installation the computer does the hard work of analyzing a complex visual field data, but the task of meaning making is left to the observer. The algorithms find face images in the video field that barely meet the requirements of a facial arrangement. The structuring of these pixels as faces, consisting only of blobs and indeterminate grain, is left to the observer.

Seeing, representation and interpretation of external phenomena has never been a matter of objectivity. Seeing is a complex activity and the perception of visual forms, aesthetic experience and cognitive interpretation are more at home with the aleatory, the misperceived and phenomena of indeterminacy than with the notion of the world as a fixed reality. It is these that drive the Sound of Silence.

Gerald Edelman describes what he calls primary consciousness as:

> . . . the state of being mentally aware of things in the world, of having mental images in the present. But it is not accompanied by any sense of [being] a person with a past and a future. . . In contrast, higher-order consciousness involves the recognition by a thinking subject of his or her own acts and affections, It embodies a model of the personal, and the past and future as well as the present. . . It is what we as humans have in addition to primary consciousness (Edelman, 2000).

Past and present predict a future predicated on subjective experience. Past experiences are related to new events, but are not necessarily rooted in cause and effect, nor are they inevitably related to the external world. They are based on what has had value or meaning in the past. One of the brain's abilities is to predict future events.[3]

The Global Consciousness Project (GCP),[4] originating from Princeton, uses random number generators to predict the future. The results seem to predict major global events some time in advance of their occurrence. The brain is different. It constantly predicts what is likely to happen, and in doing so predicts huge numbers of possible futures. Some are likely most are not, but none are disposed off until the need for action, and some may even remain as past images of the future. What we begin to build is a picture of the brain as a device for generating and interpreting the possible.

> A phenomenon of pure representation, such relativism is always at work in the appearance of events, of things as they happen, precisely because we always have to make a subjective leap in order to recognize the shapes, objects and scenes we are witness to (Virilio, 1994).

The brain's natural tendency is to see pattern in randomness, in an attempt to predict the future. If a pattern can be detected early enough its next iteration can be forecast. The human brain is constantly seeking patterns in its environment. The body/mind is built to recognize meaning and significance. Our ability for making sense comes from patterns of successful and failed perception and action, and it is this prior knowledge that forms the basis of our predicting capacity, our imagination. The body/mind is driven to make meaning, and in a world where we seek meaning, we are surrounded by signs and omens.

Psychologist George Wolford has examined the tendency of people to find patterns in random sequences, even when they are told that the sequences are random: "There appears to be a module in the left hemisphere of the brain that drives humans to search for patterns and to see causal relationships, even when none exist . . . The interpreter drives us to believe that 'I can figure this out."[5]

Sound of Silence is a generative, closed system. Random noise from a CCD camera is analyzed for patterns. An algorithm looks for patterns that match the basic geometry and physiognomy of the human face. What it actually finds are pixels on a screen forming blobs and patches of colour that have no actual relation to a real world face. They have no indexical relation to an object. They are not images of people, but another kind of image loaded with meaning, which arises accidentally, but irresistibly, from the hybrid interaction between machine and body. To all intents and purposes when these patches of pixels look like faces, they are images of faces. That such obscure images resolve themselves into faces without conscious effort, and that remain even when attending closely to them, suggests that it is paradoxically their lack of objective meaning that generates their form. It is the very ambiguity and intedeterminacy of the images that allows the brain to reconfigure them as indexical.

In Antonioni's film Blowup (1966), the central character, a fashion photographer takes some photographs of a lover's rendezvous. The negatives reveal what seems to be evidence of a murder. A succession of increasingly big enlargements reveals what appears to be a body and a gun.

The photographer

> ... sees something (or maybe not) while photographing in a London park. The swinging world recedes as Thomas becomes morbidly obsessed by the possibility he has in fact photographed a murder. Is that a body under the bushes? If not, what, then? The title's meaning becomes clear as Thomas repeatedly enlarges, studies, and reworks his negatives to locate the proof for something, any evidence of a possible occurrence that seems, as his obsession grows, to become increasingly elusive. Subsequently, the film becomes, in part, a meditation on the very nature of reality and how we deform the natural with interpretations and inflexions... any observed happening (however neutral or scientific the act of observation) is altered by the mere presence of the observer. Observation is thus never a neutral or abstract process. Certainly if there is a central informing principle to the narrative matrix of Blowup it is this: that no phenomenon is pure unto itself—especially when human emotions come into play, let alone consideration like guilt, obsession and, finally, fear (Dawson, 2006).

The black and white prints, enlarged to the point of blobs of grain are indecipherable. And yet they are compellingly open to interpretation. What exactly do the photographs index—a lover's tryst, a park, a murder, or are they merely misread? Cultural and biological compulsions form these blobs into patterns that are more and more recognizable—and in turn more and more believable. A recursive loop is formed—the more believable the images are the more they look like the subject's desires. They operate as multi-dimensional ink blots—maps of future and past memories.

> Antonioni brings to the surface that tantalizing semiotic opacity of which all of the visual arts partake but which so infrequently gets articulated. What is more, the film flaunts the contingencies of visual hermeneutics, it emphasizes the inter-relations of photography, corporeality, decomposition, and death, and, most obviously, it configures the instability, the limits, of photography's truth. The complex presence of the film's corpse pinpoints this perennial interpretive dilemma: "we are certain that we see it, but we cannot be certain that what we see is real" (Grønstad, 2004).

An ambiguous image gives rise to multiple interpretations. As we negotiate the world, we model the world by mapping our sensations to our perception of the external world. This map is dynamic and as we acquire new information, the mapping changes, suggesting that perception, like Antonioni's photographer discovered, is a matter of belief.

Asbjørn Grønstad describes Blowup as delineating:

> . . . what we perhaps somewhat conservatively would want to claim is a decisive attribute of the photographic, namely, the aleatory. This concept, I would argue, is closely connected with what the French call l'insolite; that is, a curious, incongruous element within a given totality. Both the aleatory and the incongruous-with reference to the photograph-are redolent of the punctum, though not entirely in a Barthesian sense. As you will remember, the punctum is the second "theme" that Barthes finds in the photograph and is that aspect which transcends the "unconcerned desire" of the studium to "pierce" the viewer. Though the idea of the punctum has been largely over-discussed by now, it may nevertheless be worthwhile succinctly to review its basic tenets as described in Camera Lucida (1980). Barthes himself appears to put forth an affectionist point of view with regard to the punctum, which he conceptualizes in terms of for instance a wound, a prick, a mark, a sting, or a cut. The punctum, he says, is a detail "which attracts or distresses me", hence, it is a phenomenon the possibility of which hinges on its emotional impact on the viewer. But the existential mode of the punctum is aleatoric. It is, as Barthes writes, something that simply "happened to be there", something "offered by chance and for nothing", much like the blurry outline of the dead body in Thomas' photograph (Grønstad, 2004).

Seeing is a dynamic, evolutionary and aleatory process—the eye evolves through and by the demands being placed upon it. The eye is properly seen as an outcropping of the brain, so the application of technology to enhance vision not only changes the way the world is perceived, but also the way it can be perceived, represented and understood. It changes what can be perceived. Harnessing lenses or microscopes as visual aids can be thought of as an example of cyborg vision, the technological enhancement of the body and natural behavior.

Within the realm of historic art there exist visual images which are not characterized by high definition or mimetic exactness, but function on the apparent deficiency of exactitude. The particular skills of painters like Rembrandt and Goya, and the followers of their loosely expressive style, as well as those artists who paint impressions, are the immediate transformation of material into image. This only happens at a distance. In some opposition to Antonioni's blowups, upon close examination the object of representation disappears. This gradual deconfiguration of the image was pursued extensively in the late nineteenth century and in the first decades of the twentieth century.

Like maps, paintings (and photographs) are highly coded. An observer needs to fill in the missing details or experience or they will lose their way. This is the case with every category of image. Seeing is an active and collaborative practice and it requires conjecture and supposition.

> Every picture is a picture of the body. Every work of visual art is a representation of the body. To say this is to say that we see bodies even where there are

> none, and that the creation of a form is to some degree also the creation of a body. And if a splash of paint or a ruled grid can be a picture of a body—or the denial of a body—then there must be a desire at work, perhaps among the most primal desires of all; we prefer to have bodies in front of us, or in our hands, and if we cannot have them, we continue to see them as after-images or ghosts. This is a beautiful and complicated subject, the way our eyes continue to look out at the most diverse kinds of things and bring back echoes of bodies (Elkins, 2002).

It is this very lack of body that characterizes an image. An image has no depth, no dimension at all, without a surface on which to cling. It exemplifies the absent body. This friction between absent and present bodies and dematerialized images makes us see the body in the most abandoned and empty of spaces, seeing ghosts where there are none.

N. Katherine Hayles explains:

> The contemporary pressure toward dematerialization, understood as an epistemic shift toward pattern/randomness and away from presence/absence, affects human and textual bodies on two levels at once, as a change in the body (the material substrate) and a change in the message (the codes of representation).

This suggests a new category of subjectivity, an embodied hybrid of human and machine, that privileges pattern, and its opposite noise, over the material and moves between the material and the immaterial. A new body acknowledges the absent. This body is biologically developed to see pattern and grant it significance in its dynamic mental model of its world. This is why degraded images can be recognized, why patterns can be seen in clouds, or in the noise of a television tuned to a dead channel. And it is why indeterminate images, patterns emerging from noise, can and must be reconfigured as images of things.

> ...if you cast your glance on any walls dirty with such stains or walls made up of rock formations of different types. If you have to invent some scenes, you will be able to discover them there in diverse forms, in diverse landscapes, adorned with mountains, rivers, rocks, trees, extensive plains, valleys, and hills. You can even see different battle scenes and movements made up of unusual figures, faces with strange expressions, and myriad things which you can transform into a complete and proper form constituting part of similar walls and rocks. ... look into these stains on walls, the ashes from the fire, the clouds, the mud, or other similar places. If these are well contemplated, you will find fantastic inventions that awaken the genius of the painter to new inventions, such as compositions of battles, animals, and men, as well as diverse composition of landscapes, and monstrous things, as devils and the like (Da Vinci, 2001).

Signs, marks and visible pattern have never constituted the totality of visual art works. There are always aspects of images that are invisible. There are gaps, discontinuities and ruptures in every image, such as a lack of continuity of space, incompleteness of form, figural discrepancies, omission of detail. The mere perception of painted images requires a great deal of completion on behalf of the observer.

The brain has a remarkable capacity for dealing with this incompleteness and filling in the gaps. As V. S. Ramachandran and S. Blakeslee say,

> . . . the human visual system has an astonishing ability to make educated guesses based on the fragmentary and evanescent images dancing in the eyeballs. . . . the filling in is not just some odd quirk of the visual system that had evolved for the sole purpose of dealing with the blind spot. Rather, it appears to be a manifestation of a very general ability to construct surfaces and bridge gaps that might be otherwise distracting in an image—the same ability, in fact, that allows you to see a rabbit behind a picket fence as a complete rabbit, not a sliced-up one (Ramachandran, 1998).

In extreme cases, this ability to function in the absence of a complete picture is remarkable. In studies of *blindsight* (the ability to respond appropriately to visual inputs while lacking the feeling of having seen them), patients who are effectively blind can carry out tasks that appear to be impossible without sight. Individuals can reach out and grasp an object, accurately describe whether a stick is vertical or horizontal, or even slide an object into a narrow slot. According to Ramachandran the explanation for this phenomenon is that visual information travels along two pathways in the brain. If only one is damaged, a patient may lose the ability to see an object, but still be aware of its location and orientation. This example suggests that while there is a terminal threshold beyond which a stimulus is no longer detected, we can and do respond to stimuli of which we are not conscious.

There are also areas of the brain that process information through conjecture that work on the basis of assumption or hidden knowledge about the world. It is this last process that brings us back to the sounds and images in the Sound of Silence.

When this work was first shown it was installed in Rosario, province of Santa Fe, Argentina.

On March 24, 1976, the Argentine military overthrew President Isabel Peron and remained in power until 1983. Local police departments were put under military control. The military government seized, tortured, disappeared and eliminated anyone suspected of sedition. There were 14,000 political prisoners. Government agents took another 30,000 people. The Centro Popular de la Memoria (People's Memorial Center) is a former illegal detention center in Rosario, used by the police to hold people without charging them, and torture them, under the pretense of fighting radical left-wing political subversion and terrorism. There is evidence of its use as a detention center since 1976 and until 1979. It was informally termed *El Pozo*: The Pit.

The Sound of Silence was installed in a room directly above The Pit.

Naturally enough the images and sounds that observers saw in the random noise from the camera related directly to the horrors inflicted on those incarcerated in El Pozo. The random noise was interpreted in the context of the lost and invisible bodies that had been incarcerated. Such contextual imagining is not unusual – works of art are never autonomous, but always part of a contextual continuum. Another installation occurred at a mental hospital in Trieste, Italy. Here the noise was characterized differently and different sounds and images emerged from the work.

The overlaying of information on the work brings us back to the central issues of recognition and reality in Antonioni's blowup. Can we believe ours eyes? The unavoidable textual and contextual redefinition of the subjective dictates not only what we look for, but also what we find. Such recursion leads us to imagine a world where we can see only what we believe, and can only see what we expect to see. This is a world of infinite thinness and limited dimension. Half-seen, half-lived.

It these beliefs and expectations that fuel conspiracy theories, and images of the celebrity dead that are the daily substance of the tabloids. The phenomenon of seeing relationships and pattern where there is none is properly described by the terms apophenia and pareidolia, and together they drive the interpretative dynamic in Sounds of Silence.

Apophenia is the tendency to see connections between seemingly unrelated objects or ideas. August Strindberg details the over-interpretation of a chance encounter that characterizes apophenia's sense that things are more meaningful than they are:

> There on the ground I found two dry twigs, broken off by the wind. They were shaped like the Greek letter for "P" and "y"... It struck me that (they) must be an abbreviation of the name Popoffsky. Now I was sure it was he who was persecuting me, and that the Powers wanted to open my eyes to my danger (Strindberg, 1988).

Pareidolia is a type of illusion or misperception involving an indeterminate or vague stimulus that is perceived as clearly being something. Well known examples have been natural rock formation on Mars that was seen as a face, or the seeing of animals in clouds.

While apophenia and pareidolia may be the engines of Sounds of Silence the forms that emerge suggest something else. Neither the images nor the sounds in Sounds of Silence have an indexical relation to reality, as does a photograph or a recording. They emerge from the random flickerings of pixels on a screen. They have neither substance nor form other than the chance interactions and collisions with one another. If they are to be thought of as faces and voices then they hint at an immaterial body that exists in the interstice between machine and body that exists in patterns and information flows.

Pattern tends to overwhelm presence, marking a new kind of immateriality which does not depend on spirituality or even consciousness, only on information.

> ...even though information provides the basis for much of contemporary society, it is never present in itself. ...In information theoretic terms, information is conceptually distinct from the markers that embody it, for example newsprint or electromagnetic waves. It is a pattern rather than a presence, defined by the probability distribution of the coding elements comprising the message. If information is pattern, then non-information should be the absence of pattern, that is, randomness. This commonsense expectation ran into unexpected complications when certain developments within information theory implied that information could be equated with randomness as well as with pattern. Identifying information with both pattern and randomness proved to be a powerful paradox, leading

> to the realization that in some instances, an infusion of noise into a system can cause it to reorganize at a higher level of complexity. Within such a system, pattern and randomness are bound together in a complex dialectic that makes them not so much opposites as complements or supplements to one another. Each helps to define the other; each contributes to the flow of information through the system (Hayles, 1993).

Generated in the interstice between subject and object, they are data made incorporeal flesh through the dematerializing fusion of body and machine. These are true ghosts in the machine – images born out of pattern and randomness. They are true images of the absent body. Thought of this way the work moves away from ideas about apophenia and pareidolia to suggest that real information is contained within the random noise of the work, that is always on the point of becoming. Existing not out of

> ...a reliance on interpretive presence for the formation of content, but rather in the immanent state of information without audience—the metamorphosis into the absence of an unidentifiable body that nevertheless seems, for all intents and purposes, to still exist despite its unidentifiability (Hiebert, 2003).

Using the coincidences of science, culture, technology and art, the Project examines the fundamentally contingent nature of meaning. The work of the Project has grown out of an interest in bodies in motion, worlds in flux, and in the endlessly recombinant texts and forms of our worlds, and out of an interest in the seen and unseen, the half perceived and misperceived things at the limits of our perception and in the reanimation of the lost bodies and past events that constitute this invisible world. The world is not entirely what it appears to be and that the surface of the visible world needs only to be lightly scratched to reveal the invisible worlds above and below.

These works provide a means to examine the Projects interest in systems of meaning making that rely on pattern recognition, and the relationship between meaning and the meaningful. The development of meaning in the Project's work is dependent on an increasingly, yet seemingly infinite, complex recursive and recombinant loops between meaning made and meaning found and the increasingly indeterminate limits of a super distributed and polymorphic body. In these loops the external and internal worlds are blurrily indistinct, each acting upon the other in the construction of a new selfspace forever suspended at the point of becoming. The effort to restablilize the self in this world that changes moment to moment where everything is in play, is questioned and negotiable, is unavoidably revelatory and re-problematizes current and preceding models of authenticity and resistance.

In Kobe Abe's *The Ruined Map* (Abe, 2003), the wife of the missing Nemuro remarks:

> He says a single map for life is all you need. It's a saying of his...the world is a forest, a woods, full of wild beasts and poisonous insects. You should go only through places where everyone goes, places that are considered absolutely safe.

None would enter such a world without a map. How else to tell what lays ahead for an unwary traveler? But what exactly should that map tell us? What can they represent, if the 'sovereign difference' between a map and its territory has, as Baudrillard suggested, disappeared?

Madame Scudéry's Carte du tendre[6] is a topographical allegory that functions as a map for life, as a map of emotions and of memory. Every rock, hilltop, and feature is embedded with meaning. This is a world invested with signs and symbols and psycho-geographical contours that impact the body and its agency. Filled with omens and portents this kind of geography assembles the body in the world as it predicts a future world predicated on the trace evidence of the past. Perhaps here the future function of the map is clearer: the inevitable recursive spatialization and diffusion of an emerging body through an ever nascent landscape.

> The map transformed the abstract space into concrete places and thus subjugated the earth's surface to human rule. ... transforms all concrete places into abstract places (randomly interchangeable and constructible spaces) and thus subjugates any relationship of representation to the rule of the code. The operation of a substantiating representation of space... using 'maps' has thus not become impossible, but it has only mythical functions by now (Schetsche, 2006).

Koji Suzuki's Kaoru (Suzuki, 2006) is so sure of his environment:

> ...he didn't turn around, even though he could feel someone watching him from behind. He knew what would be there if he did; the same living room, the dining room beyond it, the kitchen next to it, all unchanged.

His journey to understanding the complicated, conditional relationship between different realities and how they are coming into being is one that echoes throughout fiction, and is symptomatic of our sense of the contingent nature of reality.

The Project suggests that we should be less sure of an unchanging world. Constantly unravelling and remade by a body both acting on and reacting to its environment, we imagine a world and a body always on the verge of becoming, a transitory emanation from the glittering, noisy dance of electrons.

Notes

1. While the word representation is used throughout this paper, and invokes the right kinds of discussion, it is not properly or fully descriptive. Its traditional use as representing something already in existence, in spite of the increasing conflation of the representing and the represented, is in conflict with the core ideas of the Project. The products of the brain are rather seen as "material potentialities" simultaneously virtual and material, at the threshold of becoming. Rather than representations they are expressive potentials of material and matter—they are a kind of material hallucination.
2. For a fuller discussion of this concept see: Andy C. and Chalmers D. J. (1998), *The Extended Mind*, Analysis, 58, Pg. 10–23.

3. Professor Dick Bierman, at the University of Amsterdam suggests that in order to accommodate psi phenomena we might have to propose new types of realities, intermediate between the subjective but hardly deniable reality of consciousness and objective physical reality. His work on pre-sentiment, the apparent anomalous activation of bodily arousal preceding emotional events, suggests the body's ability to sense the future.
4. http://noosphere.princeton.edu/fristwall.page.html "A global network of electronic devices produces continuous random data sequences. Subtle patterns in the data are linked with events that cause shared thoughts and emotions in millions of people. The results challenge common ideas about the world, but independent analyses confirm the unexpected patterns, and also indicate that they cannot be attributed to ordinary physical forces or electromagnetic fields".
5. As quoted in Money Magazine, Zweig, Jason, The Trouble With Humans Why rats and pigeons might make better investors than people do. November 2000. See also Wolford G. L., Miller M. B. and Gazzaniga M. (2000), *The left hemisphere's role in hypothesis formation.* Journal of Neuroscience, Vol. 20.
6. In 1654, M.lle de Scudéry published the first installment of "Clélie", a novel of love and courtship. Her main character's knowledge of emotions is so great that she is able to draw a map of the course of a love affair.

Bibliography

Abe K. (2003), *The Ruined Map*, Vintage, New York.

Bergson (1998), *Creative Evolution.* Dover, New York.

Bohm D. (1980), *Wholeness and the Implicate Order.* London, Routledge.

Braidotti R. (2003), *META(L)MORPHOSES*, The Becoming Machine. In: *The Body: Critical Concepts in Sociology*, edited by the Aberdeen Body Group (Andrew Blaikie *et al.*), Rout, ledge, London.

Crary J. (1990), *Techniques of the Observer: On Vision and Modernity in the Nineteenth Century.* MIT Press, Cambridge.

Csordas T. J. (1997), *Computerized Cadavers: Shades of Being and Representation in Virtual Reality.* From After Postmodernism conference, University of Chicago, November 14-16.

Dawson J. (2006), *Blowup, Senses of Cinema.* January 2006, http://www.sensesofcinema.com/contents/cteq/05/34/blowup.html, accessed March 2006.

Edelman G. M. and Tononi G. (2000), *A Universe of Consciousness: How Matter Becomes Imagination.* Basic Books, New York.

Elkins J. (2002), *Pictures of the Body: Pain and Metamorphosis.* Chicago: University of Chicago Press.

Grønstad (2004), *Asbjørn Anatomy of a Murder: Bazin, Barthes, Blow-Up.* The Film Journal http://www.thefilmjournal.com/issue9/blow-up.html, accessed January 2008.

Flusser V. (2000), *Ins Universum der technischen Bilder.* European Photography, Göttingen.

Hayles N. K. (1933), *Virtual Bodies and Flickering Signifiers.* October, Vol. 66, Autumn, MIT Press, Cambridge.

Hayles N. K. (2006), *Traumas of Code. Critical Inquiry.* Vol. 33, Number 1, Fall.

Hiebert T. (2003), Unpublished in conversation with Alan Dunning and Paul Woodrow. Vancouver 2003.

Hiebert T. (2004), *In conversation with Paul Woodrow*. Calgary/Vancouver.
Higgins D. (1999), *A Dialectic of Centuries: Notes Toward a Theory of the New Arts*, BOA Editions.
Kenneth G. (2000), *The Saturated Self, Dilemmas of Identity in Contemporary Life*. Basic Books.
Kroker A. and Kroker M. (1987), *Body Invaders-Panic Sex in America*. St. Martin's Press, New York.
Kroker A., Kroker M. and Cook D. (1989), *Panic Encyclopedia*. New World Perspectives. Montréal.
Leonardo Da Vinci (2001), *A Treatise on Painting*. Adamant Media Corporation.
Massumi B. (1996), *The Autonomy of Affect in Deleuze: A Critical Reader*. Paul Patton (ed.), Blackwell Publishing, Oxford.
Massumi B. (2002), *Parables for the Virtual*. Duke University Press, Durham.
Mitchell W. J. (1994), *The Reconfigured Eye: Visual Truth in the Post-Photographic Era*. MIT Press, Cambridge.
Mead G. H. (1913), *The Social Self*. Journal of Philosophy, Psychology and Scientific Methods, 10.
Palumbo M. L. (2000), *New Wombs: Electronic Bodies and Architectural Disorder*. Birkhäuser, Basel.
Pribram K. (1997), *Holonomic Brain Theory : Recent Experimental results*.
Ramachandran V. S. and Blakeslee S. (1998), *Phantoms in the Brain: Probing the Mysteries of the Human Mind*, Quill William Morrow.
Schetsche M. T. and Temme T. (2006), *Some Brief Remarks on a Theory of Superface*. Flusser Studies 02 - May.
Sterling B. (2005), *Shaping Things*. Mediawork Pamphplet Series, MIT Press, Cambridge.
Strindberg A. (1988), *Inferno / From an Occult Diary*. Eklund T. (ed.), Sandbach M. (Translator), Penguin Books.
Suzuki K. (2006), Loop, Vertical, New York.
Tomas D. (1991), *Old Rituals for New Space*. In: *Cyberspace: First Steps*. Michael Benedikt, (ed.), MIT Press Cambridge.
Virilio P. (1994), *The Vision Machine*. Indiana University Press, Bloomington.
Weitemeier, Hannah, Klein, Taschen, Koln, 2001.

Chapter 2
About the technological *bricoleur*[1]

Tatiana Mazali

Abstract

This paper originates in the need of bringing out keys to understand the *practices* that are imposed on our bodies in the contemporary technological context. The author intends to focus on the following topics:
1) What do we mean by *Unpredictable uses?*
2) The sense of the term *Performativity*, a word that the author uses as a key to explain the relationship between body and technologies.

1. Unpredictable uses. In agreement with Alexandre Mallard (2005), the author thinks that, if we want to find a positive meaning in the practices of the techno-bodies, we have to use the paradox of the unpredictable uses. We are not expected to observe and capture the activities of users, but to somehow anticipate the trajectories of uses and their associated drifts. This approach can prove useful in the analysis of the dynamic processes involved in ICT.
2. Performativity. The term "performativity" comes from the linguistic field of John Langshaw Austin (1962). Characteristics of the term performativity, which the author will use to explain the techno-bodies of users, are: act (an act implies making to exist, so creativity in action; the act creates a step between content and form, it is therefore subversive); satisfaction *versus* truth; strength *versus* meaning. Nowadays we talk about performative technologies and performative bodies to stress the process and relationship involved. The techno-bodies imply a relationship between different parts: inside/outside, product/consumption, producer/user, and there is also a third parameter that sets the techno-bodies in action: the context of use, the environment in which they operate. The author will use the theoretical frames of two authors: Lévi-Strauss (1962), author of the fertile notion of the *Science of the Concrete* and the related *bricolage/bricoleur*; and Lev Manovich (2001), an artist and intellectual famed for its *Database Logic*.

The first aim of this essay is to find out the keys to understand the "practices" that our bodies have to undergo in nowadays technological world — practices that have become the new place of knowledge. This basic demand for clues becomes really urgent when the techno-bodies stop being an exception and grow to be our habitat and our *habitus* (Mauss, 1936; Bourdieu, 1972).

The techno-bodies, instead to be an exception, have become the rule: they are *the users*, to say it with a general term that is being applied in all modern analysis of the relationship between man and technology. Technology is an extension of man that links him with the world; it has lost its mainly instrumental function — technology allows human beings to better survive and to artificially make up for their organic flaws and deficiencies (Gehlen, 1983) — and is increasingly becoming the widespread and pervasive habitat of mankind. Since McLuhan, we are indeed aware of the close relationship between communication technologies and the human psycho-physical structure. According to him, the role of technology is to allow the individual to overcome, through a self-amputation mechanism, the functional stress induced by accelerating rhythms and increasing labour burden, both historically imposed by the money Medium and the written-word Medium (McLuhan, 1964). In spite of talking about instrument-technology (Leroi-Gourhan, 1964), or habitat-technology or even *habitus*-technology, everybody is now applying the term *user*to emphasize the shift from a "dominating" technology to a technology "dominated" by the active intervention of the user. In his relation with the "machine", the user is not a passive individual/mass anymore, but an active subject.

In this essay, the expressions *user* and *techno-body* are to be understood as synonyms. There is a general belief that the techno-bodies are extreme elements of far-away worlds — which has generated tons of "popular culture"[2] — but day after day new technological innovations prevail in our real world unbeknown to individuals or associations; compared to them, science-fiction is a mere fantasy world for nostalgic souls. The relation between man and "machine" generates real techno-bodies that are geared, on one hand, by the predictability of the uses derived from the technologies and, on the other, by the anarchy allowed by the same technologies. The ICT bring forth the difficult balance between predictability and unpredictability; they faithfully reflect the complexity which, in the new scientific theories, makes it clear that the world is not deterministic anymore and that "the possible" is now in command (Prigogine, 1996).

I shall therefore talk in my essay about the techno-bodies that are nowadays defined as *users*. I shall try to determine the actions, the movements and the forms that the use of the technologies produces. My purpose is to understand the keys of the (from my point of view extremely creative) gap that appears between "the rejected" and the average, "the excess" and the norm, "the unforeseen" and the determined: the gap that might allow the passage from the *user* to the proper *actor*.

Techno-user profiles

In the techno-social milieu of our network society some new users are emerging. They are techno-bodies and they bring along a variety of needs, quests and abilities. They have very different levels of computer literacy, cultural consumption and life style. They are not a mass audience anymore, but a segmented audience. They are socially and culturally differentiated. According to Manuel Castells definition, they are stratified and divided between the *interacting* (the slave-owners: able to choose their own multidirectional communication paths) and the *interacted* (the slaves: consumers of a small range of pre-established choices).

The information and communication technology system is constantly creating new scenarios, but can still recognise a couple of tendencies in them: the blurring of the limits between consumers and producers, the *user generated contents*, and the Cross-Media phenomenon[3] — or the *convergence culture*, to put it in Henry Jenkins' words.

In a cultural context that is constantly redefined and remediated (Bolter, 1999) by individual and personalized forms of elaboration, it is important to understand the way in which every single person, from the most innovating to the most traditionalist one, leads his own way towards signification and appropriation of the technological realm.

In the Contemporary Age, the user has an absolutely central place. Technology seeks the protection of the individual user — specific, localised, geo-located, but simultaneously connected and collective — in the same way as a medieval subject sought his monarch's protection. But what is the tangible result of the clamorously acclaimed liberty, creativity and sociality of the new technological man?

The aim of this study is to check whether the potential capacities of the new technologies actually generate a corresponding creative use among individuals or communities. As a first step, I shall analyse the figures and data from several research and monitoring surveys, carried out in Italy as well as abroad, about the use of technologies. Special attention will be given to the data related to the worldwide web, because the social and cultural weight of the Net and its practical applications have made it the focal point of a wide range of theoretical and empirical analysis in almost every field: media studies, general sociology, anthropology, psychology, linguistics and even multidisciplinary surveys on the interaction between man and computer (Di Fraia, 2004).

My method of interpretation tries to follow what Pierre Bourdieu (1972) defined as the *praxiological knowledge*: we must begin with the empirical observation of the use of technologies (phenomenological knowledge, a "record of whatever is given exactly as it is given") and with the analysis of the objective structures of those new technologies (objectivist knowledge "that catches the practices from the outside, as an accomplished fact") in order to correctly understand and describe the generating principle of the practices; our thought must "place itself at the very core of the movement of their [the practices'] accomplishment" (Bourdieu, 1972: 186).

We do have a lot of images of the Internet and of the techno-bodies that haunt it. They are delivered on an annual (sometimes even monthly) basis by national and international research Institutes, as well as by other people who carry out periodic reports and continuously monitor the web.

In the last few years, a whole new range of communication products and network applications has come into the view of the general public and the media. The users can develop an ever-increasing interactivity with the net contents, in more and more complex and sophisticated ways. Some applications that are being identified as the new frontiers of the Web 2.0 (like Flickr, Wikipedia or YouTube) allow their users to split away from the concept of mass public and to become active techno-bodies on the mediatic arena. The level of interactivity on the Net has shifted from the possibility to select and navigate among hyperlink contents following a personal path — typical of the so-called Web 1.0 — to the possibility to insert self-made contents, thus determining the forms and the substance of global communication (Jensen, 1999; Van Dijk, 2001).

In consequence there has been a disruption in the models of analysis and interpretation that sociologists where applying to study the effects of the media on the general public. It is time to update the list of the methods, the fields and the objects of investigation. All of them are facing a growingly composite and complex scenario of uses: there is a new order and relationship between the consumption styles/levels, the contents management and the new interpretations and social uses of the media.

It has become capital but at the same time more complicated than ever to understand the knowledge keys involved in the relationship between man and technology. These new usage skills of the techno-bodies and these new forms of interactivity offered by the Net, what consequences do they have? And what is the exact relationship between the innovations in technology and communication of the last years and the changing processes of the individuals — their attitude and behaviour towards content consumption and production?

In fact, according to the most recent research results on that matter, it turns out that the so-called *active users* are an indeed interesting but certainly minor percentage of the huge mass of media consumers.

In any case, all studies agree on this one point: the proliferation of the technologies that allow the *costumer-consumer* to create, together with the spreading and the cost drop of the broadband, have led to a swift increase in the number of users that produce and share digital contents on the Net. Today the quantity of information produced by the consumers is higher than the one generated by the big networks (Lyman, Varian, 2003; Saracco, 2005). According to the research *How Much Information* (2003), led by the School of Information Management and Systems, University of Berkeley - California[4], the information produced by individuals and consumers have exceeded the information produced by the Major (in the following chart (fig. 2.1), see sms and books, DVD, CD, newspaper).

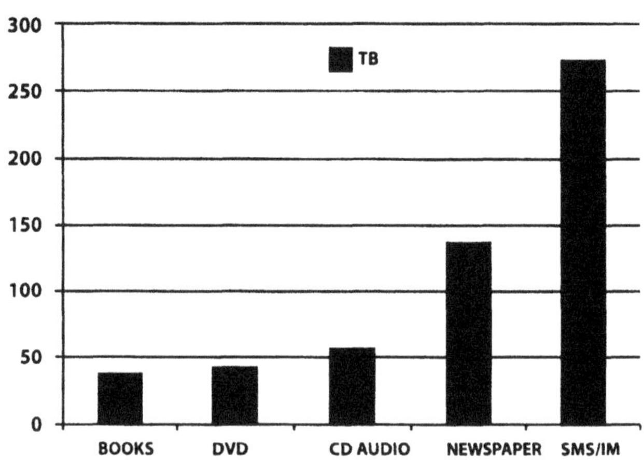

Figure 2.1 *Note: TB = terabyte (measurement for data storage capacity. The value of a terabyte is defined as one 1000 gigabytes).* Source: *Lyman P., Varian H.R., 2003, How much information?*

It is in that sense that we talk of *prosumer* techno-bodies, using a neologism coined by Alvin Toffler in 1980; the term is a contraction of the words *producer* and *consumer* and refers to the persons that both produce and consume contents, information and knowledge on the Net.

In December 2006, the famous weekly magazine Time[5] published an issue with a computer on the cover, which had a mirror instead of a screen. Chosen "Man of the Year" were all the individuals that had contributed to the explosion of digital democracy by using the Internet to share the so-called *user-generated contents* — that is, texts, images and video recordings created by the users, as the ones that have boosted such successful sites as YouTube or MySpace.

Nevertheless, the recent HITWISE study called *Measuring Web 2.0 Consumer Participation* (2007) shows that the mass of passive consumers greatly outnumbers the users that upload materials — or more precisely: *user-generated contents*. To give an example, according to that American survey, only 0,18% of YouTube visitors actively participate in its contents' creation. Besides, only 4,38% among the users of the collective encyclopaedia Wikipedia contribute to its editing. Finally, the number of visitors that post their own photographs on Flickr reaches a scarce 0,12%.

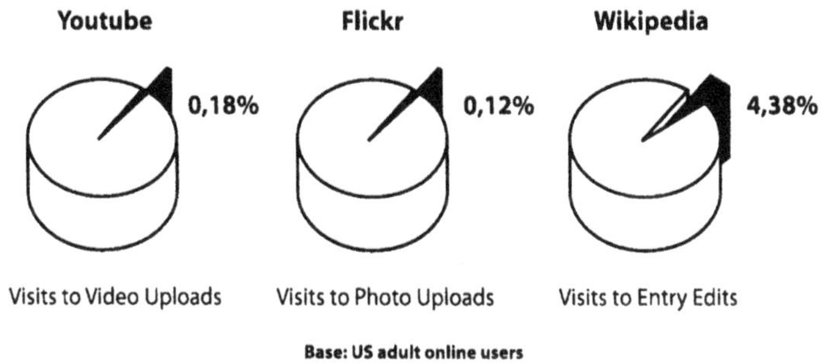

Figure 2.2 Source: *Hitwise, 2007, Measuring Web 2.0 Consumer Participation* (http://www.hitwise.com/downloads/reports/Hitwise_US_Measuring_Web_2.0_Consumer_Participation_June_2007.pdf).

The disparity between Youtube, Flickr and Wikipedia could be explained by the technical barrier of participation. Youtube and Flickr require some technical sophistication on behalf of the consumer (creating the video or photos, transferring the files to a computer, uploading the files to a computer, uploading the files to their respective websites) while Wikipedia edits require a simple mouse-click and text entry (Hitwise, 2007).

Additionally, the TECHNORATI report on *The State of the Live Web* (2007), presented at San Francisco's Web 2.0 Expo, points out that it is mainly younger people who look up words in the Wikipedia, whereas it is more often adults who are behind its contents edition. In the field of online video uploads, the less active users happen to be those of the 18 to 24 year-old range. On top of that, the percentage of active blogs is clearly diminishing: from 36% in May 2006 to 21% in June 2007.

The FORRESTER RESEARCH survey called *Social Technographics* (2007), has established a catalogue of six different Web 2.0 application users, according to the type of activity developed on various Internet social media. I shall put forward here these six categories as much definitions of the various sorts of techno-bodies to be found on the Net:

- *creators* (13%), blog and website authors who upload and share online their own video recordings;
- *critics* (19%), users who vote, write comments on the blogs and make reviews on the sites;
- *collectors* (15%), users who use the feed Rss and the tags to organise their own resources;
- *joiners* (19%), people who make a general use of the sites of social networking;
- *spectators* (33%), users who limit themselves to exploit in a passive sense the *user-generated contents*;
- *inactives* (52%), people who do not use at all (not even passively) the Web 2.0 tools and applications.

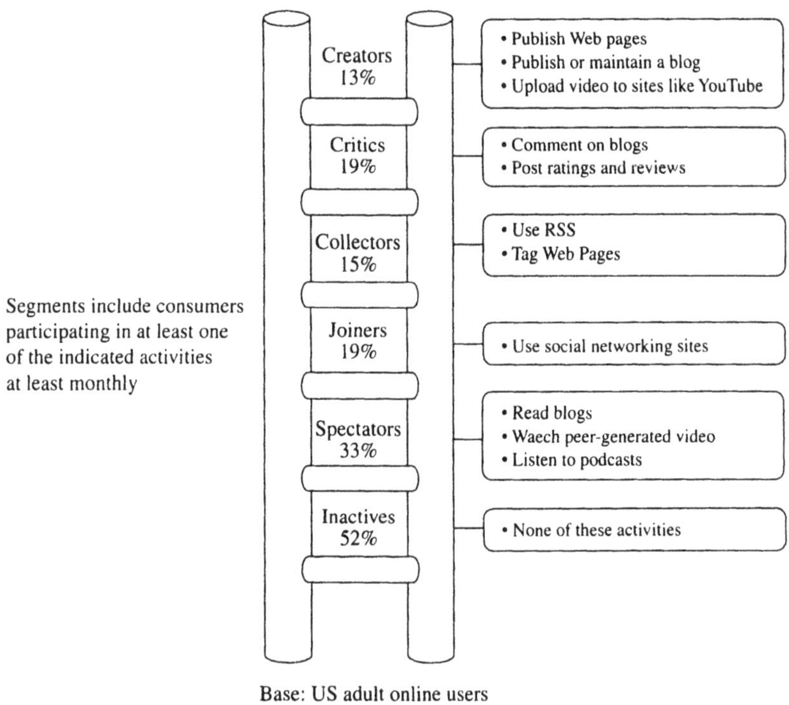

Figure 2.3 Source: *Forrester Research, 2007, Social Technographics.*

It is worth noting that, according to this last survey, only less than half of the users do really participate in the *social web*.

Many companies approach *social computing* as a list of technologies to be deployed as needed — a blog here, a podcast there — to achieve a marketing goal. But a more coherent approach is to start with them, based on what they are ready for (Forrester Research, 2007).

Another point of agreement in all the studies is the tendency to a Cross-Media communication, and in particular the convergence of the media on the web. The 15th edition of the HARRIS INTERACTIVE survey, *NetObserver Europa* (2007), states that "among the digital technologies, Internet has strongly contributed to abolish the barriers between the media contents and their support, thus favouring the convergence in the use of the media. The European internauts happen to be great online consumers of media contents"[6].

Content production by internet users, in the last 12 months, maybe illustrated with the following data: regarding males aged 15-34, no less than 24% of them have created or updated a profile on a community site or personal/professional social network. In addition, more than 12% of males aged 15-34 in the countries studied and even close to 1 in 4 in the UK have uploaded a video on a video-sharing site. And more than 10% of them have published an article to contribute to a type of website such as Wikipedia (*Note: Contribution and production of content online in the last 12 months. Source: Harris Interactive, June 2007, NetObserver Europe).*

As a result of all these empirical data, we must admit that the users' approach to the realm of technology is rather complex and ambiguous, even among the members of the new generations, born and *domesticated* within the digital world — the so-called "native digitals", to use Marc Prensky's definition (2001).

All this brings in discussion the real extent of the techno-bodies' degree of interactivity and interchange. Interactivity, understood as a dividing line between mass-media and personal media — between passive and active techno-bodies — is a concept that needs to be adjusted. As Lev Manovich (2001) has wisely emphasised, there are different levels of interaction, and each of them derives from very precise aspects of the medium and of the public/users for whom this medium is intended. To start with, there exists a psycho-emotional interactivity, typical of the mass-media and of the cinema in particular. It responds and also leads to the individual's identification with the contents of the medium itself. It is an interactivity that ends up in an emotional participation. Paradoxically, we could define it as a passive interactivity, even if it can be extremely immersive — think, for example, of cinema, where the public's absolute passivity can nevertheless correspond to a full implication in the contents. This type of interactivity is exceedingly gratifying because it does not require any effort of conscience.

The specific interactivity of the new media, on the other hand, is based of the users' possibility to actively interact with the medium's contents. It sets in motion a feed-back process, which allows the adjustment of the communicated contents in accordance with the consumer's re-action — in Van Dijk's words, this interactivity, which is at the third of the four levels, depends on the degree of control exerted by those who interact; the level of control is determined by the sender's and the receiver's possibility to change roles at any point and to exert influence on the contents (Van Dijk, 1999). This type of interactivity does bestow on the user the control over the medium, but it breaks the mechanism of identification and emotional implication. According to Lev Manovich, only a few media products manage to cover the long distance between *identification* and *control*: for instance, the videogames and the immersive virtual reality.

The different levels of interactivity correspond to different forms of implication of the body, which also means, to different techno-body morphologies.

Some media are characterised by an "emotional" interactivity and allow to reach high levels of identification. They have been defined by Derrick De Kerckhove as "tactile": we are mainly talking about television and cinema. His theory on TV's tactility is based on the consideration that television speaks to the body, not to the mind; it sets in play a prevailingly epidermic knowledge, it transmits an "experiential meaning" through submuscular stimulation. In De Kerckhove's words (1996), the submuscular activity consists in the performance of movements and actions through a sensorimotor mimicry that involves the whole body. The "meaning of television" is not the same as the "meaning of the book", it is rather closer, instead, to what the American psychologist and philosopher Eugene T. Gendlin has named "experiential meaning". Gendlin defines "experiential meaning" as the concurrence of hundreds of thousands of cognitive operations, performed by the body in a fraction of a second in response to some stimulus (Gendlin, 1964). "Experiential meaning" can thus be understood as the result of submuscular stimulation.

Under these conditions, what reigns supreme is an unconscious body that reaches high levels of emotional satisfaction.

Conversely, other media have totally different effects. The Web — in its more advanced forms, like Web 2.0 — and to some extend the mobile technologies as well, put the body in the open and constrain it to do conscious efforts. In addition, the media that reach more complex levels of interactivity, like the video games and the immersive virtual reality, produce techno-bodies that are "veterans" of a *domestication* (Goody, 1977): they have been forced to acquire some practical knowledge, based on action and movement instead of prediction and theory. In all these technologies, what reigns supreme is a body that has internalised and automated its abilities.

In such a context, mobile technologies have a field of their own: on one hand, they tend to deal with ever more natural and instinctive uses — very close to body functions — , on the other hand, they require deep attention to the interactive processes, because the mobile techno-bodies are always confronted to a series of simultaneously converging stimuli. The mobile managing of space and time requires effort and attention.

Some empirical information on younger people's techno-bodies — a range of users commonly associated with a great open-mindedness towards technological innovation — reveals that they prefer the media that allow a "passive" usage of their contents (with a low or elementary level of interactivity, like television, cinema, radio, etc.) to the more interactive ones (Gay, Mazali, *et al.*, 2008). In that sense, the Net and the new technologies turn out to be channels that do not affect the traditional media status quo — in a neither qualitative nor quantitative way; at most, they increase and accelerate the diffusion, the consumption and the permeability of the contents. In other words, we can proclaim our willingness to fully enjoy the interactivity allowed by the new media if we are somehow preserving the mechanisms of identification and emotional implication that keep fulfilling even the most innovative users.

As a matter of fact, the survey results on the real interactive participation levels on the Web 2.0 tend to show that the Net is, above all, a huge database: an instrument to access and redirect media contents, but seldom to re-edit them in a creative way.

So now the question is: do the techno-bodies really and fully perform the *bricolage* of styles, traditions, sources and points of view that the Web's new technologies offer? Where are the new uses of the techno-bodies, under the influence of the new technologies, leading us to?

Beyond the empirical data: towards the unpredictable uses

The above-depicted situation requires an approach beyond the empirical data, an approach through a model with different scenarios, not only based on the predictability of the uses of the new technologies, but also on their unpredictability. The importance that we nowadays attach to the issue of unpredictability is due to the internal changes of the new technologies — the transition from closed technologies to the ones open to reconfiguration — and to the changes in the nature and function of the *users*.

In agreement with Alexandre Mallard (2005) I think that, if we want to understand the practices of the techno-bodies, we have to use the paradox of the "unpredictable uses". Social sciences are nowadays obliged to anticipate the scenario of the applications of the new technologies. They have indeed two main reasons to work against the clock: on one hand, the steady speeding-up of technological innovation; on the other hand, the diversity and unpredictability of the applications invented by the users.

That is why the notion of "unpredictable uses" comes in handy, for we not only have the so-called creative uses of technology (which foresee scenarios that may proliferate and get "normalised"), but also "directions" that become determinative (what Mallard calls "uses drifts"). We are not expected to observe and capture the activities of users, but to somehow anticipate the trajectories of use and their associated drifts. This approach can prove useful in the analysis of the dynamic processes involved in ICT.

To sum up, when analysing the techno-body practices we shall assign a different and significant role to the users. These tend to take ownership of the ICT, thus completing the life span of a technological application, which consist in: predictability (production phase); unpredictability (consumption phase, role of users); and propagation of the "unpredictability" (multiple possible uses for the same type of device, from "pathology" to the normal pattern).

Unpredictability must therefore be understood in a positive way. It represents a new heuristic field, it requires and implies a shift in the way of considering the research methods: from observation to participation, and from phenomenological knowledge to praxiological knowledge (Bourdieu, 1972).

In the last few years, the evolution of technology has given birth to several paradigms of unexpected uses that range from the SMS's to the various systems of *peer-to-peer* data interchange. The SMS's might be the most clamorous and commented example of divergence between the expectations on a technology and its real outcome. Originally they had been planned to convey purely logistical information, but once in the hands of the real techno-bodies/*users*, they soon became a personal communication tool, operating at many levels in an extremely powerful way, and conveying also, and above all, information on emotional and interpersonal issues. SMS's have become such an obvious example of *killer application* that Howard Rheingold has coined the term "Thumb Tribes" to refer to the mobile phone techno-bodies (Rheingold, 2002).

We should notice here the increased importance of the *users* as innovators, an active role that explains the phenomenal proliferation of the *peer-to-peer* and *open source* systems: the first system corresponds to a semi-active user role — limited to sharing and exchanging — but the second one clearly corresponds to a fully active role, with the cooperation of a whole community of individuals.

These circumstances have been the premises for the elaboration of the economic models intended to spread innovation. I do think, in full agreement with Mallard, that the conditions established in such models — "the existence of incentives for users to innovate, the existence of incentives to reveal innovations, the possibility to distribute innovation at low cost and under competitive conditions with commercial production and distribution" (Mallard, 2005) — do not sufficiently explain how and why the techno-bodies make unexpected uses of the technologies, leading to equally unexpected forms of innovation.

Those economic models used to appear at moments of great effervescence related to important technological novelties — as was the case with the first forms of the Web 2.0 — during which the market was trying to identify their potential applications. Nowadays all these potentials have been fully exposed to light in a whole range of immensely wide-spread and heterogeneous applications, but they have not become totally predictable. On the contrary, the practices are still subject to high levels of unpredictability, due precisely to their proliferation and to the particular applications of uncountable individual and collective users; any interpretation becomes complex and uncertain.

In that sense, it might prove useful to read again, for instance, the above-mentioned survey results on the real degree of user participation in the Web 2.0; according to them, the highly participatory instruments do not meet a corresponding proportion of active users.

The new technologies seem to be subject to the following paradox. In the passage from exception to rule, that is, in the passage from the "start up" to the spreading stage, we do not enter a phase of predictability, but quite the opposite; we hit upon a new and stronger unpredictability, not a potential unpredictability — which Pierre Lévy would define as "virtual" — but a fully accomplished and concrete unpredictability, brought about by the practices of the techno-bodies.

It is in this "rebound" unpredictability that I see a fertile ground for a real creative appropriation of the technologies by the individuals.

It might be undisputable that the ICT usually provide a leading role to the users' creativity, but we should not naively think of a linear relation between active users, innovation and creativity. The users' unpredictability is often a secondary factor of the technological predictability, and the technologies keep on being driving forces independently of the users' own needs.

Once again Mallard recommends cautiousness, as he comments on the new "market pull" theories, which have replaced the old "technology push" ones in the analysis of technical innovation: "The ICT's domain thus appears as one of the sectors where uses on the market can precede the technological proposition. This is probably true. But it is probably also true that in many cases this innovation potential hangs on given products. Who would have said 15 years ago that there was a need for communication in mobility? [...] Pure market pull schemes are quite rare" (Mallard, 2005).

In short, the unpredictable uses are not the result of the users' free creativity, but rather the complex consequence of the itineraries that originate when technological planning and users' practices get confronted. The unpredictable uses derive from the meeting of the "strategies" of the industrial actors, on one hand, and the "practices" of users, on the other. That is why it is not enough to emphasise the user's importance, to point out the change in the consumer's cliché and to dismantle the modern concept according to which the public was a passive entity, totally conditioned by the goods and their madly intensive merchandising. In this respect, it is worth reading Giampaolo Fabris' (2003) and Vanni Codeluppi's (2000) studies on the post-modern consumer, as well as the traditional research studies on "uses and gratifications" (Blumer, Katz, 1974), which explore every single function and use of the media, stressing the individuals' and groups' ability to filter the technologies in relation to their own interests.

As a matter of fact, according to these theories a person is not a passive and inert entity, unwillingly exposed to the media contents, but an active and self-determined entity that selects, interprets and re-elaborates the media messages and instruments in line with his own specific cognitive, emotional and interpersonal needs. All these theories would not be complete without the important teachings of Michel de Certeau (1980), to whom we owe the observation of the users' creative role in relation to the artefact in itself: the user's *art de faire* becomes the pulling force of the artefact, and the latter therefore becomes the result of a dynamic process, instead of a predetermined process.

Nevertheless, it is precisely our duty to go beyond such theories and to acknowledge that the relationship between the artefact and its use — between technology and its appropriation — is constantly subject to an unpredictable complexity. And for this we may take advantage of Mallard's very helpful theoretical frame on what he calls the "uses drift": "The idea of drift evokes the fact that an innovation's use is like a boat in the ocean: it sails in the direction in which the winds blow. And the winds can blow in many, conflicting directions" (Mallard, 2005).

When looking into *use drifts*, the parameters are the following:

- the technological object (product, service, technical platform, medium): the peculiar features of the technological means taken into consideration are determining factors; they may not be the only ones, but they set the directions along which the uses can develop — in Marshall McLuhan's terms "the medium is the message";
- the user: it is essential to establish and to understand the profiles and the actions of the different users;
- the group in which the user's action takes place: due to the social and collective nature of communication, there is no other option than to contextualize any single use in its group field;
- the usage context: the mobile technologies have made it absolutely clear that the environment in which the techno-bodies develop their practices is of the greatest importance; the "environment" means the space-time coordinates that define the moment of the usage, as well as the social context and the place's "physicality" (the environment's "body" interacts with the user's techno-body and conditions it); in that sense, the techno-bodies' actions are *situated actions*.

Not a single one of this parameters can be underestimated, all of them contribute to establish the foundations of the possible usage directions for the practices of the techno-bodies. It is in the sum of all these parameters that we may find the platform from which to understand the technologies *domestication* and appropriation processes.

The technological *bricoleur*

Technology has modified our way of apprehending, our corporal structure, our environment; in the practices as well as in the theoretical approaches. It has brought closer and has con-fused the concrete knowledge (the *savage mind* in Claude Lévi-Strauss's words) and the abstract knowledge (or scientific mind); the same with the symbolic-reconstructive cognitive modality (the mind's knowledge) and the perceptive-motor modality (the body's knowledge).

The techno-bodies' environment — the place of definition and placement through the concrete movements of the action, the experience and the contact — establishes, inaugurates and strongly claims the need of a type of knowledge that shall not exclusively happen by means of a thinking process *domesticated* by the spoken or written language (the alphabetic-numeric mind). And this because, on one hand, technologies need the body to get to be known and, on the other hand, man is using more and more corporal technologies to get to know the world that surrounds him. This statement derives from the study of the techno-bodies' practices, but it is in line with the theories of many an author: Claude Lévi-Strauss (1962); Walter Ong (1982) and his formulation of the secondary orality; Jack Goody (1977) with his *domestication* model; and, in a more general way, Silverstone (1992) with his domestication theory, which highlights the role of users in innovation (the work done by individuals and communities in order to make a technology from the outside do practical work, and make sense within that community).

Having realised the importance of the technologies and their uses in the process of human thought's evolution/mutation, we nowadays consider the body as the place of knowledge of the new technological instruments.

The theoretical approach that has derived from such considerations puts forward a *new domestication*, based on the recovery of the experience of the concrete. Lévi-Strauss was the first to put it into words in his definition of *bricolage* and *bricoleur*; Sherry Turkle (1995) and, later on, Mario Ricciardi (2005) have applied these same concepts. A body that knows is, first of all, a body that feels (related to perception), that moves (related to spacialization) and that speaks (related to temporality). Knowledge is action first of all, and only then reflection. In the "place of the action", there appears and gets solved a contradiction between thinking and doing, between will and being.

But who is today's *bricoleur*? How could we apply Lévi-Strauss's concept to a propositive approach to the contemporary techno-bodies? Which ones are, today, the territories where the type of knowledge that we call *bricolage* has a real shape and presence?

For me, the territory of artistic performativity is an exemplary modern paragon of this. I am convinced that *bricolage* finds a real channel of expression in artistic creativity. Artistic performance can provide the *place* where a techno-bodied *bricoleur* may enact the passage from the *predictable uses* to the *unpredictable* ones, transforming the *user* into an *actor*.

Lévi-Strauss starting point is: "There are two distinct modes of scientific thought. These are certainly not a function of different stages of development of the human mind but rather of two strategic levels at which nature is accessible to scientific enquiry: one roughly adapted to that of perception and the imagination, the other at a remove from it" (Lévi-Strauss, 1972: 15). At present, the nature that we have to domesticate is technology. The modern technologies of communication have newly raised the question of the modes of knowledge to which Lévi-Strauss had found an answer through the concept of *bricolage* — a concept that defines a "concrete" knowledge inside scientific knowledge.

Allow me to emphasize, in agreement with Lévi-Strauss, that every form of thought has a need of "rational ordering" at the base, which will set the rules for any practice. Even the ritual practices are nothing but "classification" or "putting in order". The performative techno-bodies of the Contemporary Age build their own unpredictable rituals precisely on the rituality of the predictable technology. Again and again, they break and reconstruct the rules of the man-machine relation. They favour the uprising of new classifications, and these might be heterogeneous and arbitrary — thus joining the so-called "social tagging" activities, set in action by the social networks — but they help to establish a shared "memory".

According to Lévi-Strauss, the purpose of these contemporary rituals is the "speculative organization and exploitation of the sensible world in sensible terms" (Lévi-Strauss, 1972: 16).

Bricolage is a knowledge, an erudition, a science that Lévi-Strauss prefers to call "prior" rather than "primitive". It represents a kind of lost wisdom — of old memory of the concrete — made of body and matter. In its origins, the verb *bricoler* was applied to cheerful activities, like billiards or horse-riding, and it referred to an "extraneous movement": the precise point of contact with the matter, the very instant of avoiding an obstacle; in few words, the "now and here" that individuates and contains, in its single occurrence, the sense of the practice. Nowadays, by *bricolage* we mean the activities accomplished with one's own hands and using devious means in relation to those of a craftsman.

The *unpredictable* techno-bodies bear "extraneous movements" that are different every time and thus create difference. The techno-bodies are not professional, but they acquire knowledge through the practice.

It is important to note that the *bricolage* is a form of scientific knowledge, it is not opposed to it. Scientific knowledge can be conveyed by two different archetypes: the *bricoleur* and the engineer. How do their *modus operandi* differ? "The set of the *bricoleur*'s means cannot be defined in terms of a project, as in the case of the engineer. It is to be defined only by its potential use or, putting this another way and in the language of the *bricoleur* himself, because the elements are collected or retained on the principle that 'they may always come in handy'" (Lévi-Strauss, 1972: 18).

I personally see a clear analogy between this concept and Lev Manovich's theory of the new media. When he talks about creativity in using the new media, Manovich employs tools and notions that seem to be in perfect harmony with Lévi-Strauss's old recurring themes of mythical reflection and science of the concrete. In the essay *The language of new media* (Manovich, 2001) there are three main topics that I would like to highlight: the operation of *Selecting*, the operation of *Compositing* and the *Database Logic*. According to these, the creativity related to the new media happens at the core of a repository of predetermined possibilities and tools (all new media products, especially world wide web, are based and depend on databases of information and structures), where the *consumers* practically limit themselves to realising selective actions. That is why they are *users* and not *actors*. On the contrary, the *professional* users attain a more complex operating level, called *compositing*: they are able to creatively recombine the elements of the original database. In other words, *compositing* is an operation through which the users experiment with their own unpredictable and creative uses, whereas *selecting* corresponds to a predetermined and predictable use of the technologies.

The contemporary *bricoleur* is a techno-body that evolves between selecting and compositing: "His first practical step is retrospective. He has to turn back to an already existent set made up of tools and materials, to consider and reconsider what it contains and, finally and above all, to engage in a sort of dialogue with it and, before choosing between them, to index the possible answers which the whole set can offer to his problem" (Lévi-Strauss, 1972: 18).

Nevertheless, there is a difference between Lévi-Strauss's *bricoleur* and the contemporary performative techno-body. The difference resides in the fact of "cataloguing before choosing": today's actions on and with technology tend to shorten the time for the selecting process and include it inside the practice itself, the selection is enacted before it is decided.

The final result is analogous anyway. The act of choosing inside the "inventory" — what Lev Manovich calls the Database — contributes to the definition of a set which has yet to materialize but which will ultimately differ from the instrumental set only in the internal disposition of its parts. The selection of elements from the original set can be enriched by compositing them, that is, by their recombination. What we obtain is a definition of creativity that has to do, not with the romantic myth of the inventing genius anymore, but with the capacity to recombine and reconfigure predetermined sets of elements, instruments, possibilities, "contraptions" and "signs". The creative possibilities "always remain limited by the particular history of each piece and by those of its features which are already determined by the use for which it was originally intended or the modifications it has undergone for other purposes" (Lévi-Strauss, 1972: 19).

The elements gathered and used by the *bricoleur* are "pre-linked". To use Manovich's terms, they possess a modular structure: the new media objects have a fractal structure. They are made of minimal units of significance and function, which may be combined and recombined, structured and de-structured; they do keep their original signification/function, but they also take on new ones in their new global configurations. The elements are assembled into large-scale objects but continue to maintain their separate identities. The objects themselves can be combined into even larger objects again, without losing their independence.

Lev Manovich's modular structures are the equivalent of the elementary units in Lévi-Strauss's myth. Are then the new media a new myth — nevertheless characterised by impermanence in opposition to the ancient myths' permanence? And is this impermanence, that seems to pervade the flexibility of the communication technologies, really "impermanent"? The issue of the new technologies shows an unstable balance, not only between predictability and unpredictability, but also between impermanence (of the uses, styles, contents and "means") and permanence (the permanent effects on the techno-bodies, on their way of knowing and on their "physicality").

Unlike the Lévi-Strauss *bricoleur*, the *modus operandi* of the contemporary *bricoleur* does not come down to a mere arrangement of elements that form the same object again and again, only changing its internal composition. Whereas selecting, as Manovich defines it, follows Lévi-Strauss's line of thought, it is *compositing* that overcomes it. The compositing action can lead to tremendous changes of the original elements database; the applications of the Web 2.0 are a good example of this.

In the field of the digital performing arts in particular, the artistic action progresses through *selecting* and *compositing*, and it results in different levels and forms of configuration of the starting points (contents, means/technologies and practices). The techno-body of the digital performing arts is an engineer as well as a *bricoleur*, it develops both *modi operandi*: "The engineer is always trying to make his way out of and go beyond the constraints imposed by a particular state of civilization while the *bricoleur* by inclination or necessity always remains within them. This is another way of saying that the engineer works by means of concepts and the *bricoleur* by means of signs" (Lévi-Strauss, 1972: 19-20).

This line of thought can also be approached from the theoretical point of view of *Performativity* (Loxley, 2007; Auslander, 2003). The term *performativity* comes from the linguistic field of John Langshaw Austin (1962) and I use its characteristics to explain the artistic techno-bodies. These characteristics are: act (an act implies making to exist, so creativity in action; the act creates a step between content and form, it is therefore subversive); satisfaction *versus* truth; and strength *versus* meaning.

Techno-bodies perform acts (actions) that have unpredictable consequences (and uses). These actions are essentially processes, in other words, a part of them will always be subject to transformation, absolutely non-describable. It is indeed the performativity that places the techno-bodies of the artists in a shifting space. And this because every action that is performed becomes something else right in the instant of its performance, the non-coincidence of act and intention is unavoidable.

Nowadays we talk about performative technologies and performative bodies to stress the process and relationship involved. The techno-bodies imply a relationship between different parts: inside/outside, product/consumption, producer/user. And there is even one more parameter that sets the techno-bodies in action: the context of use, the environment in which they operate.

The artist's techno-body goes beyond the "concepts" and the "signs". Its performative actions find their meaning and their form in the very act of happening: it is through concrete action that concepts and signs come to light. The artists put themselves in an unstable balance — inside or beyond the line — and proceed by alternating the use of events and structures. In Lévi-Strauss's terms, they proceed by alternating the contingent and the necessary dimensions, they combine an external and an internal knowledge, they bring together "being" and "becoming": "Events in this sense are only one mode of the contingent whose integration (perceived as necessary) into a structure gives rise to the aesthetic emotion" (Lévi-Strauss, 1972: 27).

The performativity is linked to the event, the event is linked to the *bricolage*, the *bricolage* is linked to the new technologies and to the techno-bodies of the digital performing arts: the *bricoleur* makes/processes structures by means of combining the events (Lévi-Strauss, 1972). In the performativity and in the event, technology and art have intrinsic affinities. They are both in unstable balance between structure and event, necessity and contingency, interior and exterior. The *bricoleur* is performative; in the art of digital performance he finds fertile ground for his actions. A performance is, in fact, a thought in action. It is idea and action simultaneously. It is a processuality open to improvisation and experimentation. It often investigates and handles both reality and the imaginary visions of it, altering or even illuminating the core of human nature — body, life, love and death, individual and social identity. It is interdisciplinarity and concrete multimediality. To sum it up:

- very often a performance repeatedly destroys the codes and the common practices; it can thus be considered as an act against the specificity of language;
- a performance is an excess of energy that enhances the asymmetry between meaning and references; in terms of Jean-François Lyotard (1973) it is *Théâtre énergétique*;
- as a creative action, the performance almost never stands up to expectations: it is unpredictable in its end result and subversive in its intentions; this unfulfillment makes it basically mobile and exceptional, which Richard Schechner (1984: 177) conveys with these words: "the performative activities are fundamentally *processual*, that is: a part of them will always remain subject to transformation, and will be absolutely impossible to define"[7].

The performances with techno-bodies have the same nature as the happenings: they are subversive; they hurt the artist's body and the public's emotions; and they seek destabilization through excessive ways of expression.

Performativity is the dimension that we should try to master, because it is the binding place, the place where technologies and corporeity meet. Performativity is action, experience and movement, three words that describe very well the processual nature of the new communication technologies — not anymore instruments designed according to the standards of modernity, but places governed by flexibility and permanent reorganization — as well as the processual nature of the constitutional mobility of the body — the body does not stop, it is born, it grows, it dies, the body moves, it is never representation but eluding instant and unstoppable event.

The Modern Age and the industrial societies have "disused" the rites, the myths and, therefore, the *bricolage* — which they have converted into a hobby or pastime. The new technologies are bringing back the concept of a *concrete knowledge*; the techno-body of the *bricoleur* and his knowledge paradigm are again on the foreground, leaving the models of modernity in the far background. Nevertheless, we are even beyond the post-modernist attitude, which had made of the *bricolage/collage* its only way of action. Today the *bricoleur* is not detached from the *engineer*, and creativity is more and more the result of social actions and networks that apply unpredicted uses to reformulate the uses, the functions and the technologies.

The digital performing arts can act as a fertile ground for the premonitions and the revelations of the unpredicted uses of technology. They can be the place where to try and overcome hedonism in order to reaffirm a space and a time of one's own.

We do not use technology because it is unavoidable. We use it because it may help us conquer, through creativity, new proposals for a world that is not deterministic anymore. The goal is not to make a creative "factory" while creativity is suffocated by the *predictable uses* of technology. Neither is it to create an illusion of technological interactivity that would screen a new mass of *spectators*. The goal is to establish a subversive and anarchic creativity in order to make something positive out of the obliteration of the linear and deterministic models of the past.

I would like to conclude by giving a partial response to the questions that have lead my considerations — namely: Who is this digital *bricoleur*? Is he a modern craftsman whose knowledge is based on an knowledgeable know-how, or is he an *idiot savant* that converts the practice in a sort of direct, immediate and corporeal intelligence (Sacks, 1995; Mazali, 2007: 96)?

The difference between the contemporary *bricoleur* and Lévi-Strauss's mythical thought *bricoleur* resides in the way they apprehend the world. Lévi-Strauss refers to a world dominated by the will to classify, that is, by an ordering principle that leads to stability and permanence. The *bricoleur*/techno-body of nowadays, on the contrary, fully acknowledges the fundamental role of fluctuation and instability in any possible field of observation or action. Today's *bricoleur* has to deal with multiple choices and "limited predictability horizons" (Prigogine, 1996); he knows that certainty may be comforting, but means the negation of time and creativity.

Notes

1. The title of my essay came from a book by Bianca Maria Pirani who I thank for having suggested this title (refer to Pirani, B.M. (forthcoming), *La danza dei corpi. I luoghi sociali della tecnologia*, Roma: La nuova cultura).
2. For a study on the popular culture of our time and on the way new technologies influence and change it, refer to Henry Jenkins, 2006.
3. As to this subject, refer to the European project ACTeN. ACTeN is a EU-funded thematic network (September 2002 - August 2004). It is part of the EU Program "Information Society Technologies" (IST). In particular, "Cross-Media" can be downloaded at: http://www.acten.net/cgi-bin/WebGUI/www/index.pl/cross_media (consulted in may 2009).
4. http://www2.sims.berkeley.edu/research/projects/how-much-info-2003
5. Refer to the article *Time's Person of the Year: You*, Time Magazine: available online at: http://www.time.com/time/magazine/article/0,9171,1569514,00.html
6. Conclusions of the 15th edition of the "NetObserver Europa" analysis, available online at: http://www.harrisinteractive.fr/services/pubs/HI_FR_NetObserverEurope_V15_IT_Sep07.pdf
7. Translated by the author.

Bibliography

Auslander P. (ed.) (2003), *Performance. Critical concepts in literary and cultural studies.* New York, Routledge.

Austin J. L. (1962), *How to Do Things With Words.* Cambridge MA, Harvard University Press.

Blumler J. G. and Katz E. (1974), *The Uses of Mass Communication, Current Perspectives on Gratification Research.* Beverly Hills, Sage.

Bolter J. D. and Grusin R. (1999), *Remediation: Understanding New Media.* Cambridge, MA, The MIT Press.

Bourdieu P. (1972), *Outline of a Theory of Practice.* Cambridge, Cambridge University Press.

Castells, M. (1996), *The Rise of the Network Society.* Oxford, Blackwell.

Codeluppi V. (2000), *Lo Spettacolo Della Merce: I Luoghi Del Consumo Dai Passages a Disney World.* Milano, Bompiani.

De Certeau M. (1980), *L'invention du quotidien. 1. Arts de faire.* Paris, Union generale d'editions.

De Kerckhove D. (1995), *The Skin of Culture: Investigating the New Electronic Reality.* Somerville Press.

Di Fraia G. (2004), *e-Research, Internet per la ricerca sociale e di mercato.* Bari, Laterza.

Fabris G. (2003), *Il nuovo consumatore: verso il postmoderno.* Milano, Angeli.

Gay N., Mazali T., Monaci S. and Taddeo G. (2008), *Giovani, media e consumi digitali.* Napoli, Liguori.

Gehlen A. (1978), *Der Mensch. Seine Natur und seine Stellung in der Welt.* Berlin, Junker und Dünnhaupt.

Gendlin E. T. (1964), *Experience and the Creation of Meaning.* New York, Free Press.

Goffman E. (1967), *Interaction Ritual.* New York, Pantheon.

Goody J. (1977), *The Domestication of the Savage Mind.* Cambridge University Press.

Jenkins H. (2006), *Convergence culture: Where Old and New Media Collide.* New York, New York University Press.

Lakoff G. (1987), *Women, Fire, and Dangerous Things: What Categories Reveal About the Mind.* Chicago, University of Chicago Press.

Lakoff G. and Johnson M. (1999), *Philosophy In The Flesh: the Embodied Mind and its Challenge to Western Thought.* New York, Basic Books.

Leroi-Gourhan A. (1964), *Le geste et la parole.* Paris, Albin Michel.

Lévi-Strauss C. (1972), *The savage mind.* London, Weidenfeld and Nicolson (Orig. pub. *La pensée sauvage*, Paris, Plon, 1962).

Lévy P. (1995), *Qu'est-ce que le virtuel.* Paris, La Découverte.

Loxley J. (2007), *Performativity.* New Yourk, Routledge.

Lyotard J-F. (1973), *La Dent, la Paume.* In: *Des dispositifs pulsionnels*, Paris, Union Général d'Éditions.

Lyotard J. (1984), *The postmodern condition: A report on knowledge.* Minneapolis, University of Minnesota Press.

Mallard A. (2005), *Emergence of Unpredictable Uses? New Stakes and Tasks for a Social Scientific Understanding of ICT Uses.* In: Haddon, L, Mante, E., Sapio, B., Kommonen, K-H, Fortunati, L. Kant, A (eds.), *Everyday Innovators. Researching the Role of Users in Shaping ICTs*, Dordrect, Springer.

Manovich L. (2001), *The Language of New Media.* Cambridge MA, MIT Press.

Mauss M. (1936), *Sociologie et anthropologie.* Paris, PUF.

Mauss M. (1992), *Techniques of the Body [1934].* In: J. Crary and S. Kwinter (eds.), *Incorporations*, New York, Zone Books, Pg. 455–475.

Mazali T. (2008), *Ex-Peau-Sition: nomadic bodies and new technologies*. In: M. B. Pirani and I. Varga (eds.), *The new boundaries between bodies and technologies*, Cambridge Scholars Publishing.

Mazali T. (2007), *Ex-Peau-Sition. Il corpo ri-scritto nelle performance tecnocorporee*. Milano, Guerini.

Merleau-Ponty M. (1945), *Phénoménologie de la perception*. Paris, Gallimard.

Ong W. (1982), *Orality and Literacy: The Technologizing of the Word*. London and New York, Methuen.

Prensky M. (2001), *Digital Natives, Digital Immigrants*. In: *On the Horizon*, NCB University Press, Vol. 9 No. 5, October.

Prigogine I. (1996), *La fin des certitudes.Temps, chaos et les lois de la nature*. Paris, Editions Odile Jacob.

Rheingold H. (2002), *Smart Mobs: The Next Social Revolution*. Place, Basic Books.

Ricciardi M. (2005), *Lo schermo e lo specchio*. In: S. Turkle, *La vita sullo schermo*, Milano, Apogeo.

Saracco R. (2005), *Information and communication technologies: disruptions to look for and their impact on countries evolution strategies*. Torino, Telecom Italia Lab.

Schechner R. (1984), *La teoria della performance*. Rome, Bulzoni editore (Orig. pub. *Performance Theory*, London and New York, Routledge, 1988).

Silverstone R. *et al.* (eds.) (1992), *Consuming Technologies: Media and information in domestic spaces*. London/New York, Routledge.

Sacks O. (1995), *An Anthropologist on Mars*. New York, Vintage Books.

Silverstone R. and Haddon L. (1996), *Design and the Domestication of ICTs: Technical Change and Everyday Life*. In: Silverstone R., Mansell R. (eds.), *Communication by Design. The Politics of Information and Communication Technologies*, Oxford University Press.

Toffler A. (1980), *The Third wave*. New York, William Morrow.

Turkle S. (1995), *Life on the Screen: Identity in the Age of the Internet*. New York, Simon and Schuster.

Van Dijk J. (1999), *The Network Society. An Introduction to the Social Aspects of New Media*. London, Sage.

Varela F., Thompson E. and Rosch E. (1991), *The Embodied Mind: Cognitive Science and Human Experience*. Cambridge MA, MIT Press.

Surveys

Forrester Research, 2007, *Social Technographics Report*, http://www.forrester.com/Research/Document/Excerpt/0,7211,42057,00.html http://blogs.forrester.com/charleneli/2007/04forresters_new_.html

Harris Interactive, 2007, *NetObserver Europa*, http://www.harrisinteractive.fr/services/pubsHI_FR_NetObserverEurope_V15 _IT_Sep07.pdf

Hitwise, 2007, *Measuring Web 2.0 Consumer Participation*, http://www.hitwise.com/press-center/hitwiseHS2004/web20.php

Lyman P., Varian H.R., 2003, *How much information?*, School of Information and Management Systems, Berkeley, University of California, http://www2.sims.berkeley.edu/research/projects/how-much-info-2003/

Technorati, 2007, *The State of the Live Web*, http://technorati.com/weblog/2007/04/328.html

Chapter 3
Cyborg identity and contemporary techno-utopias: adaptations and transformations of the body in the age of nanotechnology

Marina Maestrutti

Abstract

The possibility of improving the human body through a closer relationship with technology in order to overcome the human species towards new stages of evolution is a constant element of techno-utopian visions, among others transhumanism. This projection to a radical transformation of the body - and mind - as a result of technological action is based on the concepts of adaptation, or non adaptation, of a human being to a world constantly changed by techno-science.

The belief is that not only the body has to change, but that identity is not a stable concept. This mobility in the relationship between body and identity is typical of the post-human thought, which inherits from the informational model the conviction that the biological embodiment of the human is to be regarded as an accident of history rather than as an essential condition of life.

Hybridization is therefore valued by the post-human thought as a condition which has "made" the human as he is today, and it appears as a fundamental topic in any discourse on nanotechnology, biotechnology and development of human-machine interfaces. Contemporary technologies in multiplying situations of contamination and hybridization promote de facto a horizontality in the hierarchy of life. On the other hand, to use more and more invasive and advanced miniaturized technologies - based on our biological substrate, able to communicate with him and sometimes to control their functions - opens a new range of human "performative" possibilities. The *cyborg* - the cybernetic organism, a human-machine hybridization - is generally regarded as representative, the new subject topic of the post-human age, and the two term stend to become synonymous.

This article would contribute to analyze the foundations, the challenges and limitations of this hybrid design of the body and identity in the discourse of contemporary techno-utopias.

> Humanity must not stagnate [...] Humanity is a temporary stage along the evolutionary pathway. We are not the zenith of nature's development. It is time for us to consciously take charge of ourselves and to accelerate our transhuman progress.
>
> Max More, *On Becoming Posthuman*

> The human desire to acquire new capacities is as ancient as our species itself. We have always sought to expand the boundaries of our existence, be it socially, geographically, or mentally. There is a tendency in at least some individuals always to search for a way around every obstacle and limitation to human life and happiness (Bostrom 2005).
>
> Nick Bostrom, *A History of Transhumanist Thought*

Humanity is "always" eager to improve and has "always" wanted to exceed the limits holding up the fulfillment of happiness. To Nick Bostrom, this is precisely what justifies the will to enhance defining his movement, transhumanism. The idea of enhancing physical abilities, the need to cross limits and reach a new stage of human evolution is a constant element in transhumanist thought, particularly in the establishment of close body/technology relationships. Significantly transforming the human body – and mind, especially the intellect - using technology, implies relying on the notion of human adaptation and inadequacy abilities, in a world changing as a consequence of technological developments. These perspectives are extreme. Nonetheless, they can shape the conceptual and ideological framework of research projects in the nanotechnology field, often converging with biology and physics.

The concept of human adaptation and inadequacy abilities in times of technological evolution is developed within a specific stance on the natural-artificial relationship, where the acceleration of technological evolution is opposed to the tardiness of the organic one. It would thus seem that machines has more chances and opportunities of imposing its evolutionary model than living organisms and human in particular do. Several theories will be presented before we proceed to the identification of models of bodies used to substantiate the discourse regarding human beings' inadequacy in a technological environment, increasingly considered as a "doubled nature."

For the most part, the examples will be derived from transhumanist literature, omnipresent in "metaphysics" inherent to the hypothesis on converging technologies (nano-bio-info-cogno, NBIC Convergence) which aspires to enhance human abilities and increase their power. It should be pointed out that this literature is essentially of American origin. Yet, more and more transhumanist stances are present among the European academic world, especially in England, and not completely rejected by the "weak" European version of convergence, suggested in the report *Converging Technologies: Shaping the Future of European Societies* in 2004[2]. A model of unifying bodies seems to emerge from this analysis: the idea that bodies can be conceived, built, modified and laid out like a personal project as well as a collective one.

The evolution of machines: a competitive kind of natural evolution

Describing humanity's artificial and natural dimension using definitions such as "technosphere" or "technonature" seems to result in the double process of naturalness' artificialization and the naturalization of the artificial, and a desire to abolish the dualism existing between nature and artifice. In this sense, certain authors talk about "the evolution of machines". George Dyson (Dyson, 1997), for example, expresses an amenable outlook where the evolution of machines and natural evolution are in symbiosis. The reconstruction of humanity's artificial history, from the first bit of flint to silicon microchips[3], follows an adaptive continuity, far from opposing biochemistry to electronics, and founded on two presumed methodologies: a biological point of view on the evolution of machines and a mechanical point of view on life. The co-evolution of the organic and the mechanic is a constant, characterizing the past and predicting the future.

However, a breaking point is often considered for this common process. This is what Vernor Vinge and Ray Kurzweil describe as "singularity". Indeed, there is a difference between an increase in machines and networks' intelligence and the human biological evolution. The latter is deemed slower and less capable of being integrated in what Dyson calls "global electronic intelligence."

Singularity is possible only if human beings succeed in building a conscious machine whose intelligence is superior to theirs. This contingency was often discussed and discredited (Penrose, 1980). The creation of such a machine is still in question to this day. However, if an "intelligent" computer is developed as explains Vinge, the concerted action of computer networks and their human users could lead to the creation of super-intelligence capable of increasing the human intellect. In 1993, referring to the curve of computing progress during the last decades, Vinge predicted that such an event could happen between 2005 and 2030.[4] Singularity can be considered as the end of human civilization and the beginning of a new era.

The promise of this kind of accomplishment in the future is grounded on a law meant to govern progress and change it into a regular and necessary process. As far as computing and nanotechnology are concerned, Moore's law on miniaturization is the most relevant example. Gordon E. Moore, Intel's co-founder, noticed that the amount of transistor's per integrated circuits had double every 5-6 years since 1959 and announced in 1965[5] and 1971 that the components' power would double equally every 18 months. This law seemed to make this potential future necessary and inevitable but found its limit in the size of microchips and transistors that cannot be reduced indefinitely. Yet today, nano-dimensions make it possible to "tease" Moore's law, as some researchers like to say, and push miniaturization's limits.

Ray Kurzweil, inventor, scientist, futurist and specialist in applied computing, suggests in his essay *The Law of Accelerating Returns*[6] and his last work *The Singularity is Near* (Kurzweil, 2005), to extend Moore's law in order to include other forms of calculations. He believes that the exponential growth of technological progress could be a pattern visible throughout humanity's history, even before the arrival of life on earth. According to the law of "accelerating returns", when sciences and technologies dispose of efficient and affordable computing means, they can achieve more discoveries requiring new calculation methods. Technology's role is not only about making tools; it is also about making things allowing for the conception of new tools. Moreover, progress is contagious. Indeed, progress in a field leads to progress in others. Technology generates technology, yet again according to an accelerated growth curve. Technology thus becomes the generating factor of evolution in the present phase.

Hence, it is possible to see how the constant complexification and the regular development of human-technology relationships converge towards a point of no return, a radical rupture, a "singularity" defining an entirely new stage in history. However, Kurzweil's interpretation of Moore's law has changed the meaning and the area of relevance of this law, since it could only be applied to semiconductor circuits. Kurzweil extrapolates boldly: from predictions on material components of information technology he proceeds to the spiritual performances of human mind.[7]

Then what is humanity's future in the convergence of biochemistry and electronics? Dyson warns us in the preface of his work: "In the game of life and evolution, there are three players at the table: human beings, nature and machine. I am firmly on the side of nature, but nature, I suspect, is on the side of machines" (Dyson, 1997:86). This point of view is shared by the spokesmen of a "strong" interpretation of the development of artificial intelligence. Hans Moravec (Moravec, 1999), for example, agrees with him and is convinced that man's chances of surviving in the case of an encounter with the superior specie of artificial beings are very small. The end of humanity stands for a lost battle walking down evolution's path.

The human body amid inadequacy and hybridization

Some extreme points of view on the adaptation dynamics existing among the dimension of the artificial and man, express the belief that man is *maladjusted* for a/the technosphere, maladjusted to integrate himself in it. Several models of the human body are at the basis of this belief. An old model makes reference to the theory of human incompleteness according to which man is an imperfect being capable of improving by means of culture – in particular due to his "autopoietic capacity" (Remotti, 2003:28), that is to partly build himself – going back to a long tradition in occidental thought. Indeed, this is recounted in the Prometheus myth, found in Hesiod's *Theogony* and Plato's *Protagoras*. The legend tells the story of Prometheus and his brother Epimetheus. During the creation of world, the gods assigned them the task of distributing qualities and physical abilities among all living beings. Epimetheus gives all animals a special virtue, until no good is left for man, no protective covering and no quality. To atone for his brother's mistake, Prometheus steels the secrets of fire and arts to Hephaestus and Athena, and gives them to mankind to compensate for their deficiencies. At the Renaissance, Giovanni Pico della Mirandola thematizes the link existing between natural deficiencies and the indeterminateness of the human condition, referring it on the one hand to the condition of animals and on the other to the individual potential for liberty and responsibility[8]. Descartes asserts that, unlike animals whose existence is entirely determined by laws of matter, the incomplete adhesion of the human body to these laws and the influence of the soul, allow for human emancipation and gives him means to know the world and himself. During modern times, Johan Gottfried Herder systematizes the canonic theory of incompleteness especially in his *deen zur Philosophie der Geschichte der Menschheit* in1784-1791 (Von Herder, 1966). He also defines man from a median perspective: an animal within the animals, bound to earth and material needs, characterized by transcendent capacities resulting from rationality, freedom of action and the desire for immortality in religion, which destines him to a superior form of existence. The continuity between mankind and the animal kingdom diverges when perfection is aimed as it takes on different aspects for both. Indeed, animals reach their full potential perfectly in their natural milieu whereas men can only do it imperfectly, being required by God to enjoy a kind of superior existence.

During the 20th century, Arnold Gehlen gives an anthropological point of view on the roles that culture and technology have to play to overcome human incompleteness. According to him, human biological deficiencies (lack of developed instincts, unadapted organs, lack of hair protection for the body) are compensated by technology, without which man could not live on earth. Therefore, his imperfection is compensated by his exceptional ability to find the necessary stimuli to create and act. The development of technological acts would have generated a "shift in bodily participation", favoring "superior" cognitive and symbolic activities. For Herder, this superior skill is particularly well expressed via language. Thus, highly capable learning abilities have eventually privileged the emancipation of intelligence as a factor using technology to ensure human existence (Gehlen, 1940).

The concept of incompleteness strengthened the idea that other capacities could retrieve human beings from their animal condition; make them more independent in respect to life's materiality, as a result of the development of technologies and the influence of culture. But how did we get to the point of declaring that the body, the biological support of human intelligence, is becoming obsolete?

Indeed, today's technology seems to exceed environmental and physical constraints to the extent that some humans are no longer satisfied with their body. We own a body that we no longer desire. As the Australian artist Stelarc likes to say: "*The Body is Obsolete*":[9]

> It is time to question whether a bipedal, breathing body with binocular vision and a 1400cc brain is an adequate biological form. It cannot cope with the quantity, complexity and quality of information it has accumulated; it is intimidated by the precision, speed and power of technology and it is biologically ill-equipped to cope with its new extraterrestrial environment.
> The body is neither a very efficient nor very durable structure. It malfunctions often and fatigues quickly; its performance is determined by its age.[10]

According to Stelarc, we are entering a postevolutionist era:

> In fact, I think that evolution ends when technology invades the human body [...] Nowadays, technology is ubiquitous, it is becoming a component of our body — including watches and artificial hearts; for me, it represents the end of Darwinist

> evolution as the organic development over millions of years, through natural selection. From now on, with nanotechnology, man can swallow technology. The body has to be comprehended as a "structure". It will only be possible to readjust our conscience of the world by modifying the body's architecture.[11]

As Stéphan Barron[12] noticed, the human being described by Stelarc is almost like an object submitted to technology: an organic machine interacting with technological machines. For Stelarc, the body is one of technology's spin-offs, forced to adapt itself to technological needs and design: "I don't recommend to adapt space to our body but to restructure our body. So the question is: how can we restructure a human pan-planetary physiology [...] how can we restructure a body to enable it to exist in varied atmosphere, gravitation and electromagnetic field conditions?"[13]

The artist's words reveal posthuman and transhumanist principles where the body's model is obsolete, arbitrary and hybrid.

Jeffrey Deitch, the art critic who organized the interesting exhibit *Post Human* in the 90's about the relationship between body, identity and representation, goes back to an idea shared by the majority of technophile movements linked to posthuman circles. This idea asserts that human evolution is entering a new stage that Darwin could never have predicted. According to Deitch, "Social and scientific trends are converging to shape a new conception of a self, a new construction of what it means to be human being. The matter-of-fact acceptance of one's "natural" looks and one's "natural" personality is being replaced by a growing sense that is normal to reinvent oneself" (Deitch, 1992).

Artists' actions – as well as writers' and philosophers" – focus on the transformation of the body, its disidentification and transmigration towards other configurations of expression, where genre and form no longer refer to human beings. This is the case in Orlan's *body art*, which modifies his features by means of live surgical interventions, Stelarc's declaration on the human body's final obsolescence where man-machine borders are redefined using prosthesis, implants, mechanical and technological interventions, and in Matthew Barney's appraisal of the human-animal boundary.

Of course, this concept of body-technology hybridization remains in the background. Cinema is also studying this desire to blend organic and artificial distinctions. Indeed, body-machine hybridization is present in David Cronenberg's work, and the *Alien* cycle, the cyborg figure in *Terminator 2: Jugement Day* (USA, 1991) by James Cameron and the virtuality theme discussed in the Wachowski brothers' trilogy *Matrix*, all discuss human/non-human hybridization.

Posthumans value hybridization as the condition that "made" humans what they are today. This concept is central in any discussion dealing with nanotechnology, biotechnology and the development of man-machine interfaces. 21st century technologies multiply contamination and hybridization situations: biotechnological practices (genetic engineering, creation of chimeras), the use of foreign tissues in xenografts, the use of synthetic organs and transgenesis. These technologies favor the horizontality of life's hierarchy. Moreover, the increasingly invasive use of miniaturized and advanced technologies —whose mechanism must rely on our biological substrate while conversing with it and sometimes managing its facilities— create new human performative possibilities. *Cyborgs* are *cybernetic organisms*, the hybridization of humans and machines, and are often regarded as the spokesman, the new subject of the posthuman era and both terms tend to become synonymous (Haraway, 1991).

Katherine Hayles (Hayles, 1999) covers all stages found between humans and posthumans by identifying the evaluation of the informational model – which regards biological incarnation as a historical accident rather then an essential condition of life – as the first step towards posthuman condition; the second step is to acknowledge that conscience results from evolution's process as an epiphenomenon, and no longer as the base of human identity. The body is ultimately regarded as a prosthesis that can be modified and controlled. This leads to the complete achievement of posthuman anthropology: the human-machine junction. New technologies and scientific progress have brought technology and body closer, to the extent of combining them.

Converging technologies and the stake of body enhancement

This idea regarding the obsolescence and inferiority of the body in opposition to technosphere, is found at the core of the concept of converging nano and biotechnologies. The most anticipating and utopist version of converging technologies, also known as "genetics, nanotechnology, robotics" (GNR) or "Nanotechnology, Biotechnology, Information technology and Cognitive science" (NBIC), questions the body's present state and its possible modification. Converging technologies refer to the synergetic combination of four major NBIC provinces of science and technology: nanotechnology, biotechnology and biomedicine, information technology, and cognitive science.

The report *Converging Technologies for Improving Human Performance* (Roco, 2002) edited by Mihail C. Roco and William Sims Bainbridge, brings together over 50 opinions of experts representing the government, the academic world and private sectors. It aspires to express major points of the development of convergent technologies. The concept of information is at the basis of the convergence of technologies. Yet convergence's keyword is "nano": human beings are made up of tiny elements and "The integration and synergy of the four technologies (nano-bio-info-cogno) originate from the nanoscale, where the building blocks of matter are established" (Roco, 2002:vii). Indeed, the ability to control bits, atoms, neurons and genes should enable the control of nearly everything, since it provides a comprehension of matter's informational code at all levels. The heroes of this development are the engineers, the only ones capable of understanding nature's laws and their mechanism in order to adapt them to requirements of design(cf Michael J. Heller (Roco, 2002:191)).

The prospect of convergence is an excellent argument of all discussions regarding the European nano-initiative and the American one (NNI, *National Nanotechnology Initiative*), launched in 2001 by the Clinton administration and supported by the actual Bush administration as a top-priority scientific and industrial development section.

In regards to the body, it is obvious that this new knowledge will need to be applied on human organisms. If, as Roco and Bainbridge concede, technological progress of the last decade allows for a better understanding of human biology, then the expectations of the development of nano-bio fields are more radical. A better understanding of human biology isn't enough anymore. We want to reach a new dimension and the ability to transform human biology (cf Michael J. Heller (Roco, 2002:191)). The aim of control is obvious in the project of deciphering human beings' essential mechanisms (Roco, 2002:179). This could be possible for the human body, if it is presumed that all human activities, both physical and cognitive, are entirely organized as a physical process (fig. 3.1).

Despite many promises, the prospect of convergence is regarded as very concrete and real. It seems to be confirmed and asserted as the best track to follow to accomplish a second Renaissance. As it is articulated in the title of the report *Converging Technologies for Improving Human Performance*, this new Renaissance is founded on the idea of human enhancement. The chapters' titles express it very clearly. The prospect of enhancement begins with taking the expansion of cognition and human communication into consideration, followed by the improvement of health and human physical abilities, in order to create a better relationship with

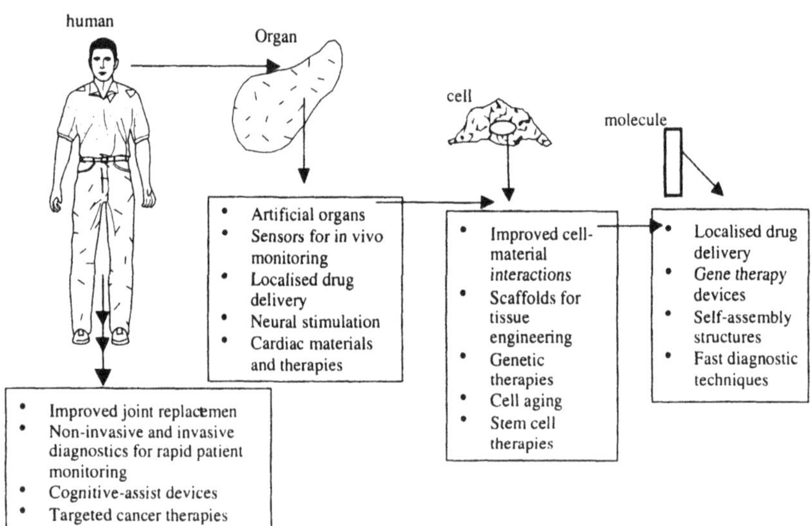

Figure 3.1 *Examples of levels for intervention of nanobiotechnology in human life extention (Roco, 2002:185).*

other individuals and society in its entirety. The report's authors are convinced that converging technologies "could achieve a tremendous improvement in human abilities, societal outcomes, the nation's productivity, and the quality of life," in the interest of individuals, society and humanity, in the long term (Roco, 2002:ix). The problem then consists in giving a sense to the notion of "enhancement" or "increase." How do the experts of this report define these notions? James Canton of the *Institute for Global Future* defines human normal and enhanced performances as follows:

For the physically-challenged the definition may entail gaining sight or mobility. For the aged, it may entail having access to one's memory. Even bolder, the definition of human enhancement may entail providing people with advanced capabilities of speed, language, skill, or strength beyond what humans can perform today. Just as plastic surgery and pharmacology have given new choices to human beings today, enhancement treatments will no doubt shape tomorrow.[14]

The military has a great need for human body enhancement. According to the DARPA (*Defence Advanced Research Projects Agency*): "With the infusion of technology into the modern theater of war, the human has become the weakest link, both physiologically and cognitively. Recognizing this vulnerability, DARPA has recently begun to explore augmenting human performance to increase the lethality and effectiveness of the warfighter by providing for super physiological and cognitive capabilities"[15] (fig. 3.2). Regarding nations and the world's welfare, human enhancement seems to be the main challenge.

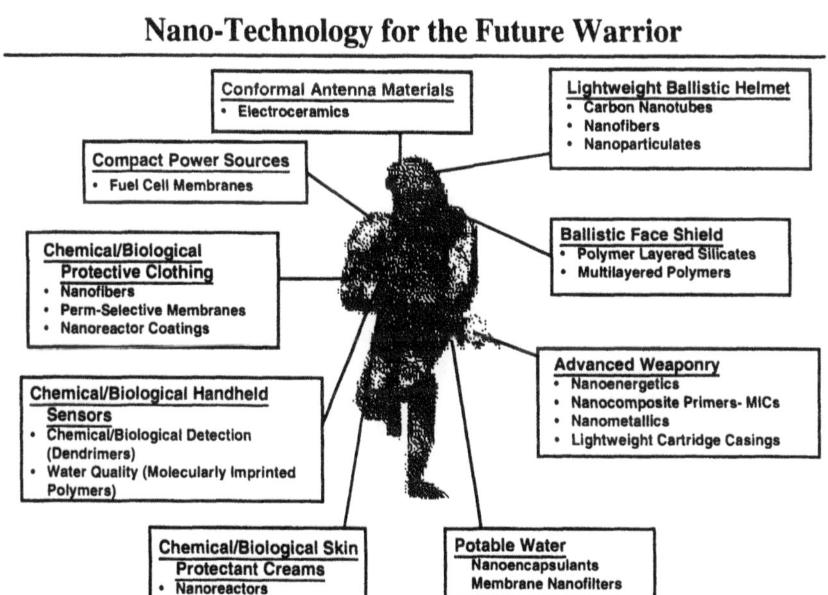

Figure 3.2 *Soldier system of the future (Roco, 2002:354).*

Notions of health and health care remain central in these discussions and emblematic of the nano and bio convergence. Thereby, nanomedecine presents itself as a different and revolutionary outlook on health care. The significance of economical investments and political strategies mobilized for the support of nanotechnology development demonstrates the extent to which the convergence of biological, medical and physicochemical research has become a priority for all industrialized countries, from both a social and scientific point of view. The public seems to become increasingly familiar with terms like "nanomedicine" and "nanobiotechnologies", promising research fields whose effectuation and expectations are so numerous and varied that it seems difficult to predict their social, technological and industrial impact.

Generally, the inclination to miniaturize devices likely to interact with the human body was welcomed as a possible improvement of preciseness in health care. Over a short period of time, developments in biological research, on the one hand, and nanotechnologies on the other, seem to offer significant solutions capable of solving certain medical issues. However, on the long run, these developments could lead to an entirely different outlook on medicine. Expectations are high, as is often the case when new technological horizons appear. For instance, the American *National Cancer Institute* regards nanomedicine as an investment of utmost importance to eradicate by 2015 the pain and mortality generated by cancer. Europe has established a technological platform to support initiatives in the nanomedicine field hoping to offer decisive solutions to serious pathologies such as cancer, diabetes, Alzheimer's and Parkinson.

In the 90's, Robert Freitas, an American researcher active in promoting and developing nanotechnologies applicable to human health, puts forward a new concept of "nanomedicine". The latter is founded on the visionary project aiming to create and realize nanorobots – or nanobots- capable of traveling in the human body and step in on cells at a molecular level.

According to Freitas, it will be possible in 10 or 20 years, to create "real" molecular machines, nanorobots capable of keeping the human body healthy and alive more efficiently then "natural" cells. Robots of nanometric dimensions will circulate in our veins and arteries to purify our blood (*respirocytes*), find awry or missing elements and eliminate toxins and pathogenic agents (*microbivores*).

Ray Kurzweil pushes the expectation even further: nanorobots will be connected to biological neurons allowing them to control our senses and feelings, create a total immersion in virtual reality and above all increase our memory abilities and form our non-biological intelligence. Connected online, our brains will be capable of exchanging new knowledge and abilities.

The shift from a healthcare notion to a notion of enhancement is achieved by the possible application of nanotechnologies and the miniaturization of chips. In his latest work *The Singularity is Near* (Kurzweil, 2005), and in *Fantastic Voyage: Live Long Enough to Live Forever* (Kurzweil, 2004), Kurzweil introduces an external transformation of the body using alimentary hygiene, as well as an inner transformation using nanotechnology. His works give a meaningful idea of the visionary predictions regarding the effects converging technologies would have on the human body and the general notion of humans, with nanotechnologies placed in the foreground. His vision is close to that of transhumanists, who regard Kurzweil as a point of reference, and insist on the need of transforming the body using any available means.

Following a movement comparable to the one responsible for the independence of sexuality from reproduction by encouraging its communicative and sensual aspect – at least in the industrialized world –, it could be possible, according to Kurzweil, to predict a similar emancipation process for other biological activities linked to sociality and sensual pleasures.

Food is a good example of that. Our digestive system and metabolism have evolved with time but do not correspond to our present situation: obesity is a symptom of this dysfunction. Our body was essentially chosen to ensure reproduction, its aim not being longevity. Yet, the partial abundance of goods, techno-scientific progress and knowledge on the organ system have enabled life extension by questioning the human status in version 1.0, as Kurzweil says using computing language.

In an upper technological level, "traditional" food could become a cultural and sensory experience by avoiding the absorption of nutritious elements and having nanobots eliminate excess substances.

Nanobots will be able to come in and out of our body easily and find nourishing substances in the environment. Version 2.0 of mankind will therefore be more independent and resistant. According to Kurzweil's prediction, most of our organs and cells should have been eliminated by 2030. Only the skeleton, skin, sexual organs, mouth and upper part of the esophagus and the brain would be left. Everything could be improved or transformed by nanotechnology and new materials: the skeleton would be stronger, more resistant and self-maintained; skin would become a solid material capable of resisting coldness and heat; treating the brain like a circuit would enable the establishment of direct communication with nerve cells, the recovery of vulnerable functions and the delivery of medicine at specific locations.

Version 2.0 of mankind would be the result of a longstanding evolution during which we were increasingly in touch with technology. At first, computers were big machines in air-conditioned rooms. Then, by dint of miniaturization, they came closer to our desks, our bags and today our pockets. Kurzweil claims that they will soon be in our brains and bodies. As of 2030, we will be more nonbiological then biological and by 2040 nonbiological intelligences will be much more powerful then biological ones.

The main argument used to justify this shift towards the artificial is that biological evolution will always be limited by the use of a finished range of materials (proteins and amino acids). But this is a slow building movement. However, it created a species that thinks and manipulates its environment and is capable of accessing its *design* and improves it: biological principles could be reviewed and changed. Version 3.0 of mankind, expected by Kurzweil for 2030 or 2040, will correspond to a total revision of the human project.

The body as project: the continuity of the modern challenge

All of these different models – obsolete, arbitrary, hybrid, modifiable- seem to be attached by one in particular: the body as "project."

In itself, the notion of project implies the idea of carrying "outside" (from Latin *projectus* and *proicere*, throw ahead), bringing something to attention. A project can therefore utterly be connected to the notion of "product" (latin *productus*, past participle of *producere*, to carry outside, impel, then bring, cause) in the sense that it predicts and sets things in a way that enables their accomplishment. The idea of project implies the act of anticipating a concrete accomplishment, the idea of an aim often deemed achieved. This is true for technological projects, relying on a type of determinism ensuring their future achievement. In the technological field projects essentially belong to engineers, thus any intervention on the body is studied with an engineering outlook. *Biodesign*[16], for example, is introduced as a new field bordering bioengineering, biomedical sciences and health technologies. A kind of ergonomics; a technological project of the body based on biorobotic engineering designs, applied to medicine and in synergy with biological sciences until the point of reaching nano dimensions.

This project can also be interpreted as one of identity construction, a theme developed in art, feminism and *transgender* movements. To be a project, the body must be privatized: the individual is responsible for its construction, reconstruction and daily redefining. According to Anthony Giddens (Giddens, 1991), bodies are the visible carriers of individual identities; hence each individual is responsible for the design of his own body. This duty of body cultivation, assumed by its owner, meets all the products on the market, tools and handbooks required for this care (Bauman, 1992). Society provides the tools and models but the undertaking of the task is the body owner's responsibility. Thus the individual becomes transformation material as well as the active subject of this transformation. He is free to pursue his project but is also defined by the sociocultural constraints, somehow suggested/imposed on him. This idea of body plasticity and freedom refers to a *culture-body*, a body entirely considered within its cultural elements, slowly growing more distant from its biological features to assume the ones characterizing it as a cultural, unnatural and worked up object. This vision of the body given by Kurzweil, is freed of all the human psychological, emotional and relational aspects of its physiological functionalities. Yet this body remains subject to a standardized social and cultural perspective, in which adaptations to ancient and new, individual and general, cultural and social body ideals are expected. This definition of the body as project agrees with the postuman vision of the body suggested by Kurzweil and transhumanist movements, and with the discussion on converging technologies. Technology has been interiorized[17] and is an integral part of identity construction, refusing any material or bodily boundaries and human or codified limits. It lives through gender, animal and human forms, what is considered normal or pathological, in both natural and artificial realms.

According to the German philosopher Peter Sloterdijk, mankind's will to act on its own nature goes back to a component of human condition: the ability to tame the world and tame oneself[18]. In the long process of "domestication of Being" attended by the movement of hominisation, technology is an essential tool allowing for human's build-up and which will continue using technologies increasingly embedded in living structures. This "anthropotechnique"[19], enabling man to be his own *artifex* up to the most ultimate and problematic point of genetic manipulations of the genome, is stigmatized by Günter Anders (Anders, 1956) who denounces the qualitative and quantitative double transformation of the intervention of modern man on nature. According to Hannah Arendt *homo creator* is destined to succeed to *homo faber* and is now capable of generating an entirely artificial "double nature" while having an effect on *homo materia*, that is humanity transformed into raw material, tamed, educated, transformed, conditioned, manipulated and eventually *created.* So despite their heavy projection towards the future, trans- and postshumanism seem to be the extension of the modern project: the conquest of death. Indeed, modernity has led to the denial of accepting death as life's ultimate outcome and to the idea that life isn't formed as a curve with an end to it. Modern times are portrayed as an arrow pointing to a single direction. Present and past are of no value: future is the source of value in modernity's build-up. The complete achievement of rationality is a utopia giving value to the present as the next step towards times to come. Modern life is therefore considered as a process with a commencement - birth -, but continually projected in the future as "endless potentiality" (Bloch, 1959) for change and creation.

Within an individual, collective and life p roject[20], the individual has a long-term aim, a work to achieve, talents to express, a world to build or change. Death is interpreted as the absurd interruption of these projects. It forces mankind to remain in the incompletion of their project and conveys the upsetting feeling of unfinished work. Thus, death in modernity is understood as "waste", an absurd loss of potentialities forever never to be expressed (Thomas, 1993).

Thinking of life as the creation of a project implies negating death as its necessary end. This outlook on death is in itself the result of a long process of changes and modifications of the individual's status in society, his relationship with the world and accordingly his viewpoint on death.

Transhumanism's great endeavor to adapt the human body to the evolution of machines can thus be explained as a renewed attempt to oppose the utmost offense made to the human almightiness, a will to take up reason's challenge – mortality – using extreme self-technologisation.

Notes

1. http://www.maxmore.com/becoming.htm
2. European Commission, Converging Technologies: Shaping the Future of European Societies, 2004, reporter Alfred Nordmann, http://ec.europa.eu/research/conferences/2004/ntw/pdf/final_report_en.pdf.
3. A continuity available in the model of the outline established by A. Leroi-Gourhan

about the evolution of knives in his work *Gesture and Speech*, Cambridge, Massachusetts & London, MIT Press 1993 (particularly in section II, *Memory and Rhythms*. Leroi-Gourhan outlines a similar functional "technical trend" uniting the Australanthropes' irregular cutting edge to the present knife. Dyson suggests a more haphazard linearity going beyond any trace of technological trends.

4. Concerning the debate about technological singularity, a number of participants created associations present on the Internet. *Singularity Institute for Artificial Intelligence* (SIAI, official website http://www.singinst.org) is the movement of Singularitarians' website. Michael Anissimov (http://www.acceleratingfuture.com/michael/blog) animates the debate on the blog: the website gives links to many personal websites and to websites of other authorities related to transhumanist and immortal movements or merely praising radical nanotechnologies. Eliezer Yudkowsky, for example, is for a "friendly AI", an artificial intelligence capable of increasing the human intellect prior to neurotechnology or genetics. He is also the author of *Singularitarian Principles* (http://yudkowsky.net/sing/principles.html). The *Acceleration Studies Foundation* l (ASF, official website http://www.accelerating.com) promotes *Accelerating Change*, a multidisciplinary conference at Stanford University on the acceleration of technological change and affiliated to the website *Acceleration Watch* (http://www.accelerationwatch.com).
5. Gordon E. Moore's first article about this is "Cramming more components onto integrated circuits", *Electronics*, vol. 38, n. 8, 1965.
   See also http://www.intel.com/research/silicon/moorespaper.pdf.
6. http://www.kurzweilai.net/articles/art0134.html?printable=1
7. For a review of the technological determinism expressed by Moore's law and the inclination to apply principles of this law in fields different from the development of semiconductors' industry, see the article written by Ilkka Tuomi "The Lives and Death of Moore's Law", *First Monday*, vol. 7, n. 11, 2002. Computing publication, website http://www.firstmonday.org/issues/issue7_11/tuomi/. Tuomi's critics, later resumed by a text clearly against Kurzweil's interpretation of Moore's law (*Kurzweil, Moore, and Accelerating Change*, august 2003, http://www.meaningprocessing.com/personal Pages/ tuomi/articles/Kurzweil.pdf), lead Ray Kurzweil's to answer in the article *Exponential Growth an Illusion: Response to Ilkka Tuomi* published in September 2003 on his personal website,
   http://www.kurzweilai.net/meme/frame.html?main=/articles/art0593.html.
8. Giovanni Pico della Mirandola, *De hominis dignitate, Discourse on the Dignity of Man* (1486), see *Pico Project* University of Bologna – Italy and Brown University, http://www.brown.edu/Departments/Italian_Studies/pico/. Man, created lastly by God, does not receive any special virtue, but is placed at the center of the world and anything given personally to each individual will be come common to him, This is a reference to the famous speech given by God to Adam during the creation: a man who does not hold a defined nature is free to follow his judgment and define his own nature (§ 4).
9. http://www.stelarc.va.com.au/parasite/index.htm.
10. http://www.stelarc.va.com.au/obsolete/obsolete.html. See also STELARC, "Prothèses, robotique, existence à distance", Arts et cognition, Marc Partouche (edited by), Ed. Ecole d'Art d'Aix-en-Provence, 1994, p. 44.
11. STELARC, "Portrait robot de l'homme-machine", interview with Jean-Yves Katelan, L'autre journal, n. 27, sept. 1992, pp. 24–41.
12. http://stephan.barron.free.fr/technoromantisme/stelarc.html. See also Stephan Barron, Technoromantisme. Arts, esthétique, vie culturelle, communication, médias, science et techniques, Paris, l'Harmattan, 2003.
13. STELARC, "Portrait robot de l'homme-machine", cit.

14. James Canton, "The Impact of Convergent technologies and the Future of Businnes and the Economy", (Roco, 2002:78).
15. Michael Goldblatt, "DARPA's Programs in Enhancing Human Performance ", in Roco, Bainbridge, 2002:337).
16. See the multidisciplinary project of Stanford University, http://www.stanford.edu/group/biodesign/index.html.
17. A set of devices is meants to to live close to our body one day: nanorobots, or nanobots, that Kurzweil mentions are an example of that.
18. See especially P. Sloterdijk, *La domestication de l'Être*, Paris, Editions Mille et une nuits, 2000 and *Regeln für den Menschenpark. Ein Antwortschreiben zu Heidegger "Brief über Humanismus"*, Surkamp Verlag, Frankfurt am Main, 1999.
19. Some authors prefer the term "anthropotechnie", see Jérôme Goffette, *Naissance de l'antropotechnie. De la médecine au modelage de l'humain*, Paris, Vrin, 2006. Goffette associates this notion to "bricolage", that is man's ability to manipulate himself, without knowing the outcomes or implications of it, but always applied for change.
20. The notion of modern project corresponds to Jean-François Lyotard's definition of "metarécits": great stories that impressed modernity – the emancipation of reason, work, the notion of human and social progress due to technological progress, and the universalistic Christian discourse. Unlike myths, these stories find their legitimacy in their will to be universally achieved in the future without looking for original foundations in the past. This is what is called "modern project".

Bibliography

Anders G. (1956), *Die Antiquiertheit des Menchen. Über die Seele im Zeitalter der zweiten industriellen Revolution*. C. H. Beck Verlag, Munich.

Bauman Z. (1992), *Mortality, Immortality and Other Life Strategies*. Polity Press, Cambridge.

Bloch E. (1959), *Das Prinzip Hoffnung*, Suhrkamp Verlag, Frankfurt am Main.

Bostrom N. (2005), *A History of Transhumanist Thought*. Journal of Evolution & Technology, Vol. 14, n. 1.

Deitch J. (1992), *Post Human*. DAP, New York, Pg. 27. See also http://www.artic.edu/~pcarroll/PostHuman.html.

Dyson G. (1997), *Darwin Among the Machines*. New York, Addison-Wesley.

Gehlen A. (1993), *Der Mensch: seine Natur und seine Stellung in der Welt*. (1940), Frankfurt am Main, V. Klosterman.

Giddens A. (1991), *Modernité and Self-Identity. Self and Society in the Late Modern Age*, Stanford University Press, Stanford.

Gottfried Von Herder J. (1966), *Outlines of a Philosophy of the History of Man*. Bergman, New York.

Hayles N. K. (1999), *How We Became Posthuman: Virtual Bodies in Cybernetics, Litterature, and Informatics*. Chicago, The University of Chicago Press.

Haraway D. J. (1991), *Simians, Cyborgs and Women: The Reinvention of Nature*. New York, Routledge.

Kurzweil R. (2005), *The Singularity is Near. When Humans Transcend Biology*. New York, Viking Penguin.

Kurzweil R. (2005), *The Singularity is Near*. New York, Viking.

Kurzweil R. and Grossman T. (2004), *Fantastic Voyage: Live Long Enough to Live Forever*. Rodale Books.

Moravec H. (1988), *Mind Children. The Future of Robot and Human Intelligence.* Cambridge, Massachussets, Harvard University Press; Robot. Mere Machine to Transcendent Mind, New York, Oxford University Press, 1999.

Penrose R. (1989), *The Emperor's New Mind. Concerning Computer, Minds, and The Laws of Physics.* Oxford University Press.

Remotti F. (2003), *De l'incomplétude.* In: *Figures de l'humain. Le représentations de l'anthropologie.* Paris, Editions de l'Ecole des Hautes Etudes en Sciences Sociales.

Roco M. C. and Bainbridge W. S. (supervised by) (2002), *Converging Technologies for Improving Human Performance. Nanotechnology, Biotechnology, Information technology and Cognitive Science.* NSF/DOC, June 2002. http://wtec.org/ConvergingTechnologies/1/NBIC_report.pdf.

Royston Searle J. (1980), *Minds, Brains, and Programs.* The Behavioral and Brain Sciences, vol. 3, n. 13, 1980, Pg. 417–424.

Thomas L.-V. (1993), *Mélanges thanatiques.* L'Harmattan, Paris, Pg. 230–232.

Chapter 4
Body becoming architecture: a continuum for the body's interface with architecture through digital networks

Chris Speed

Abstract

This paper is interested in articulating a continuum for how the body engages with architecture, and in turn how architecture is re-learning to engage with the body through the use of digital technology.

The paper identifies problems inherent in contemporary architecture, and in particular its commitment to material processes that inhibit the bodies relationship with a buildings form, structure and systems. The author maps how with the advent of networked digital systems, the walls and barriers that divide architecture and the body are de-materialising to offer new design spaces in which the separation between people and building is indistinguishable.

As a means of setting the scene, the paper begins by reflecting upon architecture's method for alienating the construction process from a place during the design and build phase. This graphic 'image' establishes the conditions in which the author begins to explore how the body (personal and social) become excluded from the production of space in typical urban contexts. The author proceeds by stepping through six practical and theoretical processes that through the increasing use of digital networks begin to transform how we understand the distinction between body-space and architectural-space.

At the end of the paper the author presents an arts / science project in which the use of digital social networks are able to both recover architectures ability to provide people with a sense of place, and the body's ability to traverse the boundaries that have alienated them from architectural spaces.

The homeless building

Whilst walking through a British city, the pedestrian encounters many opportunities to reflect upon themselves and the environment around them. Walking the city becomes an "enunciation" (de Certeau, 1984:111) through which we are able to express our place in the hustle and bustle of the city matrix—economic, political, structural and physical (de Certeau 1984:94). That is, until it you find some spaces that are trying not to be there. Hidden by huge walls in the middle of a street that run the length of a field, that are too high to look over and too dense to look through. I am making specific reference to the large scale urban building sites that are surrounded by a running wall of wooden boards usually painted white, grey or a single colour. Over the past twenty years a highly tuned professional process has led to the development of these walls that obfuscate a site from its context. Gone is the ugly wire fencing that allowed people to see into a building site, and in their place is an opaque wall with illustrations of what will be behind the wall when it is taken down. These 'vacuums' in the space/time fabric of the city become interesting because they provide a rupture in the relationship between the 'social body' and architectural space.

The practice of making new architectural projects invisible from an environmental space so as to allow designer and engineers the opportunity to quarantine a project until it is ready for introduction in to society, is a deeply anti-social practice. This temporal separation from society designed to hide a new building is a conscious act that limits the process of design and construction to take part in any existing or new opportunities for social discourse. Whilst the practical deportation of a building site has obvious health and safety benefits, its impact on the human geographical fabric of the city is akin to Certeau's critique of a vacation as it prevents the stakeholders in the community taking part in the construction process:

> "Only a rationalized cell travels. A bubble of panoptic and classifying power, a module of imprisonment that makes possible the production of an order, a closed and autonomous insularity—that is what can traverse space and make itself independent of local roots". (de Certeau, 1987:111)

The choice to isolate a building as it is being built at a local address is comparable to making it 'homeless', as all ties are cut between itself and the social and spatial context within which it is being built.

How the building lost its sense of place

This apparent constructed barrier between the time-based process of building and the social-spatial context that surrounds it exemplifies a critical problem that architecture is now trying to work out for itself; how to recover an architectural design process that supports its users personal development for a sense of place. In many ways this is a result of a architectures historical embracing of technology that initiated what can be indentified as the Modern movement, and that resulted in a shift away from architectural processes that intimately linked us to a sense of place. The embracing of technologies and the movement to modernism "free[d] people from the limitation imposed on them by family or clan or by their village community, offering them unheard-of options and often material improvements as well" (Heynen, 1999:15). In doing so, there was an ensuing loss of particular certainties such as the sense of 'being' that Heidegger sees as being intrinsically tied to the concept of 'dwelling'. Through the mapping and subsequently the commodification of space, Heidegger suggests that the Enlightenment separated us from

an intrinsic connection to land and, in doing so, we lost our sense of 'dwelling' (Heidegger, 1971:155).
By tracing the term 'building' back to the German phrase 'buan', which is related to "I am", Heidegger concludes that building and dwelling are closely bound to the concept of being. For Heidegger, dwelling does not stem from building, but the reverse; building comes from the need for shelter and a sense of being in the environment.

> "The nature of building is letting dwell. Building accomplishes its nature in the raising of locations by the joining of their spaces. Only if we are capable of dwelling, only then can we build." (Heidegger, 1971:155)

However, as Modernism concentrated upon the machine and the development of the city for a space for living, an insurmountable distance was created between the metropolis and a place to dwell. Subsequently, as dwelling receded, building took on a new aggressive form in which modern homes such as bungalows and apartments constituted new living environments and the attraction of the city led to migrant populations and homelessness.

> Under modern conditions the world has become impossible to live in; modern consciousness is that of "the homeless mind," and foreigners and migrants provide a model for the experience of every individual in a modern, mobile, and unstable society." (Heynen, 1999:17)

Modern architectural values then questioned the nature of dwelling as something that could not be qualified simply through structure and attachment to a place, since the human continuity with 'nature' and space can no longer be taken as a given. During the machine age Architects and designers began to embrace technology because it was considered intrinsically opposed to nature, and provided new systems for shelter through materials that distinguished themselves from organic matter, further rendering us 'homeless' (Heynen, 1999:14). Examples of this include Le Corbusier's concrete Marseille apartment block on stilts, suggesting a rejection of the soil beneath it, and Buckminster Fuller's post-war Dymaxion house, that made use of aircraft manufacturing technologies that replaced hand-made, wooden dwellings. Architecture's role in reconciling our relationship with the world remains in discussion and provides the differences in theoretical and practical approaches to designing forms and structures.

More recently, in an effort to overcome the tension between modernity and dwelling, contemporary architects have written about the complexity of the architectural process at a conceptual level, involving narratives from literature and the arts as well as critical theory and politics. Post-modern architecture can be understood as an attempt to reconcile the dilemma as it rejects science and invests in cultural imagery as a resource for representing the value of the vernacular and exposing our lack of faith in modernism. Mimesis and the use of apparently naturally inspired systems becomes a framework for strategies involving the form, materials, plan and experience of a building. Of the most significant architects of recent decades, Rem Koolhaas and Daniel Libeskind have incorporated complex historical and contextual references into their work. These architects would argue that they have extended the service of providing dwelling by addressing the complex nature of our relationship with history, space and place. However, due to the scale of many of these projects and the modern process of constructing them, the buildings still struggle to escape the dilemma of 'homelessness'. The detachment that the pedestrian feels as they encounter large scale architectures and the prevalence and ever increasing density of the walls that hide building sites seems to demonstrate this further.

Since it is unlikely that urban buildings are going to get smaller and clients budgets become more relaxed, strategies for addressing the problem have come not only from the desks of architects, but from the people on the ground who negotiate around the walls that divide body and space. Getting over walls, breaking and entering spaces is nothing new, but a recognition of social and cultural practices that attempt to transgress spaces and reclaim an engagement with a place, are part of a continuum in which digital technology is dissolving the distinction between body and architecture.

The next six parts of this paper describe a short but diverse spectrum of spatial practices that demonstrate analogue approaches to dealing with the wall that divides body and architecture, through questions and theoretical strategies for dealing with material processes, to the dissolution of the barrier through the explicit use of digital systems.

At this point it is important to declare my fluid use for the term 'body'. In discussing the body in the context of its relationship with architecture it seems appropriate to use it to describe both the individual and the group. As the narrative proceeds and introduces more digital technology, any distinction becomes irrelevant as networks extend our body and social systems make it impossible to separate us from a crowd or social body.

Recovering our sense of place within architecture

Using the body to reclaim architectural space

Skating the wall

A demonstration of a physical social practice that intentionally attempts to negotiate exclusive spaces is skateboarding. Iain Borden, uses the 'pastime' as a means of accessing Lefebvre's models within *The Production of Space* and extrapolates its concerns through its birth, growth and engagement with space and society. Skateboarding operates on a multitude of different levels as supporting the body's dialogue with space, originally adopted as an extension to surfing in the 1960's it quickly began to use swimming pools and other large architectural forms to explore the extraordinary thrills experienced by the body as it is spins, jumps and flies from surface to air. However after an initial investment into purpose built skate parks, from their lack of upkeep and mass closure by the end of the 1970's a desire emerged from individuals to explore the street as a space for play, and skateboarding became an interrogation of the spatial and social rules of the city.

Often perceived as anti-social in their manner due to their performance at a different pace to that of pedestrians and their occupation of public spaces, skateboarding can be seen to be transforming the social contracts of the urban environment. Borden uses these practices to articulate a process of redefinition of space that he suggests is occurring as skaters try define a space for themselves within the powerful forms of control and jurisdiction of the city. He describes the breadth of skate boarding's dialogue with space and how acts and processes support a powerful conceptual and physical intercourse with space.

> "Skaters produce an overtly political space, a pleasure ground carved out of the city as a kind of continous reaffirmation of the notion that beneath the pavement, lies the beach." (Borden, 1996:86)

Skating also engages with the transformative nature of objects in space, their scale, materiality and functional properties as benches become objects to 'grind' across and fire hydrants become opportunities for 'ollie-ing' over. Through the suppression of skateboarding, an extraordinary language of control has emerged through the signage and design of surfaces and furniture that highlights the acceptable codes of practice within space that councils and local governments enforce; No Skateboarding signs, studded pavements, alternately raised wall heights. Plymouth City Council residents ranked the banning of Skateboarding and Rollerblading within the highest priority bracket of its survey published in March 2004.

> "Skateboarding is then a kind of unconscious dialectical thinking, an engagement with the spatial and temporal rhythms of the city, skateboarders using themselves as reference but then rethinking the city through their actions." (Borden, 2001:228)

At this point the permeability of our wall, or any material fabric that purports to uphold a given authority has the potential to be transformed or dismantled under particular circumstances. Skateboarding, through its transformation of the street into a beach, or the bench into a ramp, redefines spatial territories through the conscious use of the board and body, skateboards have the capacity to reconfigure the social potential of a space by treating the materiality of a space differently.

Physical synthesis between body and building

Becoming the Wall

A second technique for dealing with our wall is to become part of it. Eve Dent is a live performance artist who uses herself and photography to develop 'body installations' with architectural structures and spaces "in an exploration of the threshold between bodily self and material thing" (Phillips M. *et al.*, 2004:41). The 'epidermic' dialogue with the environments that she engages, expresses a great deal of the tensions between body and space. Fitting her body behind, underneath and inside the fabric of a room, parts of her body become completely hidden to the audience, whilst other parts are starkly visible.

> "In the creation of my body/architectural hybrids I seek to animate the hidden poetic life of a space and reveal the ambiguous nature of the body, its familiarity and foreignness, its beauty and horror, our fears of being overwhelmed by the void within and without. Through my live installations, perceptions of structural and spatial relationships are fundamentally altered, blurring the boundary between animate and inanimate, subject and object, built structure and flesh, through acts of union and rupture". (Dent, 2006).

Each of Dent's works demonstrate a literal engagement with the place in which she has chosen to work. Dent cites Caillois's essay 'Mimicry and Legendary Psychasthenia' first published in the Surrealist Journal Minotaure in 1935, as an influence to her work. A powerful text that uses animal mimicry to explore the close tensions between bodies and their surrounding environments, but one that also reminds us of how science had began to change how we considered space in the early twentieth century.

> "It is with represented space that the drama becomes specific, since the living creature, the organism, is no longer the origin of the coordinates, but one point among others; it is dispossessed of its privilege and literally no longer knows where to place itself. One can already recognize the characteristic scientific attitudes and, indeed, it is remarkable that represented spaces are just what is multiplied by contemporary science: Finsler's spaces, Fermat's spaces, Riemann-Christoffel's hyper-space, abstract, generalized, open, and closed spaces, spaces dense in themselves, thinned out, and so on. The feeling of personality, considered as the organism's feeling of distinction from its surroundings, of the connection between consciousness and a particular point in space, cannot fail under these conditions to be seriously undermined; one then enters into the psychology of psychasthenia, and more specifically of legendary psychasthenia, if we agree to use this name for the disturbance in the above relations between personality and space". (Caillois, 1935)

Through Dent's work the wall begins to become plastic and negotiable, as she mutates and attempts to become part of a building material. Once the audience recover from the shock of seeing a body within a wall, her stillness leads them to begin to believe that the body is wall. After all, the body is no longer operating as a body might anymore; it does not move, it does not communicate, it certainly doesn't dance nor make noise. On perceptual level Dent is changing her substance to being closer to the architectural fabric of a space than the social body that engaged in communication and exchange. The reduction in communication and physical activity as Dent 'becomes' the wall, gives us clues as to the role that our wall plays in the city. The wall doesn't do anything, it can't it because it has no agency, but the nature of its composition does obscure onlookers from listening and seeing the source of communications that are behind it.

Readdressing architectures relationship with technology

Dematerialising the wall and the body

In order to understand the next step in our continuum toward an integrated model between body and architecture we must first reflect upon architectures condition in the early twenty first century. Having developed the ability to construct large scale complex buildings so successfully through its adoption of modern technology, architectures relationship with science shifts as the modern project began to falter toward the latter end of the twentieth century. In his analysis of Post-Modernism Jameson describes how architecture began to develop an aesthetic relationship with itself as cultural, economic and social tensions lead to scepticism in science. Such a position rendered the industry incapable of using technology in an informed manner and subsequently prevented it from fully updating its technical processes. In particular this has meant the retention of the Cartesian model for describing space that is in conflict with a social and cultural experience of place. A context that ultimately limits Architecture further in responding with material strategies for recovering the loss of a sense of place that Heidegger yearned for.

Jameson reflects upon Mandel's three stages of machine power through the technological revolutions of the past 150 years: 1. Steam power, 2. Electric and combustion power, and 3. Electronic and nuclear power. In describing the lack of representational capacity that the computer has in its grey 'outer-shell' compared with that of the "racing car whose hood is adorned with great pipes, like serpents of explosive breath" (Marinetti, The Futurist Manifesto, 1909) he reminds us that the PC is a machine of reproduction, demanding an entirely different "capacity for aesthetic representation" (Jameson, 1991:37). Architectures use of computers to draft and render space, contribute to his subsequent proposition that post-modern architecture celebrates a 'sublime' form of this technology to support the development of an eclectic, image-conscious landscape that disregards itself from any hard science that had underpinned its modern activities.

> "Architecture therefore remains in this sense the privileged aesthetic language; and the distorting and fragmenting reflexions of one enormous glass surface to the other can be taken as paradigmatic of the central role of process and reproduction in postmodernist culture". (Jameson, 1991:37)

Subsequently, space and time—in a post-modern, culturally dominant context—become highly distorted currencies. Not only is there a closing of the 'critical distance' that provided the modern producer and theorist a space to 'stand outside', but Jameson suggests that 'postmodern bodies are bereft of spatial coordinates and practically incapable of distantiation' (Jameson, 1991:48). As a series of strategies for understanding post-modernism, Jameson demonstrates how the breakdown of time as a Cartesian currency, and architecture in its creativity, results in buildings in which function is subordinate to form. This reconfigures the modernist tenet that form should follow function[1] and posits a shift from absolute time to extreme relativity.

In closing his analysis of post-modernism Jameson describes the conditions that place Cartesian systems under massive stress for use in any cultural activity. Jameson suggests that "space-postmodern hyperspace has finally succeeded in transcending the capacities of the individual human body to locate itself, to organize its immediate surroundings perceptually, and cognitively to map its position in a mappable external world" (Jameson, 1991:44). Maps have become redundant in their existing form and he references the work of physicist Heisenberg who developed the post-Newtonian 'Uncertainty Principle' as an indicator of their limitations to provide truths (Jameson, 1991:52). Despite these conditions in which Cartesian systems are at breaking point, the tools and methods of architectural practice remains entirely based on Descartes' coordinate system. Although the context of an actual building site and the laying of bricks may not require a more complex model of physics, digital space and its mathematical constitution is absolutely where architects should be able to exercise a post-Cartesian design that situates the body closer to the map.

Following architecture's distancing from science and the adherence to Cartesian representations of space, questions such as 'what is architecture?' and 'what's the matter with architecture?' (Gunkel, 2001:73) have begun to emerge. Caught in this minor crisis is my third coordinate; Ben Nicholson who through his own architectural development offers a methodology in which our wall changes its density. Nicholson trained as an architect in the 1970s but failed to obtain the professional qualifications. When the internet emerged as a social, technological, commercial medium Nicholson identified the term architect as being used in a whole new context—one that extended the practice of people who 'make and organise space that anyone can enter' (Nicholson, 1999:79) in the actual world, to those that do the same for people in the virtual. Comparing our experience with navigating through actual buildings to that of digital websites, Nicholson constructs an opportunity for architecture to rethink its methodological practices. "The hyper-visitor has just crashed architecture's party, to un-tech the well manicured, encyclopaedic nature of experience" (Nicholson, 1999:79). Referencing the non-teleological form of the internet where "time forward, backward and time around are given value", (Nicholson, 1999:79) he constructs a context for architecture to reconsider how it has become a victim of its own methods.

His position is best understood by Loaf House, his design for a dwelling in Chicago. Although it began as a proposal for an actual building, it ended up being entirely virtual. Nicholson's journey through a design phase and analysis of the function of the building led him further and further away from understanding his job as someone who should specify bricks and mortar, but to someone who should specify an enclosure that "permits one to occupy space, engage in various relationships and generally enjoy the view" (Gunkel, 2001:79). In dematerialising the building, he pushed the concept of what it is to be an architect to a limit at which the material that is used to define the practice dissolves to leave a critical practice.

Nicholson's method is useful because he steps outside of the architectural continuum that has perpetuated a Cartesian method and proposes something radically different, but which remains as architecture. Gunkel uses Nicholson's 'blasphemy' (Gunkel, 2001:77) in opposing the materiality of architecture as a means of describing the discipline's crisis in dealing with computers as drawing tools and not as networked social systems. "What's the matter with architecture, therefore, is that it both resists its self-dematerialisation into the digital bits of cyberspace and is, at the same time, unable to solidify its dedication to materiality in opposition to such dissolution" (Gunkel, 2001:83). In an article to support a new form of digital architectural practice, Nicholson uses a runner bean as a metaphor to unpick the disciplines Cartesian dilemma that resists embracing the body as a part of a design process. The runner bean's reproductive potential, and connection to currency and measurement, represents a reconciliation of Cartesian duality through its embodiment of 'measure and chaos' (Nicholson, 2002:13). Nicholson suggests that the metaphorical digestion of the bean which terrified Pythagoreans because of the bean's miniature image of life (when dissected and looked at closely they appear to have human reproductive organs), would allow the field to move on from trying to retain the dichotomy between material practice and de-material theory, and social dynamics as an index of the organism. "Architecture is ready to make a place that is beyond measure and materiality, where anthropocentricity gets cracked once and for all time, severing the corporeal shackle" (Nicholson, 2002:13). Nicholson's approach exposes the inhuman application of Descartes' coordinate systems and provides an opportunity for a new type of architecture to emerge that integrates human agency into the definition of spaces.

At this point in the narrative our wall has the potential to be something that is no longer hard and solid that we can skate across or attempt to melt into, but something else that is closer to being a part of a network. If used poorly it inhibits communication, but if used sensitively has the potential to support navigation.

Building is a social process

Talking through the wall

Recalling our experience of the street and encountering the walls of construction sites that disengage a space from a place, we might realise that we are also excluded from the lives of those who work on the building site: the bricklayers, plasterers, steel workers, carpenters and plumbers who are imprisoned out of time. In contrast De Certeau would want us to acknowledge that although the space is separated from the pedestrians time and space, behind the walls there is an actual space where relationships that constitute the 'job' of building the building are being played out. For the workers who may know little about the urban context, and are drafted in to work side by side, often next to strangers, the wall suggests that the construction site is their home and the outside may as well be the prison in which they have no sense of home.

> "Everyone goes back to work at the place he has been given, in the office or the workshop. The incarceration-vacation is over. For the beautiful abstraction of the prison are substituted the compromises, opacities and dependencies of a workplace. Hand-to-hand combat begins again with a reality that dislodges the spectator without rails or window-panes. There comes to an end the Robinson Crusoe adventure of the travelling noble soul that could believe itself intact because it was surrounded by glass and iron." (de Certeau, 1987:114)

So the wall isn't dense to simply visually impede our 'seeing' of what's next on the city skyline, but it can also be understood to be a control measure to prevent any 'hand-to-hand combat' that may spill out on to the street. However, building sites are complex spaces that involve a great many people, doing many different things and working for different amounts of time, all of whom need to communicate at different levels to ensure the building gets built (Lee and McMeel, 2007:72). Social networks in these environments are highly dynamic and inconsistent, with some groups who will be familiar with each other, others who may work for a very short time on-site and within a specific area that requires limited contact with others, and others who oversee the project. These circumstances force individuals who have to understand how the building is put together at a very basic level to use a variety of communications systems to exchange knowledge.

> "Information sketched on a plasterboard wall, regarding the position of insulation and the air gap required, can be right beside the location where the detail is applicable, and it is relevant to the location until all the insulation and plasterboard is erected, at which point it will become redundant and will be subsequently plastered over. Similarly, it transpires that often on a building site when vital information is required, staff do not follow the sanctioned channels but rather seek a path of lesser resistance and reach for their mobile phone. The mobile, which site regulations deplore or even outlaw, offers an irresistible immediacy but also an irreplaceable and untraceable informality. Information can be accessed exactly when needed, and it can also be discussed, negotiated and translated without the constraints implied by documentary precision and permanent records." (Lee and McMeel, 2007:72)

Although the evidence of pencil marks on plaster remain within the confines of the building site, the mobile phone conversations traverse the wall and reach out to other spaces across the social network of the construction industry. Site managers will never know how many employees on their site are communicating and 'working' on another building site at the same time. But what we learn from this, is that people on the inside of the wall, can and do make it 'over the wall' through mobile communications, albeit above and beyond the jurisdiction of the site.

At the fourth stage in the continuum, the wall has now become quite permeable as telecommunications and social networks that support the building process, ignore its height, length and breadth. Such phone calls make a mockery of the Cartesian systems that started off as representations of the space, as the tolerance of actual building materials relies upon human decisions to make the building work. What seems particularly interesting at this point is to consider the building site behind the wall as a large social organism, that in order to build itself relies upon substantial tendrils that reach across to other bodies in order to put itself together. Such is the dense mesh of tendrils that in fact it may be hard to distinguish when the building actually starts and ends behind the wall.

Social digital architectures

Architecture as social navigation

The previous two coordinates in this continuum for the body's interface with architecture have introduced and demonstrated how architectural practice is struggling to make use of digital systems based upon the limitations of their Cartesian foundation for describing space. The next coordinate presented offers a platform for recovery for architecture, and a context in which the body is recognised and becomes architecture.

As architecture continues to insist on using computers to transcribe drawings of space with X, Y and Z coordinates, the public have capitalised upon the networking aspect of them. From its early beginnings as a networked reference tool, the internet has adapted according to what it was good at; being a distributed community. Peer to peer computing, the rise of instant messaging, voice-over-IP phone services and collaborative web services have all contributed to a socially driven agenda for that internet that is now being re-branded as Web 2.0.[2] In order to understand how architecture could incorporate this shift in utilising social systems and recover its relationship with the social body, it is important to look at the extent to which architecture considers the subject.

The social agenda for architecture is implicit in its practice, as Hillier identifies in his introduction to The Social Logic of Space (1984). Associating 'social' with the "decoration, embellishments, or even modifications of shape" (Hillier and Hanson, 1984:1) he complements the practical purpose of design with the further capacity to affect culture:

> "Invariably, artefacts are both functional and meaningful. Insofar as they are the first, they are of practical use; insofar as they are the second, they are of primarily social use, in that they become a means by which cultural identities are known and perpetuated." (Hillier and Hanson, 1984:1).

Interested specifically in how architecture is arranged to support good social practices, Hillier binds the task of 'ordering space' in buildings to the 'ordering of relations between people'. However Hillier returns us again to the architectural problem of materiality; namely, that while its impact upon people is rather obvious, it's very hard for architectural theorists to talk about the "sociology of buildings". Describing this as a "permanent disability", he suggests that it is easier for architecture to talk about styles and appearances than to "manufacture a socially relevant discourse out of these surface properties" (Hillier and Hanson, 1984:2). Consequently architecture, while aware that it is highly relational in its effect upon people, concentrates on what it is trained to discuss; the plan and the look of buildings once they are constructed. This associates Hillier's 'disability' with the central concept of limited models of time within architecture. That is, the design and construction of buildings is separate from its audience until the building site wall is taken down and the doors are opened, and the subsequent publication of analyses of a space that occurs through images and through words (Hillier and Hanson, 1984:3). For Hillier, the experience of a building is a four-dimensional relational system that is subject to change according to social and environmental conditions. Neither the design and building stage, nor the publication of architectural texts that follow and celebrate the buildings arrival in architectural lineage are able to support a relational, time-based context for understanding the social complexity of an architecture.

Ben Nicholson's context for a new type of architecture, and the networked space that supports communications between construction workers is located by definition within networked spaces. In the distributed and asynchronous space of the internet, architecture has a context in which a relationship between 'form' and 'social' can be entwined.

> "Localities become disembodied from their cultural, historical, geographical meaning, and reintegrated into functional networks, or into image collages, inducing a space of flows that substitutes for the space of places. Time is erased in the new communication system when past, present and future can be programmed to interact with each other in the same message." (Castells, 1996:406)

Castells chastises architecture and describes it as a "failed act" since it is has been incapable of a "mediated expression of the deeper tendencies of society" (Castells, 1996:448). He uses human geography to support his 'space of flows' model, which is derived from post-modern conditions and the rise of the network centred society. Castells identifies a change in the social production of space in which "we may be heading toward life in parallel universes whose times cannot meet because they are warped into different dimensions of a social hyperspace" (Castells, 1996:448). This 'schizophrenic' condition is similar to that of the difference between early architectural visualisations of the original interpretation of cyberspace (William Gibson's parallel world that has subsequently been depicted through cinema and virtual reality spaces) and that which has emerged as the highly social form of the internet in which websites now rarely use spatial metaphors. While deeply concerned about the split models of space that are emerging Castells finds an opportunity for reconciliation in the distributed nature of the network that allows itself to support new social structures:

> "Since networks are multiple, the inter-operating codes and switches between networks become the fundamental sources in shaping, guiding, and misguiding societies. The convergence of social evolution and information technologies has created a new material basis for the performance of activities throughout the social structure. This material basis, built in networks, earmarks dominant social processes, thus shaping social structure itself." (Castells, 1996:502)

Castells' new social 'material' provides the opportunity for digital architecture to overcome Hillier's 'disability' and offers mainstream architecture methods for embracing a social complicity. In addition, conceiving the social network as a new form of material is similar to Nicholson's de-material approach to architecture, but in addition through the recognition of social networked computing, it retains the ability to still create places, which Nicholson could only do through theory.

What then, are the technologies available for digital architecture to construct a 'social material'? Of the many approaches to developing contemporary Web 2.0 technologies that profess to be social networking services, a specific similarity to that of architecture and its socio-spatial dynamic can be found in the field of *social navigation*. Introduced in 1994 by Dourish and Chalmers, this term was conceived as a way of delineating between the types of navigation that can be found on the internet; social, spatial and semantic (Dourish and Chalmers, 1994). Identified at a point when the internet had yet to emerge as a "consensual hallucination" or as a semantic web,[3] the subsequent development of social navigation systems has survived while virtual reality has struggled. Originally described as "navigation towards a cluster of people" or "navigation because other people have looked at something" (Höök *et al.*, 2003:4) it has since broadened to encompass a wide variety of methods that involve social interactions across the internet.

As a specific field within Computer Human Interaction (HCI) social navigation is not concerned with the design of buildings. By definition, its attributes are closely matched to the socio-spatial dynamic that Hillier establishes within architecture. In his second book Hillier constructed a definition of society to support architecture's role in constituting it. In a complex set of relations, he proposes that society should be understood to be an abstract concept consisting of 'genetic' inter-social arrangements that are realised in space-time. By arguing that society has no material qualities, architecture becomes the materialisation of social dynamics at particular points in time.

> "Although society is this dematerialised genotypical thing, the means by which it is projected through time is anything but dematerialised. On the contrary, while the material form of society at any moment of time is not that society, it is the means by which that society is transmitted into the future." (Hillier and Hanson, 1996:402)

Society's material expression then becomes architecture, although "buildings are not maps of interaction. They are maps of the social genotypes of human interaction." (Hillier and Hanson, 1996:403). For Hillier, architecture operates as the contingent 'phenotype' for social interactions. He defines it as contingent in order to qualify its limitations across time, since all buildings are incapable of responding once built. This qualification is the critical difference from the modernist programme for an architecture that assumes the role of providing permanent 'phenotypes' for society (Hillier, 1996:403).

The socio-spatial dynamic that Hillier establishes is extremely similar to that of the objectives of social navigation in which the journey of the user in digital space is informed by the social environment that they find themselves in, which remains affected due to their interactions and choices.

> "What emerges from this is a picture of a reflexively populated information space. By 'populated', I mean that it contains not just information, but also people who are acting on that information, and who can see the effects of each others' actions and exploit that information in managing their own activities. By reflexively populated, I mean that not only does the structure of the space have an impact on the action of the users, but the action of the users have an impact on the space." (Dourish, 2003:291)

The value of social navigation in its similarity to Hillier's role of architecture, clearly lies in the socio-spatial dynamic that ties people to space in a highly relational framework (Chalmers, 1999:62). Both in built architecture and social navigation "people are active participants in reshaping the space" (Höök *et al.*, 2003:8). However, it is for the new forms of architecture that will emerge through ubiquitous and locative computing that social navigation offers the most opportunity. While Hillier recognises built architecture's limitations in responding quickly to social change, the time-based nature of digital space provides architecture and the mobile body with the technical frameworks to support a spatial and temporal dynamic beyond the ability of built architecture. Consequently social navigation appears to represent a form of Castells' 'social-material' which is based upon post-Cartesian understandings of space.

Social navigation as a concept and design practice within 'digital' architectures allows our two protagonists: body and architecture, to melt into a co-operative system in which walls are the response to movements of people, dynamic but supportive of travel. Whilst social navigation at this point does little in contesting the physical barrier of the wooden wall, we can begin to consider a platform in which architecture and the body can negotiate new forms of space in which technology can facilitate a co-operation in the definition of a new place without the need for barriers or the making of a building temporarily homeless.

Transubstantiation

Changing the substance of body into architecture, and architecture into body

In his use of the car as an 'update' of Certeau's *Walking through the City*, Thrift identifies digital technologies as offering a more complex substrate for enabling communications to become part of a negotiation with space in what he suggest may be a form of 'transubstantiation'. It is this the final coordinate that is at the extreme end of our continuum that projects a situation in which body and architecture are synthesised and their distinction is useless as one defines the other.

The car complimented with satellite navigation, air conditioning, musical soundtracks and a figure-hugging seat provides a very personal interface with a city, one that predisposes the driver to allow the car to become an extension of his/her body. Once driving, we find ourselves expressing a series of characteristics that indicate a deep embodiment of the car including: the charged emotional state in which we engage with others, communication techniques using lights and movement, and the 'tactics' that allow us to navigate spaces by reading the 'gestures' and actions of others (Katz, 2000). More relevant to us, Thrift extends the cars qualities even further by suggesting that the combination of embodiment, digital enhancement and communication is capable of 'producing a new, more fluid transubstantiation' (Thrift, 2004).

> "I believe that the sheer scale and sophistication of what is happening now amount to something quite different: a studied extension of the spatial practices of the human which consists of the production of quite new material surfaces which are akin to life, not objects, and thereby new means of bodying forth: new forms of material intelligence producing a new, more fluid transubstantiation." (Thrift, 2004:49)

Thrifts adoption of the term transubstantiation is particularly interesting because it offers for him a framework within which the material limits of the car are superceded by the extended and embodied manner in which we engage with them to move through space. The Roman Catholic and Easter Orthodox doctrine from which the term originates, refers to the capacity for the bread and wine in Communion to become in substance, but not appearance, the body and blood of Jesus Christ. For Thrift this process is becoming paralleled as the use of network and communications technology is increasingly providing a substrate across which we are able to exchange the substance of the pavement that Certeau defends, for the pedals, steering wheels and buttons that become our interface with the street. It is clearly not the metal or plastic that we touch that performs the miraculous transformation, but it is within the complex relationship that is emerging, as digital systems become part of our negotiation with the social body and space.

Thrifts demonstrates the potential for digital systems to extend the social negotiation with a space through the car. Complex as modern cars may be, this provides a precedent that we can invest some faith in to allow us to overcome our wall that is obscuring our building site by using similar methods. Latour's model of the social through Actor Network Theory is also placed within a sensitive model of space and reminds us that the fabric of society is not in the walls or materials, but the passage of relations across a network;

> "Society is not the whole 'in which' everything is embedded, but what travels 'through' everything, calibrating connections and offering every entity it reaches some possibility of commensurability. We should now learn to 'hook up' social channels like we do cable for our televisions. Society does not cover the whole any more than the World Wide Web is really worldwide." (Latour, 2005:242)

With this in mind, we can understand that a critical reason why the use of a temporary wall to temporarily erase a place is fundamentally destructive is because it hinders social networks to be accessed by anyone else other than those working on the site. Latours' call to 'hook up' social channels can be interpreted as a desire to see the extension of a communications substrate across which social networks can further grow.

> "information technologies allow us to trace the associations in a way that was impossible before. Not because they subvert the old concrete 'humane' society, turning us into formal cyborgs or 'post-human' ghosts, but for exactly the opposite reason: they make visible what was before only present virtually." (Latour, 2005:207)

From Thrifts use of technology as an enabling characteristic of transubstantiation between car and street, to Latour's assertion that space is simply the residue of social relations, we can begin to put faith in network technologies to provide a medium through which the body (social and individual) are part of the definition of architecture. Thrifts focus upon the car as the initial context in which a transubstantiation between body and space begins, is the mere beginning of how as individuals we will shape our streets. Already mobile phones are adopting the same navigation technologies as cars to assist pedestrians as they navigate a city, complemented by social navigation and networking software, browsers, and of course the ability to make phone calls, our bodies and the architecture around us will be on the brink of a tremendous shift in substance.

Summary

This paper has used problems within architecture to explore how the body interfaces with space. Before looking forward to the opportunities offered by digital technology to allow the body individual and social, I have attempted to sketch out the resistance of the materials involved that at present prevent any deeper intercourse between architecture and the body. I have used six practical and theoretical examples to describe the breadth of extreme strategies that people have gone to in order to engage with architecture. From skate boarders that place the jurisdiction of space under stress as they use their bodies to 'ride' across architectures, to Eve Dent who seeks a route of lesser resistance to become part to the wall. Ben Nicholson is used to identify architectures obsession with materiality as its primary tool for producing space, and his relinquishing of it in favour of the internet introduced the socio/spatial capacity of digital systems which were extended through Lee and McMeel's observations of mobile phone use across building sites. Social navigation as a design approach within HCI, was described as a technical platform in which the distinction between the networked participants and the architectural structure become dynamic. This offered a theoretical springboard to describing the potential of digital mobile systems, exemplified in Nigel Thrifts use of the car, to offer a form of 'transubstantiation' through which the body and architecture become intimately connected.

The paper has recognised three of the 'modern' legacies that remain part of architectures design and construction process that limits the way in which the body can engage with urban places. First was the limitations that the subject has in its understanding and provision for a comprehensive social dimension to architecture. Primarily articulated as the loss of 'shelter', and later through its inability to see itself part of the social networks that make up society as a whole. Second was its adherence to Cartesian drawing systems that retain a split model of time and space and offer limited modes for representation through which the body (individual and social) can find engagement with architecture. Finally the question of material was addressed and architectures struggle in overcoming it as its primary tool to 'make places'.

By lacing the examples with some of architectures problems, I hope to have constructed a narrative in which the reader appreciates the complex relational dialogue between body and space. Whilst it is true digital systems are changing how we understand the body, we have also to work hard with architects and designers to develop new methodologies that take responsibility for supporting the dialogues across the two subjects, so as to develop a language in which the body and architecture can grow without barriers. As an example of how this difficult aspiration may be considered, I make reference to a project that attempts to provide a platform upon which the architecture of a specific building at the University of Plymouth, and the social body that substantiate its function, are presented in a dialogue.

Case study of architecture as body, body as architecture through the use of digital networked systems

Arch-OS a substrate for documenting social activity

> "The Arch-OS combines a rich mix of the physical and virtual into a new dynamic architecture. Arch-OS uses embedded technologies to capture audio-visual and raw digital data through a variety of sources which include: the 'Building Management System (BMS) (which has approximately 2000 sensors in the Portland Square development); digital networks; social interactions; ambient noise levels; environmental changes. This vibrant data is then manipulated and replayed through audio-visual projection systems and broadcast through streaming Internet and FM radio." (Phillips and Speed, 2003:1)

The Arch-OS (Operating Systems for Architecture) infrastructure at the Institute of Digital Art & Technology, represents the early signs of a 'substrate' across which the social construction of space may be informed, and the space between body and architecture dissolved. The Arch-OS system has been used to present representations of social systems, and its exposition of network and environmental data has revealed the invisible conduits through which we construct social relations. Whilst many of the projects that have used the Arch-OS data have engaged with traditional themes found within art, they have all at 'different' times connected 'different' sensors to 'different' displays that explore the multiple associations between humans.

Precisely from an understanding of Arch-OS as a tool for evidencing social exchange and human activity, Wilfried Hou Je Bek launched the blog www.urbanxml.com which documents the growth of RSS feeds that trace human activity across the world. More strategic in its efforts to support architectures 'environmental' communication across digital networks is Haque's and Leung's Extended Environments Markup Language (EEML) which constructs protocols to support collaboration and exchange.

The perpetual and relentless streaming of Portland Square data to the internet over the past five years seems to demonstrate a potential for similar systems to operate not just after a building is open, but before (from pre-existing social and environmental networks) that operated on a building site, to during (as the building takes shape and as people can use networks to overcome the barriers to the site), and after the building is re-introduced to a space (through the growth of further connections from occupants). In doing so the construction of the building has the potential to become secondary to the sustaining of social networks that cross the space, and subsequently reduce the impact of the 'new' buildings detachment from society.

With systems such as Arch-OS offering interfaces to communications on and around a building site, suddenly our wooden wall may not be so opaque. As we become smarter at driving through the city and relying even more on digital systems to help us negotiate socio/spatial networks we can anticipate engaging with conversations and offering advice to people not just behind windscreens but also behind walls. In the same way the steel shell of a car has become irrelevant in the interface between the street and us, the running wall that surrounds a construction site will become less of a barrier between to the building that is being built. No longer need the building be isolated from a social and environmental context, instead the construction of a new building can be an extension of existing social networks.

> "Once the whole social world is relocated inside its metrological chains, an immense new landscape jumps into view. If knowledge of the social is limited to the termite galleries in which we have been travelling, what do we know about what is outside? Not much." (Latour, 2005:242)

Notes

1. The phrase that was adopted by the American architect Louis Sullivan informed his methodologies in the design of skyscrapers and buildings in the early 20th century. Coupled with Adolph Loos' claim that ornament is crime, these two principles massively informed the work of Le Corbusier, Gropius and van der Rohe who defined the modern architectural movement which post-modern architecture has since sought to undermine.
2. Web2.0 refers to the second generation of internet technologies and systems that allow people to collaborate and share information. Systems such as wikis, recommendation sites and social networking services all offer technical mechanisms that allow shared authorship and indexing. These qualities are understood to be intrinsically different to the Web1.0 model which was based on publishing and not sharing.
3. The Semantic Web is a project by Tim Berners-Lee to create a universal medium for information exchange. Using standardised markup languages and processing tools it makes it possible to link pieces of information together to enhance their meaning when searching for them. The Semantic Web is often described within the same context as Web2.0, with many of the contemporary applications being part of the transition to a full Semantic Web.

Bibliography

Borden I. (1996), *Beneath the Pavement, the Beach: Skateboarding.* Architecture and the Urban Realm, In: Borden, I. (ed.) *Strangely Familiar: Narratives of Architecture in the City.* London, Routledge.

Borden I. (2001), *Skateboarding, Space and the City: Architecture and the Body.* London, Berg.
Castells M. (1996), *The Rise of the Network Society (Second Edition).* Oxford: Blackwell.
Caillois R. (1935), *Mimicry and Legendary Psychasthenia,* trans. John Shepley, October, Vol. 31, (Winter, 1984), Pg. 16–32. The MIT Press.
de Certeau M. (1984), *The Practice of Everyday Life.* Trans. Steven Rendall. Berkeley: University of California Press.
de Certeau M. (1987), *Practices of Space.* In: M. Blonsky (ed.) *On Signs.* Oxford, Blackwell.
Chalmers M. (1999), *Informatics, Architecture and Language.* In: Munro A. J., Höök K. and Benyon D. (Eds.) *Social Navigation of Information Space.* London: Springer.
Dent E. (2006), *Eve Dent.* [online] *Axis Online Reseource of Contemporary Art.* Available from: http://www.axisweb.org/ofSARF.aspx?SELECTIONID=15517 [Accessed 20 June 2008].
Dourish P. and Chalmers M. (1994), *Running Out of Space: Models of Information Navigation.* Short paper presented at HCI'94 (Glasgow, Scotland).
Dourish P. (2003), *Where the Footprints Lead: Tracking Down Other Roles for Social Navigation.* In: Höök K., Benyon D. and Munro A. J. (eds.) *Designing Information Spaces: The Social Navigation Approach.* London: Springer.
Gunkel D. J. (2001), *What's the Matter with Architecture.* In: Hill J. (ed.) *Architecture - The Subject is Matter.* London: Routledge.
Heidegger M. (1971), *Building Dwelling Thinking.* In: *Poetry, Language, Thought..* Trans. Albert Hofstadter. New York: Harper Collins.
Heynen H. (1999), *Architecture and Modernity.* Cambridge. MA: MIT Press.
Hillier B. and Hanson J. (1984), *The social logic of space.* Cambridge: Cambridge University Press.
Hillier B. (1996), *Space is the Machine, A configuration theory of Architecture.* Cambridge University Press.
Höök K., Benyon D. and Munro, A. J. (eds.) (2003), *Designing Information Spaces: The Social Navigation Approach.* London: Springer.
Jameson F. (1991), *Postmodernism; or, The Cultural Logic of Late Capitalism.* London: Verso.
Katz J. (2000), *How Emotions Work.* Chicago, IL: University of Chicago Press.
Thrift
Latour B. (2005), *Reassembling the Social: An Introduction to Actor-Network-Theory.* Oxford, Oxford University Press.
Lee J. R. and McMeel D. (2007), *"Pre-Ontology" Considerations for Communication in Construction.* Ontologies for Urban Development 2007, Pg. 169–179.
Nicholson B. (1999), *Click it, Click it Good.* In: Cook P. and Spiller N. (eds.) *The Power Of Contemporary Architecture.* London: Academy Editions.
Nicholson B. (2002), *Beans, Beans, Glorious Beans.* In: Spiller N. (ed.) *Reflexive Architecture.* London: Academy Editions.
Phillips M., Speed C., Montandon A. and Dent E. (2004), *'read/write/fold'.* Performance Research, Routledge. Vol. 9, No. 2, 'On the Page'.
Thrift N. (2004), *Driving in the City.* Theory, Culture & Society, Vol. 21, No. 4-5, Pg. 41–59.
Phillips M. and Speed C. (2003), *Arch-OS v1.1 [Architecture Operating Systems], Software for Buildings.* 5th International. CAiiA-STAR Research Conference and Workshop. Consciousness Reframed July, 2003.

Chapter 5
Resistances of the body. The presence of the human body in the automated world

Caroline Moricot

La technique, c'est l'action humaine qui réussit
André Georges Haudricourt
La technologie science humaine,
Paris, Maison des sciences de l'homme, 1987.

Abstract

The place of the human being in automated technical devices is an age old question that leads us to the definition of human in these worlds, where the automats often replace the activity of men, but cannot manage without their presence, not so much in terms of technical functioning, but rather in terms of giving sense to the function. The human being is in the middle of a tension, that he has to manage, between a tendency of the automation which aims to draw him aside and a daily management of the systems which, simultaneously, makes his presence irreducible. This paradox of automation is symptomatic of the definition of human in these large technical systems. Shaped by this paradox, the body is sometimes considered as a restriction to the technical potentialities of the device (that justified its existence to assist, then in some instances replace the body), sometimes like an irreplaceable resource. This chapter will explore, through the model situation of piloting planes, some of the postures of resistance of the body : the resistance (in terms of robustness) of the body as an element of distinction and hierarchisation of the group ; the body practice and the question of change ; the capacity to resist its sensations and to have faith in the information presented on cockpit screens.

Introduction

What is the place of the human being in automated technical devices? What is, in particular, the place of their body? This age old question (which always resurfaces with the installation of new systems) leads us to the definition of human in these worlds, where the automats often replace the activity of men, but cannot manage without their presence, not so much in terms of technical functioning, but rather in terms of giving sense to the function. To be interested in the bodies of operators acting in these large technical systems[1] would imply envisaging the significance of their carnal presence as an element of understanding the functioning of these devices. Because human is a being of flesh, his relation with the world is not only intellectual or operational, but also sensitive and symbolic and these various dimensions resonate on each other.

I suggest building a reflection on this question by starting with the "model" situation of piloting planes. The following analysis is based on a series of studies carried out during the last fifteen years in the civil and military aviation[2] field. The discussion will not place emphasis on the distinctions between these two worlds, we will only highlight their specificities where it will be useful to complete our socio-anthropological point of view on the articulations which can be tied into these singular circles, between the question of the technique and that of the body.

The plane belongs to the class of objects defined at the same time by a high level of automation and the required presence of highly qualified operators. The pilots supervise the systems, place them in the right context, they recover many deviating situations, in short, they adapt in real-time an "ideal" scenario which unfolds on the screens of the cockpit. Their intellectual presence is still necessary while their body is increasingly constrained with immobility. The human being is thus present "despite everything": he is in the middle of a tension, that he has to manage, between a tendency of the automation which aims to draw him aside and a daily management of the systems which, simultaneously, makes his presence irreducible. This paradox of automation is symptomatic of the definition of human in these large technical systems.

Shaped by this paradox, the body is sometimes considered as a restriction to the technical potentialities of the device (that justified its existence to assist, then in some instances replace the body), sometimes like an irreplaceable resource. The first "resistance of the body" underlines its weaknesses (the body is considered as a constraint with which it is necessary to compromise) while the second makes the body a place of resistance to total automation (the body such as that which remains when all has been automated) because it is the place of knowledge and collective memory.

The advent of a new generation of airliners in the Nineties was the subject of sharp controversies in the civil aviation sector: automated devices entered the cockpits on a large scale to optimize, under the control of the crew or independently, the management of the flight (see table 1). Researches which I led during this period were aimed at better understanding, using the concept of appropriation, how the human, put in a new situation, rebuilds a representation of the technical object, of his work and of himself in a context where automation is increasingly present and where safety constitutes an absolute prerequisite (Moricot, 1997). How, step by step, does an unknown environment become familiar and on what premise is this familiarity based? What are the significant stages of this process and what are the preconditions? Familiarity with the objects comes from a comprehension of their use which is derived, amongst other things, from the senses. In fact, fundamentally, *objects are appropriable when one can recognize oneself in them*. In other words, a pilot cannot endorse a plane which does not return to him his identity of pilot. This is the essential issue of any technical innovation: the appropriation is the inscription in a pre-existent culture of innovation. It is not a simple adaptation of the human to the technical environment, it is much more than that, it is *the capacity to operate in a relationship of familiarity with things*. Visual, auditory and tactile familiarity constitute as many indices which do not necessarily coincide with the criteria clarified in the functional rules or the regulations of use (Thévenot, 1994). There exists an intelligence of the body which makes it possible to move in reality. The body is the mediator of our relationship with the world, our tool of knowledge of the world as the phenomenologists say. It is, in particular, a tool for mediation between the regulated and impromptu (Sauvageot, 2003). This intelligence of the body which is capable of trickery, innovation and creation, is basically sensitive and flexible (on the model of the *metis*), it organizes, in such a case, what one can call a resistance of the body.

I propose to explore in this chapter some of the postures of resistance of the body: the resistance (in terms of robustness) of the body as an element of distinction and hierarchisation of the group; the body practice and the question of change; the capacity to resist its sensations and to have faith in the information presented on cockpit screens. I will employ, each time, examples drawn from my fieldwork in the civil and military aviation sector.

The plane, an automated world

I will use the term automation here in reference to an ensemble of systems which show characteristics of power, complexity, autonomy, decision. The devices, which are part of the process of automation, actually do not constitute a homogeneous category as one can think of it a priori. The Autopilot of the 60s and the Flight Management System of the glass-cockpits have admittedly only very few things in common. From the point of view of the technique they implement, the interface they use, and to a larger extent, the perspective of the human they suggest (and of his place in the process of work), one can see that these systems present a large variety of situations. Moreover, the addition of automatisms does not produce systematically an automat, or even an automatic unit. We are interested here only in one type of automatism, that which, is characteristic of the glass-cockpits, puts in scene calculators, i.e. automatisms based on a computerization of the process. The question which arises, from a socio-anthropological point of view, is how to understand better the way these techniques are used: what are the practices and representations established around and with them?

We can, with the aid of Nadine Sarter and David Woods, define these systems according to various criteria: their autonomy, their authority (in terms of decisional capacity) and their visibility seem to constitute three characteristics of the glass-cockpits (Sarter and Woods, 1995). Autonomy refers to the capacity of these systems to deal with some long and complex sequences of actions without intervention of the pilot once programmed and "activated". According to Sarter and Woods, these automated systems have also a high level of authority which is conveyed for example in the concept of "protection of the flight domain" with which the Airbus has been equipped with since the A-320 model: the automatism is able to detect a flight configuration endangering its safety (a stall for example), to prevent it and to restore a safe configuration acting, if necessary, against the will of the pilot. As regards the visibility of the systems, their capacity of being "observed" by the users, this does not depend so much on the number of the screens and the quantity of information available but rather on the way this information can be localised and used effectively by the operators.

The improvement of the performance of calculators in the middle of the Seventies has touched the working world and, on a larger perspective, society as a whole. Their generalized use on board has transformed many aspects of the crew's work (including its composition) and denoted, it is generally thought, a new stage of the automation process. The passage of this stage is attested even in the vocabulary used to name these planes: new planes, planes of the new generation, new cockpits, and glass-cockpits. The term glass-cockpit was initially used to designate planes equipped with cathode screen display systems. Glass screens, flat screens, and square and colour screens present and synthesize information formerly disclosed on tens of small dials. These screens depict the situation in which the plane is: primary control of the flight trajectory (PFD Primary Flight Display), navigation (ND Navigation Display), monitoring of engines and fly-by-wire (ECAM Electronic Centralized Aircraft Monitor). A flight management system (FMS) was added to this new method of information presentation, a kind of super Autopilot, supported by a whole array of interdependent equipments and subsystems. On the most recent planes, the flight control devices comprise, moreover, a system of protection against any deviation of the flight domain. The glass-cockpits thus associate automated devices of connected management control systems with an electronic display of the interfaces.

What place is given to the body in an automated process of an activity like the one of piloting a plane? How, more generally, do we manage human presence in a technician world where the project is basically an autonomous technique? A quick historical detour[3] will enable us to better understand the tension which has gradually been built around these questions. The presence of human on the aircraft has become paradoxical: it is at once suspect and irreducible. Because the worlds which one says automated are nevertheless those which we do not know how to manage without the human who continues to give sense to the technical activity. The process of automation places the human in the middle of a tension between a tendency (that of the automat), which tries to distance his body, and a requirement (that of the appropriation which makes it possible to operate in the everyday life) which summons its body to be in the action. We initially will reconsider the postulates, the issues and the constraints of automation, an old process in aeronautics.

Historical detour: a progressive sidelining of the body

The use of automated devices is not a new occurrence in aeronautics. The principle of automation was present and almost constitutive in the birth of aviation and its commercial development. Ferdinand Ferber affirms since 1904 his intention to manufacture an "automatically stable" machine. On the question of balance in flight, Octave Chanute posed, at the end of the 19th century, the problem in these terms: "Of all the problems to be solved, it is in my opinion the most important and the most difficult one... It would be much simpler if a flying machine had to move only according to one constant angle of incidence and in calm air. (...) but the flying machine must go up and go down, like a bird, and must take into consideration the swirls, the movements and the gusts of wind. A bird manages this by constantly modifying its gravity centre; like an acrobat, it balances itself instinctively. The problem is much more complicated for an inanimate machine which requires, if possible, an automatic and permanent balancing."[4]

Of course, these automatisms, Chanute and Ferber are speaking of, are of the reflex type (and not management as those which one knows today). One will observe that on the other side of the Atlantic, this question of stability does not constitute a major concern for the Wright brothers who will have been precisely criticized for the unstable disposition of their plane, the "flyer". According to Emmanuel Chadeau such a configuration resulted from a deliberate choice. Chadeau reports that Orville Wright explains himself in his memories that having decided to favour the easy handling of the plane, he and his brother counted on the effectiveness of their system of control surfaces to compensate, by piloting, this instability of the plane. In other words, it was the human responsibility, rather than that of the automatic devices, to maintain the machine in balance. We do have to recall here that the Wright brothers had been bicycles manufacturers (Chadeau, 1991).

The place of the body is thus already the object of a controversy: several prospects are still possible and open in this beginning of the 20th century. On both sides of the Atlantic Ocean, the question of automatism is considered differently thus already reflecting the two visions of the role of the human on an aircraft. The comparison is undoubtedly risky, but one cannot prevent oneself from comparing this controversy to the one which took place about twenty years ago, between Boeing and Airbus and which in the same way thus presented visions of cockpit automation and of the place of the human rather differently from each other.[5] Automation is a technical, a social and also a cultural process (Batteau, 2001). From the beginning of the thirty's, the concept of an increasingly automated flight up to the disappearance of the pilot (in any case as a human being, seeing, thinking and feeling) is present. The process of automation starts between the two wars and under the impulse of the commercial development of aeronautics: trusting more and more to their instruments, the aviators will conquer little by little the night and the bad weather, they will lengthen their operating range to reach the poles and the equator (Petit, 1993; Edwards, 1988). In 1929, Doolittle, in the United States, accomplishes an entire flight (from take-off to landing) without visibility, that is to say by using instruments allowing him to rebuild reality. The first autopilots appear and, with them, the first automatisms in the cockpit of the plane. The Second World War contributes greatly to the development of aviation, by starting selection and training for a great number of pilots (the success of the missions still depends largely on the capacities and the skill of the pilots in this particularly hostile environment).

Remarkable in this process of instrumentation and of automation of certain tasks, is the way safety is already constantly in the foreground; it is always in the name of safety that innovations are carried out, and when the human being is isolated from certain type of action, it is because he is perceived as a danger![6] Commandant Mioche uses very explicit terms during the first air security congress in the thirty's: " Manual piloting is as irrational as possible and there is no doubt that in the not too distant future all planes not destined for acrobatic flights will be controlled automatically (...). It is moreover the absolute suppression of piloting faults and the masterly solution for flights at night and during foggy weather; blind flying, it is generally accepted, is only an imperfect automatic piloting, where the indications given by the instruments are transformed into movements of the control surfaces by the whimsical servo-motor which is the pilot."[7] The valorisation of the piloting fault, "the freak" of the pilot's body sensors, i.e. their instability, are indicated here to affirm the solution of automation as the best, and only possible way. The body is disqualified.

Each innovation has its intentions... As concerns the plane, the automation of certain tasks, left before to human initiative, is meant of course to discharge him of unrewarding routine and repetitive actions; it is also supposed to limit the risks related to the loss of vigilance, due to physical or mental tiredness. Automation became indispensable and still is in aeronautics (but this field is certainly not the only one) within the safety conquest and, in this perspective, human is often placed in comparison to the automat. For example, it is the automat that will execute the landing in the case of bad weather because it is considered to be more reliable.[8] On the other hand, it is the pilot who maintains responsibility, and who has to compromise with the machine which is not always easy to realize.

Harry Collins proposes the concept of "machine-like action" to point out some habits of human activity which consist in imitating the machines: "If we imitate machines in our acts, then the machines will be able to replace us in our work, but, this work must already be of mechanical type so that machines can be installed. Those who want to replace the human at work by machines have first to redefine the meaning of "work" so that it can be carried out in a mechanical way; the process of deskilling intervenes at this point. (...) Intelligent machines are useful as tools. The only error to avoid is to believe that a tool is an actor. It is easy to say: "A computer is better than me in arithmetic." But the epistemological signification of such a statement is identical to the meaning of: "A hammer is better than me to nail." To confuse computer and arithmetic is of the same order as to confuse the hammer and the one who holds the handle of it" (Collins, 1992).

The course of an automated flight sequence (a landing for example) supposes that the crew adopts a "machine-like" posture in phase with the automated procedure that takes place under the control of their eyes but without intervention of their hands. However, the crew remains responsible for the good progress of the sequence and must constantly be able to retake the plane manually in order to counter a problem. This posture, which can be described as "responsible supervision in real-time", is contrary to a machine-like posture, nevertheless, it must be superimposed upon it. This example shows very well the tension in which the crews exist, tension between two contradictory and yet simultaneous requirements, two postures of the body which do not mobilize the same resources. Quite spontaneously, in order to maintain their level of proficiency, the pilots continued to retake the plane in manual mode as often as possible. This act can be interpreted as a resistance of the body which is well aware of the impossibility of remaining skilful without practising. *"I have the feeling to have lost out in terms of pure piloting* — confides a pilot — *I would not be able do the things again as well as I did before. Training is essential"*. Another pilot adds: *"I disconnect the automatisms to keep this skill, it is still possible."*[9]

Harry Collin's concept of "machine-like action" refers to the one of body-machine expounded by David Le Breton. His critical analysis of Cartesian dualism aims to explore the idea that "living is subordinated to the machine's model and exhausts itself entirely" (Le Breton, 1990, Pg. 75). Does the rational exclude the sensitive? No, responds Antonio Damazio categorically to this injunction of Descartes. In a work entitled *Descartes' error, emotion, reason and the human brain* (Damazio, 1994), he shows the contrary. He states, on the basis of a detailed study of some clinical cases, that it is the absence of emotions and feelings which precisely prevents the human from being rational.

The project of the automat

Le joueur d'échec de Maelzel is a short-story written by Edgar Poe in 1836.[10] He relates an authentic story of deception, which involved making people believe that an automat capable of playing chess victoriously actually existed. Before measuring itself against a player, the machine was exhibited and its mechanisms revealed in order to "prove" that there was nobody hidden inside. It must however be pointed out that the appearance of the automat was that of a man, a Turk in fact, with rolling eyes, smoking a pipe, i.e. setting a rather spectacular body on the scene. Poe denounces this trickery setting forward the following reasoning: Maelzel's machine cannot play chess because no machine can. At the end of a long deduction, he arrives at the conclusion that there is a man who contorted himself in the machine, which was the case. All the stories of automats are not trickeries, but all carry in them the myth of device *acting like* a perfect and invincible man.[11] The field of data processing which is a form of automation (in terms of organization, management and transmission of information) will certainly fall into this category.

Adopting both an historical and a sociological point of view, Philippe Breton recalls a history of data processing by updating the imaginary on which it was built. He stresses the important part of this imaginary as being an "acute conscience that the human being is imperfect and that the solution to this problem must be sought in the techniques" (Breton, 1990, Pg. 157). If the computer is compared to the brain of man, it is initially with the aim of imitating it, of simulating its reasoning, and also trying to exceed it. The discourse which thus accompanies the innovation will be based on the topic of the imperfection of the human. It is this same set of themes that accompanies the automation of cockpits' today. This evidence of human imperfection reinforces, per contra, the (supposed) idea of perfection of the machine (in terms of regularity, power, constancy...). The process which aims to legitimate the comparison (on all levels) between humans and machines is at the heart of the project which cybernetics will set up and diffuse from the Forties.[12] One of the principal postulates of cybernetic thought rests on the supposition that the human brain is transposable with the computer. Norbert Wiener is very explicit when he writes: "the impossibility of telegraphing from one place to another, the model of a man is probably due to technical obstacles, (...). It does not result from an unspecified impossibility of the idea itself" (Wiener, 1962, Pg. 128). The test of Turing, *Can a machine think*, aims precisely to demonstrate that the human and the machine can produce similar behaviours. It is noted that the body is excluded here from the definition of behaviour since, during the famous test, the body is not present (in order not to betray the human or mechanical nature of the answer). One can conclude rather cynically with Herbert A. Simon that: "All that leads us to the idea that the intelligence, embedded in a human being, a dog or a computer, can be represented by a system capable of dealing with symbols" (Demailly, 1986). The body remains widely absent from the cybernetic way of thinking, which is built, one could say, against the body. *One can consider this moment as the founder of a posture of resistance of the body.* The flesh appears to be, indeed, one of the main obstacles to the development of the projects of cybernetics. The body, because of the irreducible thickness of its flesh, continues to be there, but as in deferment (remember that for Wiener the obstacle is only technical and provisional). This "humiliation of the flesh", to use a term of Ivan Illich, is the result of a depreciation of the body and of sensitive experience, characteristic, according to him, of the technical and industrial civilization that deploys a type of rationality, which in order to better functionalise reality disembodies the "real world".[13]

This is why, using this logic, the human and machines are constantly in competition. This is the conclusion of Harry Collins when he writes: "If we regard ourselves as machines, we interpret our differences compared to the mechanical ideal like a sign of human frailty rather than as a sign of creativity", adding that this vision of the world and of the place of the human in this world seems quite internalized, so much so that "while we become increasingly familiar with machines, we cure their failures without even noticing it" (Collins, 1992, Pg. 292).

The paradox of automation

This question of the physical imprint of humans in the technical systems, which has been shown to be both a difficulty and a necessity, brings us to a better understanding of what I propose to call the paradox of automation. The addition of automatisms does not necessarily produce an automat, even if this project is present in the legitimization of a comparison between the human and the calculator, with the idea of the perfection of the machines. Undoubtedly, a radical situation of total automation would be easier to manage, but as far as air transport is concerned, as in many other fields, it seems that the prospect for the automat is ruled out for the moment (even if, according to some, this preclusion is provisional). It follows from this situation that the process of automation of planes has put men in a deeply ambivalent situation. In fact, it appears that *because of the increasing presence of automation, the presence of humans is progressively necessary*. Moreover, if this presence is paradoxically more necessary, it is not always facilitated: the body is kept apart; the pilot is all the more inclined to be over-confident that the devices generally function without any problem. However, the presence of humans, related to the management of the automatisms, cannot be satisfied to be passive because it is not only a question of carrying out a supervisory role: it is necessary to anticipate, adapt, communicate, possibly disconnect to retake manually, and especially to determine the reliability of the information on the screens. In some ways, one asks at the same time more and less of the pilot of a glass-cockpit plane.

Thinking about the question of human presence in the automated devices infers the consideration of various modes of presences. Among these various presences, we will stop more particularly on the posture of resistance which invites us to imagine the presence of human like an active, a committed presence. This posture does not give a complete overview of human activity in the automated world; it elucidates that which enables us to highlight some of the tensions which cross the large socio-technical systems.

Defining resistance as a mode of regulation of large technical systems

The encyclopaedia devotes several pages to the concept of resistance which covers a great number of fields: physics, biological, mechanical, historical or legal to name but a few. The action to resist means the capacity to stand firm while reacting, to show endurance, to defend oneself, to support without weakening sufferance, tiredness. It expresses the idea not to yield to an effort in the opposite direction and, by extension, it describes the force which cancels the effects of a destructive action, as in the resistance to diseases for example. It also refers to the idea of protection and one talks in biology about vital resistance to indicate the properties which living entities posses to react to all causes of destruction, in order to protect themselves. The law proposes to distinguish between a passive resistance, an active resistance and an aggressive resistance. In French history, Resistance means the action taken against the German occupation during the Second World War. On the issue of the resistance of materials, it is question of the distortion of bodies, of the elasticity necessary for the robustness of the matter. The concepts of friction, effort and fatigue are present. Fatigue, in particular, is the subject of a detailed treatment in various categories illustrated in mathematical equations. We turn finally to a last meaning of the term, a more moral one, which describes the capacity not to yield to something attractive: to resist temptation. To resist in order to restrain, to slow down, or stop a process of which one can perceive the dangers belongs to the register of the instinct of survival of the collectives, social survival. The human is still there because he makes the technique understandable, he establishes the link between the various elements of the device (much like the conductor, as Simondon says, who directs, but does not play an instrument himself) (Simondon, 1989).

Being opposed to the movement or the action of automation because the definition of human that is proposed leaves an increasingly reduced place to the body, can appear as an act of resistance that some qualify as resistance to change, thus stigmatizing this posture by reducing it to a corporative conservatism. I propose, on the contrary, in a Durkheimian perspective, to consider the forms of resistances enacted by humans—and in particular body resistances—in the technical systems like a mode of social regulation which allows the protecting of oneself from a form of anomy which we gravitate towards: the desire of an automated world which promises to relieve the troubles of the social world, but which by the process it establishes, ends up eliminating the human. The technical solution to a problem which questions a society in all its dimensions promises a miracle which cannot be achieved. The dissatisfaction which results, far from questioning the project, nourishes it and strengthens it in an almost exponential way. The anomy which threatens our society is due to its incapacity, or its difficulty, to control this ever growing desire that is never appeased. *Resistances of the body in the technical systems such as aeronautics, can be considered like spaces of regulation* and therefore as protection against the anomy which threatens and tempts us. A collective, intuitive regulation which functions more on a tactical mode than on a strategic one, to use the dychotomy of Michel de Certeau who proposes, in *L'invention du quotidien*, to distinguish between strategy and tactic to explain the daily practices. Whereas the strategy is based on the calculation of the power struggles which supposes on the one hand to be able to isolate a subject from its environment and on the other hand to be able to draw a line around a circumscribe, i.e. to manage a relation with a distinct externality, the tactics must constantly play with the events to make "occasions" of them. Political, economic or scientific rationality was built on this strategic model, states de Certeau, whereas many daily practices are of tactical type. "The tactics cannot count on a boundary which distinguishes the other like a visible totality. The tactics have as place only that of the other. It penetrates into it, in fragments, without seizing it in entirety, incapable of keeping it apart. It is not supported by a base which capitalizes its advantages, prepares its expansions and ensures independence according to the circumstances. (...) What it gains, it does not keep it" (De Certeau, 1990, Pg. 46). While automation grows on a strategic mode, the resistance of the body uses the tactical resources to protect its territory.

From the studies of Marcel Mauss, we know that engaging one's body in an activity is nothing natural, it's a *technique of the body*, i.e. a social action which is the subject of training for its best effectiveness taking into consideration what a society valorises at one point of its history (speed for example). The postures of the body are thus learned postures, generally within a collective. This collective memory of the body knowledge, inscribed in the body of everyone makes it possible to know oneself and to be recognized, to cooperate in work and to be located in the hierarchy of the group. The expressive engagement of the body in the ordinary acts of work is an essential condition of the co-operation, points out Nicolas Dodier.

Because of its presence within the technical systems, the body is placed in position of resistance. If one agrees to summarize the previous comments in simplistic terms, one can propose to consider automation, in progress for several decades, as a process which initially sought to relieve the body of the operators of a certain number of repetitive and tiresome gestures, before seizing certain much more "noble" actions. On an aircraft the pilot is, however, still there and for the moment his presence is considered as being essential. However, this presence has been modified: much less physical than at the time of the Boeing 707, and much more intellectual and cognitive in terms of supervision. Has the body become too cumbersome, i.e. of excessive presence? (*"surnuméraire"* according to David Le Breton, 1990, Pg. 72) Can one be present at an activity without ones body? Considering the body ipso facto in position of resistance brings to the idea of a body obliged to reaffirm the need for its material presence and to find the means of doing it. What would a technician world be without human presence? If one projects oneself in such a situation, the question which arises is the one of the conditions of the rupture in the mode of human presence in the technical system. To be there without its body, an old dream of ubiquity, is this still presence? The path that is traced by the use of the drones (planes without pilot on board) prefigures the stakes raised by such a situation: firstly, space decorrelation; because the place unit is not necessary for two soldiers engaged in a theatre of war, they can be thousands of kilometres away from each other and still make the same war. Some, of course, are more exposed to the risk of dying on the battlefield than others who, in front of their screens, are incapable of invoking the honour of the combatant anymore. Being in the middle of the action, building it, modifying it but at the same time, not being there. The bodies of these new combatants are so much protected that they are absent, definitively evicted and, it seems, profoundly isolated. These assumptions will have to be checked in the years to come.

We will now review some of the postures of resistance of the body by drawing on observations and interviews realized with airline and fighter pilots.

First posture of resistance of the body: the testing of its robustness

The first form of resistance that will be questioned refers to a definition of resistance close to the one given in physical sciences: resistant is that which is opposed to an action or a movement. The act of resisting thus means to oppose a force to another in order not to endure the effects of an action. On the other hand, the resistance of a material means its hardness, its firmness and its solidity. In botany, the resistance of a plant is its aptitude to support bad weather. Applied to people, resistance means the physical quality by which one easily supports tiredness or deprivations and which makes it possible to support an intense or prolonged effort. A person can be resistant to tiredness, which is synonymous with endurance. It will be noticed that these characteristics are employed in a rather positive meaning. In the familiar language, a resistant character signifies firmness, force, solidity and tenacity. In this sense, resistant people are often perceived as being indefatigable, robust and solid.

The history of aviation can be read, from this point of view, like a history of resistant bodies or more precisely of bodies which were subjected to the resistance test. The first pilots had to face the cold and bad weather; they had to control machines which they had to dominate, physically and forcefully, for long periods of time. The difficulty of the task, its extreme dimension, and the hostility of the environment in which they were constrained to evolve, all contributed to and guaranteed the development of the exceptional character of the author of this exploit, it is the time of the pioneers, aces and heroes, the time of Guynemer, the time of air-mail, Saint-Exupéry, Mermoz and Guillaumet... The myth is built starting from the adventures of these aviators ; it reports their force, their endurance, their courage, and sometimes their sacrifices.

However, strength was not the only quality expected of a pilot at that time, it was necessary to feel the plane, to form a unit with it, to develop a particular sense of balance (piloting with the buttocks). It has previously been seen that the process of automation, already present in the early stages of aviation, took part in a progressive sidelining of the body, in particular in pure piloting. This process however did not prevent some body "prowesses" from being recognized, developed and maintained until today, they entertain a corpus of body techniques shared by pilots. These two opposite processes can be the subject of a common lecture; they should probably be understood together. Let us substantiate this with an example taken from the world of the fighter pilots in military aviation. The constraints of the flight often lead to suffering, one pilot reports: "*During combat, it hurts. Automation does not remove the body constraints, those of the third dimension. One can have a higher comfort level, but the constraints, they are clear.*" The continuation of the interview even shows that the automation, which compels the pilot to become a manager, can contribute to adding more importance to the constraints weighing down on the body: "*the combat will stop because the pilot arrives at his limits; the plane in maximum power keeps all its potential. I remember a combat here with young people where the winner was the one who held longest under 9 G. At the end of 40 or 50 seconds under 9 G, he had the same phenomenon of exhaustion that you can have at the end of a 15 km race.*" Suffering is indeed one of the components of this profession (which is also a profession of pleasure and these two terms must be thought of together) a suffering related to the performances of a technical device which constrain the human to always "play" with the limits of his physical capacities, to be at the edge of fainting without yielding to it, and to move his body without wounding it.

One understands from this extract of the interview that pilot training is the occasion to test, i.e. to experience (both in the sense of putting to the test and in terms of feeling) their resistance. To test for themselves (to feel the effects on their body, to know how to anticipate them, to evaluate them as far as possible because these effects are variable according to the days, or more exactly depending on the physical state of the pilot, it is thus necessary to be permanently attentive) and for the group because it is one of *the means of being distinguished, i.e. being positioned in the hierarchy of the group*. The question of testing the limits is pivotal to this process of resistance, since it is necessary to push the body as far as possible, including to the limits of danger (without succumbing because then the limit would be crossed). This practice is exhilarating and some pilots pursue it to perfect their training. It places the individuals in what Nicolas Dodier proposes to call the arenas of skill, inviting them to show, by their body ease, their capacity to carry out "technical prowesses". "The arenas of technical skills distribute the individuals between the poles of dexterity and awkwardness, strength and weakness, experience and ignorance, calm and nervousness, courage and pusillanimity" (Dodier, 1993, Pg. 120). In the account made by this pilot of the combat where the one who wins is the one who is able to resist longer than his "adversary" under a very high load factor, one can easily understand the spirit of competition, admiration and respect which animates the members of the group in these structuring moments for the collective, because everybody finds a place there.

From this first posture of resistance of the body, one will retain the primarily positive aspect, valorising for the human (aspect noticed at medical examinations during aptitude tests, or valorised by doing collective sports), and structuring for the group, symbolized by the clothing accessories of the fighter pilots —anti-g, helmet, maewest, etc— centred on male values (with the obligations of virility who are the corollaries).[14] The human being is, from this point of view, a fully fledged actor in the technical system. His robustness is incarnated in the figure of a young, sporty, durable body. His force and his endurance will make of him a team-member who one will be able to count on. The hunting squadrons completely integrate this dimension and the place which is granted to the body remains surprisingly visible. Fieldwork observations reveal that the implementation of an advanced sophisticated technical object, the most powerful plane in the French air force, expression of modernity, operates on a daily basis, almost at every moment in a reference to the oldest tradition (that of the knight and fencing combat). The credit which one continues to grant to the tradition aims at persuading members of the group to adhere to a vision of the world (a defensive ideology according to Ch. Dejours, 1993) which integrates the very difficult work conditions of this profession and which makes them, as well as the fear associated with this type of work, acceptable. The tradition engages all the members of the group, it binds them to each other in a concept of the world which it proposes and which is gradually revealed to the beginner during his initiation. The whole of the community regenerates itself with each ritual expression. Thus, the modernity expressed here in its most exacerbated forms (speed, advanced techniques, automatisms...) is accompanied by a social system of relations based on tradition (rites, initiations, oral transmission...). This traditional dimension of the life of hunting squadrons, must be taken into account, not only from a purely ethnographical, or even folklorist, point of view because it also brings to light the very heart of this profession: flight and its very special conditions. The tension between tradition and modernity is posed like a fundamental characteristic of this social organization; it even seems that one reinforces the other. The daily presence (but not trivialized) of danger feeds this social construction. One will notice in particular how much the tradition parades the body (bathed at baptisms, walked in a carriage from the plane to the squadron, patched...) as much as modernity makes it motionless, unobtrusive, constrained (G-suit, helmet, straps, very reduced space of the cockpit). Roland Barthes, in his *Mythologies*, had already highlighted this immobility when discussing what he called "the jet-man" (Batrthes, 1957).

The squadron building walls are adorned with paraphernalia, such as photographs, flags, medals, writings, objects or even animals, which chronicle its history. A set of rites is observed to mark the important moments of the life of the squadron: for instance, the arrival of a new member to the squadron, whether this is a young pilot who just left school or an experienced pilot who comes to take the command of the squadron, will be marked by his baptism, i.e. an immersion in a container filled with oil, water and a mixture of various other ingredients from the planes. The men themselves wear various escutcheons on their flight suit indicating: their squadron membership, of one of the three Flights; missions they have taken part in; and their war name (a nickname that they generally did not choose and which is given to them by the group). These escutcheons made of embroidered wire are often brightly coloured and are meant to brighten up the khaki flight suits. If the flight suit is indeed a uniform, the escutcheons worn like a coat of arms allow for distinctions to be made between the pilots in a very subtle way; the codes are often accessible only to the initiated (for example, the French flag escutcheon worn on the shoulder is donned only by those who have served on detachment abroad). One can note a relative freedom (and often humour) in the drawings and the words embroidered on the flashes which follow changes in fashion and are thus renewed rather frequently. The tradition (such is the word employed by the actors themselves) is omnipresent (one members of the squadron is appointed annually to this task) and even if it does not nowadays have the operational function it held before, it continues to regulate part of the social life of the squadron. In a very classical way, the tradition exerts a function of integration and adaptation: it is a question of giving rise to, maintaining, and even awakening the adhesion of the pilots and the navigators not only to the particular work conditions of this profession, but also to the moral values of the group —the ideology of hunting squadron members. This aspect is taken over by the collective through the different rituals which are accomplished. The place that the institution grants to the body in this process of tension between tradition and modernity reveals its importance and can therefore be considered as one of the keys to the comprehension of this world.

But the body of the fighter pilot is not constant: *"I know that I have a very good resistance to the load factor, now, I go on a drinking spree, a stag party for instance, and that happened to me. They make you fly the following day, I had asked not to fly, but they make you . . . At 2G, I had the impression that I was going to faint, whereas usually, I go up to 7 or 9"*. It is in particular to this inconstancy of the body that the constancy of the technique will oppose, its regularity, and its predictable mathematical resistance to tiredness. The mythical figure of the cyborg will come in the Sixties to illustrate this vision of a "perfect" body, i.e. robust, constant, obedient and soulless. The figure of Cyborg incarnates the project of a body-machine. The analysis proposed by Peter Galison underlines that two simultaneous movements allowed such a project to be carried out: on the one hand there is the process of dehumanization of the enemy, half human, half machine ("a cold-blooded machine like opponent"), which supposes the engagement of new means to fight him, and on the other hand the process of fusion binding the pilot to his machine up to the point of the elimination of any distinction between the two; thus, the components of the system are blended to the point of unity. Robustness, constancy and endurance thus characterize the cyborg, project of a half man, half machine "being" which is capable of overcoming physically and psychologically the most difficult situations, like war. Of an unequalled power, indefatigable and soulless, this warrior is designed with the measurement of his supposed enemy, and to defeat it (Galison, 1994).

Second posture of resistance of the body: body habits and the question of change

The question of the senses constituted a central issue in the controversy on cockpits automation which animated the world of civil aviation at the end of the 1980s. There were frequent confrontational exchanges of points of view (often antagonistic) between airline pilots and the engineers, which were on occasion quite brutal (Scardigli, 2001). One thus saw emerging, explicitly around the question of the use which the pilots make of their body and their senses, a reason for tension between an automation tending to sideline the body, and an appropriation of the tool which uses the body like one of the principal dynamic forces of its implementation. The type of presence of the body, or its negation, forms part of the definition and the role of the human in an automated environment. In order to illustrate this idea more concretely, let us listen to a captain describing his profession: *"The profession of a pilot consists of 85% thinking and 15% feeling. To feel, means to have muscular, physical and visual sensations; but, deprived of this 15%, one could not pilot"*. Indeed, one of the most virulent criticisms expressed by the airline pilots about the "new generation" planes is directly related to this question of the sensorial feed-back of information. The pilots complain of *"not feeling anything anymore"*, that the noise in the cockpit is so loud that it covers the sound of the engines, and they deplore the immobility of the throttles, which represents for them a loss of an essential information in comparison to "before" where they "felt" the throttles moving, going forward or backward, indicating a push or a reduction of the engines.

Let us start with an anecdote to illustrate the ways in which the body sometimes maintains the habits of sensitive piloting, i.e. with its senses. I observed, during that period, a number of newly qualified glass-cockpit airline pilots. At one of these observations, I remember an instructor who, during a simulation training session, was irritated by the captain who kept his hand on the throttles, even in automatic mode. Although the instructor pointed this out to the captain several times, the captains' hand always returned on the throttle, a little bit like when one drives a car, the hand is often in contact with the gear shift lever. This gesture, which made sense on a traditional plane (controlling the push of the engines by touch), was maintained by habit because undoubtedly his hand continued seeking this information which was, from then on, denied to him. Marcel Mauss, in *Les techniques du corps*, showed the influence of this education of the body and its effects on the way one walks and swims, and the way in which the English and French troops dug during the First World War (which resulted in an adaption of the spades, because each different way of using it fitted with a different type of tool). He even quotes in example that there are "positions of the hand, at rest, suitable or improper" (Mauss, 1980, Pg. 368). It appears from now on "improper" to pose the hand on the throttle, apart from some specific moments like take off for example, where it would be "improper" not to do it!

I was then working on the automation process and was collecting statements on the intellectualization of work symbolized by the "managing pilot". I however witnessed, at the same time, in a much more abstract and diverted way, the reappearance of the question of the body and the relearning of the senses in when contending with new machines. The pilots complained a great deal about what they then described as a "loss of the senses" because of the increasing presence of automatisms. Undoubtedly it is more justified, with hindsight, to talk about a loss of their sensory habits because the process of appropriation, whose premises were perceptible at the same time showed clearly that one cannot pilot without ones body and senses. The new practice is often set up in an opposition to the sensations (Kaufmann, 1997). The body becomes disobedient and the feeling of arduousness increases, the habit looses its force as long as the integration of the object is unable to return a feeling of comfort.

The hand of this pilot, which inevitably returned, without reason, on the throttle lever, is testimonial of the many ways in which bodies slipped into technical sets to ensure and optimize its functioning. The memory of the body led him to go on seeking in this gesture a significance the machine refused. The information concerning the push of the engines must now be read on screen. The expected immobility of the hand refers to the immobility of the body in general. To change one's practice is always difficult and Mauss describes the method he himself uses to swim, by spitting out water, and of his inability to modify his practice although he recognizes it as being useless. We will reconsider with the third posture another aspect of this difficulty of giving up one's sensations.

Is this a form of resistance to change? Perhaps, but the assertion contained in the concept of resistance to change, of a conservatism and a supposed indocility of the operators, taught us very few things in the end, except for the brutality of value judgments. Because a gesture is a technique of the body, it is carried by a set of representations and social practices which give it a sense, a sense shared by a social group or a professional community. The imposed transformation of habits thus supposes to question, at least partially, the significances which are associated to them. These gestures, like all body techniques, are natural in the eyes of their authors. This naturalness of gesture is central to its effectiveness. The dismissal from ones mind of its training is the guarantee of its success. The modification of this gesture is all the more delicate; and that is what the body expresses. Mauss uses the term of *habitus* to describe this process inviting us to treat the memory of the body like a specific memory, with its own mechanisms of conservation, stabilization and integration.

Third posture of resistance of the body: to know, in certain circumstances, not to listen to its senses

This third posture of resistance of the body might appear paradoxical given the issue I have tried to develop in this chapter. It personifies another facet of the body presence in technical systems: that of a weakened and inconstant presence which, in certain circumstances, one needs to protect. For a fighter pilot, to fight his own physical sensations is sometimes, in fact, a need to ensure his survival: "*It is necessary to be extremely wary of ones senses on a plane because the environment is hostile and you might have a completely wrong feeling about something which will kill you. For example, you have the feeling to be on ones back in the clouds and you are perfectly aligned, therefore you go back on the axis and at that moment you crash. That is called vertigo. It is an intellectual combat to keep the arm from moving, to prevent it from going to correct what the body feels. It is necessary to believe in the instruments such as one sees them there, one learns that very early on. That still happens to us, it is very simple to have vertigo when you fly in patrol with other planes very close to yours, you have vertigo in the clouds, you have the impression that the patrol is like that (he demonstrates with his hands), and, above all you must not move or change anything because you are likely either to crash, or to go into your leader. That happens in the clouds, under instrument flying. If the instruments say that the plane is in such a configuration, that the flight is ok, it is necessary to trust them. The first pilots who flew inside the clouds took along a goose on their aircraft because the goose always keeps its head upright."* The pilots are familiarized very early on with the dangers of sensory illusion. They know that it is sometimes necessary to distance themselves from the body which can betray them. Their expertise lies in the identification of these moments, the capacity to discriminate between the good and the bad feelings. Despite the high level of automation, planes today remain very sensitive to decidedly fine levels of body engagement. Humans go through an apprenticeship of "best" tactile contact with the machine, by virtue of which they discover for themselves the reactions of the plane to their actions and learn how to prevent such consequences. To test these feelings in ones body and to recognize them are two decisive moments of the process of appropriation. They take their signification in relation to each other and must be thought of together. Feeling means here the idea of the body forming a unit with the object, which then ceases to be an object as it belongs to the body of the pilot. To recognize is part of the process of *objectification*, of the sidelining of the object, and of the rebuilding of this distance (Merleau Ponty, showed the need of detaching the object of the body, to treat it like an external thing in order to establish a relation of comprehension with it). The concept of *career* used by Hughes and Becker in a very different field describes well this apprenticeship of the recognition of the feelings, as well as the classification between the good and the bad ones, and finally the way in which the criteria of classification are transmitted by the group, in a social context. The young pilot or navigator is trained by his peers within the hunting squadron. He gradually learns how to control increasingly complex situations during the flights, and follows at the same time essential training on the perception of effects. This concept of "career", in the interactionist sense of the word, clarifies this relearning of the senses which pilots accomplish. All this allows for a better understanding of the way in which fighter pilots, in a risky environment where the use of automated devices is generalized, manage to be successful in their tasks.

The body, what remains when all has been automated!

The body in automated worlds is a "discrete" body. One saw in the case of airline pilots that automation very gradually led to an obliteration of the body. This obliteration was concretely reflected in the way in which they were "constrained" to give up a type of skill linked to an intimate knowledge of the machine and which showed their capacity to break the opacity of the object. The body was thus in the posture of a presence "despite everything" (Yves Clot) which supported a "tactical" appropriation (de Certeau).

The posture of resistance of the body in the world of aviation is initially a protection against the dangers related to flying. Piloting training is mainly founded on the tracking of all the indices which could announce dangers and on the actions (an ensemble of gestures which it is necessary to be able to achieve as reflexes) which are appropriate to prevent it. We have seen that, in a fighter plane, the body can be "misled" and that it is thus necessary also to know how to evaluate this trap. To protect ones body from the risk of death (an accident) constitutes a first type of resistance of the body, basically related to the instinct of survival. To resist, signifies also to protect its body from wear, from useless tiredness (resistance as a technique of the body), important aspects if one is to last in this profession.

The few postures of resistances of the body reviewed above are very different from each other, they each illustrate a singular comprehension of the technical activity of piloting a plane and show the complexity of the body presence of humans on board. Because the human body is present (reactive, adaptive, and charged with collective memory, i.e. robust to the unexpected in ever changing situations), it is, by the force of the habit it is carrying inside, and by its analytical capacity which enables it to differentiate between good and bad feelings, still considered to be, for the moment, one of the best guarantors of safety in large technical systems like aeronautics. The posture of resistance of the body seems here in particular to be a collective means of controlling the desire for a total automation, an anomic desire, masterminded by the human being against himself.

Gaspare Mura, professor emeritus of the Pontifical Urbaniana University of Rome, is the professor of philosophical hermeneutics at the Pontifical Lateran University and of the Holy Cross University of Rome. He was the Director: of the International Institute for the Study of Atheism at the Pontifical Urbaniana University; of the Urbaniana University Press and of the journal of philosophy and theology Euntes Docete. Prof. Mura was also literary editor of the Citt Nuova editors of Rome, for whom he edited important works of philosophy and directed numerous collections and series of philosophical works. His brilliant editorial carrier performed was awarded in 2007 by the papal honor of Commentator of the order of St. Sylvester. He is currently President of the Academy of Humanities and Social Sciences(A. S. U. S.) in Rome, which is founded by several professors with the objective to promote interdisciplinary dialogue between the humanities, social sciences, empirical sciences and the science of religious convictions. He is also a consulter to the Pontifical Council for Culture. He asset numerous scientific publications of philosophical and theological and his own philosophical thought known internationally as the veritative hermeneutics. Among his writings: Hermeneutics and truth, history and problems of philosophy of interpretation (1990, 1992), (this work received the Culture Award of the Presidency of the Council of Ministers); Think the word, for a philosophy of the encounter (2001); Introduction to hermeneutics of the truth (2005); The religious phenomenon today: tradition, change, denial (2001); Holy Text and religions, compared Hermeneutic (2006).

Notes

1. Following Alain Gras who introduced the Large Technical Systems theory in France. Alain Gras, *Grandeur et dependence*, Paris, PUF, 1993, *Les macro systèmes techniques*, Paris, PUF, 1997. See also Berward Joerges, Large technical systems: Concept and issues, in Renate Mayntz & Tom Hughes (eds.), *The development of large technical systems*. Frankfurt, Campus, 9-36, 1988. Tom Hughes, *Network of power*, Baltimore, J. Hopkins Univ. Press, 1983.
2. Caroline Moricot, *Des avions et des hommes, socio-anthropologie des pilotes de ligne face à l'automatisation des avions*, Lille, Presses Universitaires du Septentrion, 1997. Caroline Moricot, Gérard Dubey, *La polyvalence du Rafale ou l'objet total. La relation entre une technologie nouvelle et ses utilisateurs*, Paris, Les documents du Cessd, 2005. Caroline Moricot et Gérard Dubey, *Pilote de chasse: tradition et modernité à l'épreuve de la polyvalence des avions*, In: Franҫis Gresles, (dir), *Sociologie du milieu militaire*, Paris, L'Hamattan, 2005. The fieldwork has consisted in observations and inteviews. The analysis related to airliners pilots are based on 70 interviews and about the same number of observations in flight (between 1990 and 1994). A quantitative inquirery has also been conducted during the same period: 900 pilots answered the questionnaire. The analysis related to military pilots are based on 30 interviews of about one to two hours each. The observations was realized during the preparation and the debriefing of the missions.
3. This practice of the historical, fictional and especially anthropological detour was theorizated by Georges Balandier, *Le Détour, pouvoir et modernité*, Paris, Fayard, 1985.
4. Octave Chanute, *Progress in flying machines*, 1894, extraits reproduits dans *Les cahiers de science et vie*, n° 1, février 1991, Pg. 60.
5. This controversy was expressed in particular in the design of the handle, i.e. the interface with the control surfaces: its size, its place in the cockpit, its capacity to restore the movements of the plane, etc made, during the beginning of the 90's, the object of interminable discussions in the civil aviation community.
6. All large technical systems were not unanimous on this point. For example, the history of railway shows different options with regard to safety. Gilles Mardon relates that, on the contrary, in the railway circles, "the engineers were fiercely opposed to the automatisms in the name of safety"; adding that in the beginning, "railway safety rests completely on the human being—sight, hearing, vigilance, endurance—and on the operational rules which distribute tasks and responsibilities". That seems very surprising when knowing the continuation of the story and the advanced automation of the TGV cockpits where, from now on, the very reduced role of the human, consists in, when all is normal, expressing his presence to the automatism while pressing on a pedal... cf. Gilles Mardon, "L'enjeu de la sécurité ferroviaire", in *Culture Technique*, n. 11, sept. 1983, Pg. 273.
7. Quoted by Alain Gras et Sophie L. Poirot-Delpech, *Grandeur et dépendance*, Paris P.U.F., 1993, Pg. 172.
8. "A decisive step in the automatisms was crossed with Caravelle, the first plane able to land in an entirely automatic way in Third Category conditions. Therefore, it had been necessary to replace the pilot by automatisms for delicate piloting in approach, for the operation of rounded and touching wheels, and for the behaviour speed by control of the push. Thus, it was proven that the automatisms made in a repetitive way and in safety conditions better than the pilots, which the latters admitted with reserve. A part of them did not accept this seizure of the automatisms on what they considered as their prerogative: the landing in difficult conditions." Jean Pinet, *Automatismes et*

facteur humain dans l'aviation de transport, communication au Mastère Navigabilité des Aéronefs, ENAC, février 1995, Pg. 4. This example of the automatic landing calls several remarks: for the first time it comes to a complete and complex sequence of flight proceeding under the monitoring of the pilot, but without his own intervention; as Jean Pinet underlines a little further in the text, the realization of this performance required of the crew "a perfect respect of a very elaborate distribution of the tasks and a perfect discipline of execution". This question of the procedures, the coordination of the crew, its communication, the principle of cross-check, etc still remains today closely related to that of automation; finally this example underlines the macro-systemic aspect of automation: indeed, it is possible to carry out an automatic landing of third category only if, of course, the plane is equipped with this instrument, the pilot is qualified for this type of approach, and also provided that the runway possess the necessary equipment (which is not the case of all the French airports). In other words, this technical progress could be carried out only jointly with the realization of infrastructures on the ground. It is what Alain Gras and Sophie L. Poirot-Delpech try to show in detail in the two chapters of "Grandeur et dependence" (1993) devoted to air transports.

9. In the results of the quantitative part of my research, it could be noted that 59% of the pilots feared to lose part of their skill because of the automatisms (800 French airline pilots had answered this survey carried out in spring 1993, the complete results can be consulted in Caroline Moricot, *Des avions et des Hommes*, Presses Universitaire du Septentrion, 1997). These results are completely comparable with those obtained by the English researchers of the Royal Air Force Institute of Aviation Medicine: 40% of the pilots declared that their manoeuvrable skill had been degraded because of automatisms (Mr. James, A. Mc Clumpha, R. Green, P. Wilson and A. Belyavin, *Pilots attitudes to flight deck automation*, R.A.F. Institute of Medicine, 1990-91). In the same way, E. Wiener's researches in 1989 showed that nearly 50% of pilots declared being concerned with a possible loss of their skill due to too much automatism. In addition, in the same investigation, nearly 90% prefering to take control of part of each flight manually, to maintain their skill. In his observations, Wiener has notice, like us, that pilots disconnect the automatisms and retake the plane manually, in certain phases of the flight (E. L. Wiener, *Human factors of advanced technology ("glass-cockpit") aircraft transport*, Ames Research Center, NASA, June 1989).
10. Edgar Allan Poe, *Histoires grotesques et sérieuses*, Paris, Gallimard coll. Folio, 1978, Pg. 104–142.
11. A hundred and fifty years later, the machine will really win the chess game!
12. The reflection on the relationship between machines and intelligent human behaviour developed in the 1940s. Philippe Breton relates the way in which the explanation of these associations was going to be sought in the idea that "the behaviour" could be an independent reality. It was first of all a question *of legitimating the comparison between human and machine*. Three researchers (N. R. Wiener, A. Rosenblueth and J. Bigelow) had developed what they called the "behavioural study method". "This method consisted in privileging, in a study of any kind of phenomenon, natural or artificial, its behaviour, that is to say, the modifications that this phenomenon undergoes because of its relationship with its environment". Because for the behavioural method, which is modelled on mathematics, "there was no other reality than the one of the relations between the phenomena". (...) "While concentrating on the behaviour of the objects, independently of the physical nature of the bodies which compose them, one made it possible to make a comparison between any kind of "objects", and, in particular, between the human and the machine. Philippe Breton, *Une histoire de l'informatique*, Paris, Seuil, 1990, Pg. 142. See also Céline Lafontaine, *L'empire cybernétique, Des machines à penser à la pensée machine*, Paris, Seuil, 2004. Maurice

Merleau-Ponty really understood this project, contained in cybernetics, of negation of a specificity of the living which made it possible for the machines to become its equivalent: "the artifice is denied and thought like nature", *La nature, notes de cours du Collège de France*, Paris, Seuil, 1995, Pg. 215.

13. Ivan Illich, *La perte des sens*, Paris, Fayard, 2004. Daniel Cerezuelle proposes a very lighting analysis of the thought of Illich, Charbonneau and Ellul in a paper written in 2004, entitled *La technnique et la chair*, http:// agora.qc.ca.
14. Cf. the analysis of Pierre Boudieu, *La domination masculine*, Paris, Seuil, 1998. The recent presence of women pilot in squadrons modifies this obligation.

Bibliography

Baladier G. (1985), *Le Détour, pouvoir et modernité.* Paris, Fayard.
Barthes R. (1957), *Mythologies.* Paris, Seuil.
Batteau A. (2001), *The Anthropology of Aviation and Flight Safey.* Human Organization, Vol. 60, n.3.
Bourdieu P. (1998), *La domination masculine.* Paris, Seuil.
Bouvier P.(2000), *La socio-anthropologie.* Armand Colin.
Breton P. (1990), *Une histoire de l'informatique.* Paris, Seuil.
Cerezuelle D. (2004), *La technnique et la chair.* http:// agora.qc.ca.
de Certeau M. (1990), *L'invention du quotidien.* Paris, Gallimard.
Chadeau E. (1991), *Voisin contre Wright ou l'analyse d'un duel.* In: *Les cahiers de Science et Vie, les grandes controverses scientifiques*, n. 1, février 1991, Pg. 60.
Chanute O. (1991), *Progress in flying machines*, 1894, extraits reproduits dans Les cahiers de science et vie, n. 1, février 1991, Pg. 60.
Clot Y. (1995), *Le travail sans l'homme.* Paris, La Découverte.
Collins H. (1992), *Experts artificiels.* Paris, Seuil (Artificial Experts, MIT Press).
Damasio A. R. (1994), *L'erreur de Descartes, la raison des émotions.* Paris, Editions Odile Jacob.
Dejours C. (1993), *Travail, usure mentale.* Paris, Bayard.
Dodier N. (1993), *Les arènes des habiletés techniques.* In: *Les objets dans l'action*, Raison Pratiques, n. 4.
Edwards E. (1988), *Introductory overview* . In: E. L. Wiener and David C. Nagel (ed.), *Human Factors un aviation*, Academic Press.
Galison P. (1994), *The ontology of the Enemy: Norbert Wiener and the Cybernetics Vision.* Critical Inquiry, n. 21.
Gras A. et Poirot-Delpech S. (1993), *Grandeur et dependence.* Paris, PUF.
Gras A. (1997), *Les macro systèmes techniques.* Paris, PUF.
Hughes T. (1983), *Network of power.* Baltimore, J. Hopkins Univ. Press.
Illich I. (2004), *La perte des sens.* Paris, Fayard.
James M., Mc Clumpha A., Green R., Wilson P. and Belyavin A.(1990), *Pilots attitudes to flight deck automation.* R.A.F. Institute of Medicine.
Joerges B. (1988), *Large technical systems: Concept and issues.* In: R. Mayntz & T. Hughes (eds.), *The development of large technical systems.* Frankfurt, Campus.
Kaufmann J.-C. (1997), *Le cœur à l'ouvrage.* Paris, Nathan.
Lafontaine C. (2004), *L'empire cybernétique, Des machines à penser à la pensée machine.* Paris, Seuil.
Le Breton D. (1990), *Anthropologie du corps et modernité*, Paris, PUF.
Mardon G. (1983), *L'enjeu de la sécurité ferroviaire.* In: *Culture Technique*, n. 11.

Mauss M. (1980), *Les techniques du corps.* In: *Sociologie et Anthropologie*, Paris, PUF. (1ère édition du texte en 1935).

Merleau-Ponty M. (1995), *La nature, notes de cours du Collège de France.* Paris, Seuil.

Moricot C. (1997), *Des avions et des hommes, socio-anthropologie des pilotes de ligne face à l'automatisation des avions.* Lille, Presses Universitaires du Septentrion.

Moricot C. et Dubey G. (2005), *La polyvalence du Rafale ou l'objet total. La relation entre une technologie nouvelle et ses utilisateurs.* Paris, Les documents du CESSD.

Moricot C. et Dubey G. (2005), *Pilote de chasse : tradition et modernité à l'épreuve de la polyvalence des avions.* In: Franşois GRESLES, (dir), *Sociologie du milieu militaire*, Paris, L'Hamattan.

Petit E. (1993), *Histoire de l'aviation.* Paris, PUF, Que sais-je ?

Poe E. A. (1978), *Histoires grotesques et sérieuses.* Paris, Gallimard, coll. Folio.

Sarter N. B. and Woods D. D. (1995), *Strong, silent and out of the loop: properties of advanced (cockpit) automation and their impact on human-automation interaction.* Cognitive systems engineering laboratory, The Ohio State University, Columbus.

Sauvageot A. (2003), *L'épreuve des sens.* Paris, PUF.

Scardigli V. (2001), *Un anthropologue chez les automates.* Paris, PUF.

Simon H. A. (1986), *Sciences de l'intelligence, sciences de l'artificiel.* In: A. Demailly et J. L. Le Moigne Presses Universitaires de Lyon.

Simondon G. (1989), *Du mode d'existence des objets techniques.* Paris, Aubier, (1ère édition 1958).

Thevenot L. (1994), *Objets en société ou suivre les choses dans tous leurs états.* Alliage, n. 20-21.

Wiener E. L. (1989), *Human factors of advanced technology ("glass-cockpit") aircraft transport.* Ames Research Center, NASA.

Wiener N. (1962), *Cybernétique et société, l'usage humain des êtres humains.* Paris, Editions 10/18.

Chapter 6
Social bodies and locative technologies

Gianni Corino

Abstract

This chapter reflects on the emerging field of locative technologies and attempts to place it within a theoretical framework within social and urban studies, human computer interaction and through the presentation of recent artistic practices. The analysis looks how socio-spatial systems such as the "agora" (and other sites of social interaction) played a role in the constitution and development of society. Findings will be evaluated against the emergence of contemporary public spaces and their relation with the social that through the proliferation of mobile, wireless and ubiquitous computing offer new platforms for public discourse and innovation in the application of update new media technologies. Critically framing the implications of location awareness and context based technologies, the author will benchmark the findings with social approaches that re-address the focus and meanings of the "social" and "space" using Manuel Castells' and Bruno Latour's concepts and definitions. The question underlying the chapter is: what relations are there between locative technologies and the social body? Examples of contemporary artistic practices will be used as evidence of this discourse, drawn in particular from the artworks of artist such as Beatriz da Costa, or in projects like GPS drawings.

This chapter reflects upon the emerging field of locative technologies and attempts to place them within a theoretical framework of social, cultural, and urban studies. It can be considered this against the background of the recent proliferation of mobile, wireless and location awareness technologies that have formed a new platform for *public discourse*. A further attempt is to contextualise locative technologies with the field of human-computer interaction.

To develop the argument, I first present a history and definition of locative technologies, and consider their main purpose in the market and their use in the setting of art. Critically framing the implications of location awareness and context-based technologies, I also benchmark the findings with social approaches that readdress the focus and meanings of the social and space using concepts and definitions from scholars of different contemporary sociological areas in relation to the concept of *public sphere*. Following this, the chapter describes a contemporary digital art practice known as locative art, illustrating the previous discussion.

Major aim at the end is to define them more precisely in order to move towards understanding the social implications of such technologies.

This chapter arouses two fundamental arguments: first, emerging social bodies; second, the relationship between locative technologies and social bodies, their nature and, as I said earlier, how they can constitute a *public sphere.*

The main concern of the discussion here is to show how the *public sphere*, far from being something defined permanently, can be *embedded in, produces* and to some extent *embodied* by places, objects and people. These three elements enter as social bodies or agents in the public domain of social, political, environmental, ecological, and cultural debates and policies. This contribution could be considered a *re-negotiation* or adaptation of the original concept *public sphere* into a new context where new agents are involved and new strategies of political and cultural actions arise, as some locative art practices and interventions show. According to the "classical" sociological definition of *public sphere* two elements are important to know: first, the participants and second, the terms of their involvement to the public discourse.

In Jurgen Habermas's original concept of the *public sphere* and further revisions by scholars belonging to different disciplines - media studies, cultural studies, philosophy - something never questioned is the kind of participants of the discussion, the ones who makes the *public sphere*: people with a critical understanding, reasoning and communication capabilities. In this chapter it is argued that from now on other participants should be included, social bodies that could be formed of humans, things or places. Not only do locative technologies contextualise the political, cultural or general societal issues in respect of their origins but also with wireless communications they give agency to objects, people and places in a way that allows them to actively and directly participate in the public discourse.

What are locative technologies? Anyone, anywhere, anything

Locative technologies are a complex field of different technologies which boundaries are not strictly defined. According to Nicholas Nova, an engineer and researcher among the earlier who pointed out the prominence of location awareness in our contemporary society, they are:

> those that use geographically defined or location-based information in order to create new types of content or services, through their ability to locate, track, map, visualise and attach information to physical location. Locative technologies deals with physical location as well other contextual cues. (Nova, 2004)

Locative technologies provide a bridge between two worlds, stressing the tension between digital world of data and physical world. As a result they open unpredictable developments in cultural, social and political arenas.

A new digital morphology overlaps the real one - a *data sphere* enacted wirelessly by information and communication technologies. This hybrid morphology is the one I'm interested to map and explore.

In the area of Human Computer Interaction the field to which this approach belongs is called Ubiquitous Computing. In the words of Martin Weiser, the evangelist of this computation philosophy, "ubiquitous computing is the age of calm technology, when technology recedes into the background of our lives". As the name suggests, it argues for a computational ubiquity that provides a more "natural" way to interact between human-computer, utilising notions of how we interact with the physical world. This line of thought provides a different approach to computing science, based on the idea that computational processing power can be embedded in the world, in places and objects instead merely in a computer. *Ubiquitous computing* is about the ubiquity and invisibility of information; it is therefore harder to map it and also to track back the traces left in our social life.

As part of this approach, locative technologies enable information to be tied to geographical space. This reconciliation allows not only tracking the implications but also to contextualise the information in its original context. The challenge in notions of space and place in a mixed reality ecosystem is enhanced and enacted by mobile and wireless communication technologies; the idea is that space and place are again meaningful in this new technological perspective. The mapping of this digital morphology could be considered a new frontier enabled by locative technologies.

Through their ability to track, locate and collect data, these technologies create a memory, physically or virtually, attached to the place, object or body. They can then communicate, elaborate, interact with the data and, at last, visualise the information in different representational ways. All these capabilities work in the real-time.

Locative technologies include not only specific location-positioning tools like GPS (Global Positioning System)[1] but also wireless communication technologies typical of digital mobility. The wireless cloud around us includes telecommunication system at different geographical scale: global, local and personal. The global level is the one used by mobile phones like GSM (Global Systems for Mobile Communications), GPRS (General Packet Radio Service), UMTS (Universal Mobile Telecommunications System), along with that there is the local level of the wireless internet with technologies such as WI-FI (Wireless Fidelity), WiMAX (Worldwide Interoperability for Microwave Access) and finally the personal level with Bluetooth and RFID (Radio Frequency Identification), along with ZigBee.

These are a group of heterogeneous technologies used in combination by researchers, artists and within industry, more and more frequently appearing as part of our everyday life. Except GPS, which is a pure locating and tracking technology, all the others above are communication technologies that by means of their technological infostructure allow also localisation and tracking in space, in short or long range, outdoor or indoor and at different resolutions. The common qualities of these technologies are their wireless nature and the fact that are often embedded in numerous communication devices (phones, computers, PDAs...), in places (such as airports, libraries, shopping malls...) and in objects (books, food products, clothing...).

Most modern mobile devices are able to access the Internet allowing information to be stored and retrieved from remote databases, as well as they start to embed location aware capabilities.

On the other side of localisation and tracking technologies stands the information system enabling the visualisation and mapping of the information. Visual mapping technologies, in the jargon of the trade, are known as GIS (Geographical Information Systems)[2] and are used to visualise and map space and data layer of geographical information.

A way to classify the different location technologies has been proposed by Bruce Sterling, a science fiction writer and design journalist, who, moving from GPS the Global Positioning System as a reference, defines the rest of these technologies (Wi-Fi, Bluetooth, RFID, ZigBee) as Local Positioning System (LPS). This is an umbrella term he has conceived for short-range technologies that could communicate among each other, with people and environments. The Local Positioning Systems are at the basis of what has been defined as the *Internet of Thing*.

The concept of *Internet of Things*[3], first introduced by the 2005 Report of the ITU (International Telecommunication Union), summarises and makes evident "the vision of a fully interactive and responsive network environment" enabled by digital technologies. This is not a prediction of science-fiction writers or the creation of a digital artist or the vision of a new media sociologist, but is rather an image based upon market analysis. Employees ID badges, food and goods in supermarket chains, sky passes, books catalogued in libraries are just few examples of presence of *Internet of things* in everyday life.

The *Internet of Things* not only looks at locative technologies but also at the convergence of mobile and wireless communication systems with location awareness technologies—a market, according to the ITU report, that by 2007 had penetrated 33.5% of the overall population of the world, and developed revenues of USD 1.5-1.8 billion by 2004 for RFID devices.

In 2005, Bruce Sterling wrote a foreseeing book called 'Shaping Things", where he reflected on the historical trajectory of objects and on the role of design and designer in our society. The evolutionary trajectory starts as the artefact moves to machine, to product, to gizmo and end up with the neologism *spime*.

"[A] 'spime' is something that, through pervasive RFID and GPS tracking, can track its history of use and interact with the world." (Sterling, 2005)

The qualities of a *spime* shown in comparison with the previous objects are its memory and its agency. Similar evidences of changes had been noted by others researchers and writers like the journalist and sociologist Howard Rheingold. In his book "Smart mobs" (2003) there is a chapter dedicated to "The Era of Sentient Things," where the dialectic between atoms and bits is discussed through describing ongoing researches in the field of augmented reality and in identifying the *sentient things*, something very close to the concept of *spime* by Sterling.

Julian Bleecker, the director of the Near Futures Lab at the University of California, has set last variant of the same phenomenon in 2007. He reviewed the concepts of *Internet of Things*, Sterling's *spime* and Rheingold's *sentient thing* introducing the idea of *blogject*. In *A Manifesto for Networked Objects — Cohabiting with Pigeons, Arphids and Aibos in the Internet of Things*, Bleecker shows how *blogjects* are interconnected together and communicate with the people.

Bleecker described the *blogjects* according to three characteristics:

- Blogjects track and trace where they are and where they've been;
- Blogjects have self-contained (embedded) histories of their encounters and experiences
- Blogjects always have some form of agency — they can foment action and participate; they have an assertive voice within the social web.

The first characteristic stresses the idea of spatial awareness, or where objects are in a space and time. The second points to the history of their social life and the context of usage. Like in spime object extend their nature by means of those new properties but what I want to point out from the quotation is their active social role. The third refers to which kind of agency they can have. Different terminology, same approach to computation in a ubiquitous and pervasive way.

It is the reference to an assertive voice, seen as a peculiar quality of their agency, which is interesting because it could affect or stress the basic definition of sociological concept of the *public sphere*. The next section expands upon this assertion by explaining why locative technologies, the Internet of Things, *Sentient Things, Spimes or Blogjects* are becoming ever more relevant in our society.

Public sphere and social bodies

The concept of *public sphere* has been such a powerful idea that over the years it has caught the imagination and drawn the attention of scholars from many different disciplines. It might be useful at this point to look at Habermas original definition before moving on with the analysis. I'm referring to this concept as a key reference point and framework to trace the implication of the locative technologies. I join the debate about the *public sphere*, not to engage critically with the basic definition or evidence but referring to it as a key term to express the public domain, I therefore refer mainly to some recent interpretation of it by other scholars.

> By "public sphere", we mean first of all a domain of our social life in which such a thing as public opinion can be formed. Access to the public sphere is open in principle to all citizens. A portion of the public sphere is constituted in every conversation in which private persons come together to form a public. They are then acting neither as business or professional people conducting their private affairs, nor as legal consociates subject to the legal regulations of a state bureaucracy and obligated to obedience. Citizens act as a public when they deal with matters of general interest without being subject to coercion; thus with the guarantee that they may assemble and unite freely, and express and publicize their opinions freely (Habermas, 1989).

The *public sphere*, as it is stated in this extract from 'The structural transformation of the public sphere : an inquiry into a category of bourgeois society", should be *open* to all *citizens* who want to act in the *public*. When they *come together*, they form a single body or, as Habermas said in a later intervention, a *public body* (Habermas, 1989).

Habermas, in another passage, defined the *public sphere* as a *body of individuals* and as a *discourse*. From the perspective outlined here, the term body refers both to physiological level of the body, and also to process of grouping that creates class, community. Agency, citizen empowerment, and participation are matters of the *public sphere*.

Can this argument be reinvigorates, renegotiated and reconfigures in light of shifts occasioned by computational and technological changes?

The technological issue is at the core of the renegotiation of the public domain. In "Rethinking the Public Sphere: a Contribution to the Critique Actually Existing Democracy", Nancy Fraser argued that Habermas view is no longer up-to-date since it fails to reflect contemporary socio-technical conditions, especially if we consider how ICTs (Information and Communication Technologies) have changed how people interact.

In a way, also Habermas later reviewed his former definition asserting that "today newspaper and magazines, radio and television are the media of the public sphere". (Habermas, 1964), attributing relevance to the medium and so to the technology used in shaping the public domain.

Starting at this very point, new digital media and in particular the Internet has reshaped the dimension and structure of social groups, towards a decentralised, globally networked interaction between individuals. Each media as Habermas confirmed affect the way we act and interact together so the way the public sphere is enacted.

In the early years of Internet Rheingold was involved in the same attempt of understanding how the public sphere is produce through that medium and he stated that:

> virtual communities are social aggregations that emerge from the Net when enough people carry on ... public discussions long enough, with sufficient human feeling, to form webs of personal relationships in cyberspace (Rheingold, 1994).

Virtual communities attempt to discuss political, societal, and cultural issues; and, rather than Habermas' focus on class distinctions and face-to-face interactions, they draw our attention to community itself–fragmented virtual communities, global networks of people "coming together" by means of the Computer Mediated Communication.

When we think of the kind of sociality that mobile communications facilitate, again Rheingold is the one who anticipated the mobile scenario as he did before with Internet with the definition of *smart mobs*.

> Smart mobs consist of people who are able to act in concert even if they don't know each other. The people who make up smart mobs cooperate in ways never before possible because they carry devices that possess both communication and computing capabilities (Rheingold, 2002).

Again, a dynamic, nomadic sociality comes into focus, a hybrid structure of social interaction, face-to-face and virtual, not merely virtual community alone but social network, not just class but mob, a dynamic and always changing agglomerate of people made aware through and by the technologies they use. An example, almost a classic, is the collective action done via SMS messages in order to coordinate a public revolt in the Philippines that eventually brought to the President Estrada resignation.

The first new social body as a public body is therefore the *smart mobs.*

In a mixed reality ecosystem the *body of individuals* loosens its borders and the individual shares agency with other agents such as places and objects. Until now, "objects" and "things" have been conspicuously absent from the field of agency.

> 'Things" in the pervasive Internet, will become first-class citizens with which we will interact and communicate. Things will have to be taken into account as they assume the role of socially relevant actors and strong-willed agents that create social capital and reconfigure the ways in which we live within and move about physical space (Bleecker, 2006).

Julian Bleecker, in order to describe "things" in the pervasive Internet he created of the term *blogject* to stress the idea that interconnected "things" can create their own blog. A blog is a very personal and sometimes private form of journal, a creation of human beings and nowadays a shared form of *public discourse*. Blogs are certainly one the of the most widely used forms of public opinion, openly accessible in the online environment, disseminating thoughts, ideas and to some extent making culture; they constitute the closest form of *public sphere* possible to find in an online environment. In the USA in 2007 more than 12 million people out of a total Internet population of 147 million Americans maintained a blog; according to data published by the Blog World Expo, 8% of the American internet population produces a blog. They make public their ideas, emotions, opinion, and they watch, comment on, and attempt to control what is occurring in the world.

The ability or the possibility to express freely and directly is the key elements of a conversation and on Internet this happens through a blog. When object can take this position they actually achieve a role usually played just by human beings. Bleecker claims that things eventually will become "first class citizens," able, through their blogging, to act and participate in public discourse. In this way, objects can be thought to aspire to citizenship. It is this that creates the second social , the *blogject.*

If people and objects play their role in the public sphere, then it may be worth considering how and where their interactions take place. Stephen Graham, an urban sociologist, claims "we are moving towards a society of enacted environments" (Graham, 2004), while Rheingold uses the term *ambient intelligence* to describe a similar concept. Both ideas, considered in relation to the digital ecosystem, pertain to information that is attached to a place and can be accessed through mobile wireless technologies. "Cities" (Graham, 2004), writes Castells (Castells,

2004), "do not disappear in the virtual networks. But they are transformed by the interface between electronic communication and physical interaction, by the combination of networks and places."

When a place, much like an object, can create memory (contextual data and relational data), it communicates and interacts and has the agency to embed and embody issues of public interest. Here there is the third social body, smart places.

Actors of the discourse, the social bodies, can be considered people, objects and places.

This is the reality of social bodies: networked individuals, objects and places that interact and communicate. As citizens, the social bodies are political, ecological and environmental, social and cultural entities. This should be the starting point of a discussion of the meaning of the social and its relevance for political, societal and cultural analysis, where locative technologies have begun to emerge and allow to map a territory that is otherwise disappearing into the background of our lives.

The enactments of social bodies: locative art/locative media art

In the title of this section two new terms are introduced and used as alternatives: locative media and locative art. Locative art or locative media are practices that make a creative use of the all range of wireless and location awareness technologies discussed until now. In this section I use alternatively the terms locative art and locative media because there is no real difference between the twos and most of the time they are used as synonymous. An element of distinction could e the kind of outcomes, more artistic in the first case more industry or business oriented in the second case. The second term differs slightly from the term of locative technologies used earlier is central to the topic of this chapter as exemplification of social bodies enacted by locative technologies. As with "locative technologies," locative art, referred sometimes as locative media, is a very broadly definition. Toward these ends, locative art makes use of the breadth of technologies introduced earlier in the chapter.

Locative art has another relevant characteristic: it usually privileges the urban context. Recent locative art pieces explore several relations among space, body, object, information and technology, sometimes in continuity with the history of art, sometimes in dialogue with scientific data or by exploring new technology as a potential means of expression. Examples are given below.

So far practice-based research has proved useful in investigating the implications of certain technologies, to explore their possible uses and to track their social, political and cultural effects. This entails enactments of possibilities, of the not-yet-there. Artistic and creative practices help to describe the framework in which social bodies manifest themselves in new public sphere. Art, in this case, points out, explores and maps the boundaries between physical and digital information spaces, at the frontiers of data spaces.

Locative art, as with other forms of practice in art history, has the ability to show in a creative and personal view those possibilities and potentialities. To some extent locative art carries on the utopian effort of public art. It creates art that embraces roles typically pertinent to other spheres of human action, sometimes embracing an utopian view other times an activist role, like being directly involved in political, social and cultural action.

The first survey and categorisation of locative art was produced by Drew Hemment (Hemment, 2004). Hemmet claimed that there are many sub-categories within locative art: moblog mapping, bio-mapping, locative mobile gaming, geo-graffiti or geo annotations, audio-graffiti or audio-annotation, locative performance, GPS drawing. This list could now be updated to with those projects mapped by Julian Bleecker in his "Manifesto for Networked Objects", since the art practices he termed blogjects are undoubtedly part of the locative art or locative media trend.

Many projects belong to the sub-category of geoannotation, which includes further subcategories such as mobile blogs (moblogs), audio and video mapping; key applications in this area are Geograffity, GeoNotes, Urban tapetstry, Area Code, SoundMapping and [murmur].

In geoannotation projects digital information is linked or attached to a specific place and can be left and retrieved by users by means of mobile devices enabled with location awareness features. Walking in the proximity of a location in which the information has been attached allows the retrieval of the content. Masaki Fujihata's 1992 work entitled "*Field-Works*" could be considered as an ancestor of geoannotation. It provided "a series of projects which reconstruct collective memories in cyberspace as a kind of video archive by using position data captured by GPS and moving images captured by Video".[4]

"[murmur] is one example of the use of geoannotation and is defined as *a non-profit, community-based, locative oral storytelling project.* Started in Toronto but then resituated to other large cities (Dublin, Edinburgh, San Paolo, among others),

> [murmur] lets people hear history as it was lived, complementing the "official" history of the city and its neighbourhoods. In experiencing [murmur], people can develop a new intimacy with places: once a story is attached to a place, that location has a new significance and can no longer be arbitrarily overlooked or forgotten.
>
> Once heard, these stories can change the way people think about that place and the city at large. These are the stories that make up our urban identity, but they are kept inside of the heads of the people who live here. [murmur] brings that important archive out onto the streets, for all to hear and experience.[5]

The aim these projects, as Hemmet notes, is to allow a "collaborative authoring or collective memory" (Hemment, 2004) of places. In theory, this develops not only personal or micro narratives related to a specific place but strengthens a sense of community in the residents who collaboratively contribute to the history of the place. The projects create a history carved in the digital material of the location. Pursuing this, they explore ways in which a collective memory can be settled and attached to a place, adding a timeline to the space beyond its physicality. The place reveals itself through the recollection of inhabitants and visitors who contribute privately or publicly to construct its history. It allows the public to experience familiar places from alternative perspectives. The work has a decentralising effect because it allows social spaces to empower communities, and it opens them to private narratives.

Stressing the idea of memory embedded into a place this project gives an historical and narrative agency to space that in my opinion announce the growth, the social body of smart places.

A different kind of approach can be found in mapping projects focusing of the use of GPS and GIS technology to map movements in the space. *GPS drawings* by Jeremy Wood are illustrations created by users through GPS. They are documents of journeys captured using Global Positioning System (GPS) technology. Travels are orchestrated and improvised generating maps of trajectories.[6] Apart from the original outcome of Jeremy Wood's artistic piece, I'm here forced to imagine how relevant non-representative maps are becoming in revealing the uses of space. Further development has to be expected from this GPS graffiti. In other similar (albeit more complex) works, like those by Teri Rueb and Andrea Wollensak, these implications are more clearly investigated. In Rueb's "The Choreography of Everyday Movement", everyday movements of dancers are tracked in a large city. The resulting outcome, in the words of the author, envisions, as a topographical mapping, the culturally inscribed nature of our everyday travels. Using GPS, the project seeks to render visible our movement through the city, revealing socio-political and poetic patterns of traffic flow.[7]

Works such as these provide a direct way to access the urban body and the everyday private consumption of urban space. These provide data that directly reveals emotion, social interaction, cultural and political actions, inscribed in the real world like tracks in the snow. As Wilson writes, discussing Andrea Wollensak's "*Global Positioning Series #4*" (1995), "they show the implications of new imaging and cartography techniques on our sense of place" (Wilson, 2002).

The project "*In Transit*" by Amy Balkin traces San Francisco taxi-cab movements. What we see on the project website[8], as an outcome, are traces over space and time. Traces on the past can bring insight on the future, history of traces can enlighten intention and recurrences, and from these lead us to deepen our understanding of public issues. "*In Transit*" is part of a larger project curated and promoted by the Exploratorium in San Francisco, called Cabspotting: cabs are a vehicle for artists, writers and researchers to explore, observe and experiment. They are an urban ethnographical tool visible through the website that collects all the data and all the projects under this umbrella. Cabspotting traces San Francisco's taxi cabs as they travel throughout the Bay Area. The patterns traced by each cab create a living and always-changing map of city life. Hinting at economic, social, and cultural trends that are otherwise invisible, it reveals "*Invisible Dynamics*". What is collected from taxi rides are statistics, demographics, conversations, and projections. Traces can tell us not only where we are, but also allow us to obtain different perspectives on how physical space is occupied. These can yield fresh insights, possibly with important consequences that shape decisions about how we behave in physical space.

There are several projects worth analysing that provide examples of the Internet of Things and the realm of *blogjects*, in particular two very similar projects created by Beatriz da Costa called "*The Pigeon that Blogs*"[9] and *Air*.[10] "*The Pigeon that Blogs*" is a flock of pigeon equipped with environmental sensors GPS/GSM technologies that find and track toxic patterns in the atmosphere. According to Da Costa, in acting as a sensor, the pigeon becomes a form of citizen who participate[s] actively in the environmental ecological discourse.

Data is collected during the flight of the pigeons in a specific city and at a specific time schedule, charts of which are then formed and uploaded to the website of the project. The pigeons, until now, have been released in many different cities helping to monitor the presence of carbon monoxide in the environment.

Evocatively a pigeon, through its flight, can talk to us and be part of an environmental discourse in which it is directly involved, much like human beings. Da Costa's "Air" provided similar outcomes but through different agents (using human beings instead of pigeons) finding similar concerns and concepts as the original project. In the website description of the project, AIR is described as "a public, social experiment in which people are invited to use Preemptive Media's portable air monitoring devices to explore their neighbourhoods and urban environments for pollution and fossil fuel burning hotspots".

An ecological problem that seems distant from our responsibility becomes very much closer and motivates the participant through a real awareness our participation in the public domain of environmental policy.

The final project to be discussed in this chapter not only helps to stimulate a public discourse but provides an example of an agent of awareness and warning. Tripwire[11], by Tad Hirsch (2006) from the Smart Cities Group at MIT's Media Lab,

> is a site-specific installation responding to the unique relationship between the Norman Y. Mineta San Jose International Airport and downtown San Jose, CA. Custom-built sensors hidden inside coconuts are hung from trees at several public locations to monitor noise produced by over flying aircraft. Detection of excessive aircraft noise triggers automated telephone calls to the airport's complaint line on behalf of the city's residents and wildlife. Documentation of noise incidents is archived for later analysis. Turning a coconut into the sense "agent" is a clever way to make the sensor have more affinity to the surroundings.

The intersections between art, activism, and technology play a significant role in public awareness. In this chapter, I have discussed the impact of locative technologies and how artistic interpretations open up horizons not foreseen, how new technologies are incorporated into everyday life, how they are turned from 'disembedding technologies' into embedded media practices, and how technology may be able to disappear into our everyday life.

There are issues excluded from the present discussion because marginal to the main topic but still relevant to the broader overall argument about locative technologies. These include aspects such as surveillance, privacy, issues of information sustainability, and design issues, to mention just a few. It is not possible to exhaust those aspects in the length of a chapter where the main focus was to reveal the how locative technologies empower and enact new social bodies. By means of a sociological excursus and review on emerging technological realities it has been show how it is possible produce a public sphere involving into the process new kind of actors.

The locative art projects presented with this chapter aid in mapping the unknown relations between social bodies and show their relevance to the public domain, pointing the way toward new possibilities for collective action and community organisation.

Notes

1. The big player in outdoor locating systems is GPS, originally a military system based on satellite technology for outdoor tracking, that has subsequently been released for civilian usage. Developed in 1970s for the American Navy and Air Force, it allows an absolute and precise measurement of the three coordinates: latitude, longitude and altitude. Nowadays it is an end user technology - in fact it's easy to buy standalone GPS devices such as car navigators, PDAs enabled with GPS as a personal navigation system or a GPS module for a camera to locate pictures with the geographical coordinates. It has been used widely for outdoors sport, surveillance, security, transportation, etc.
2. Mainly, GIS systems are large geographical databases collecting, storing, retrieving and visualising complex geographical information both in real time or on demand. The interest of the public has been caught by GIS systems such as Google Maps and Google Earth, online mapping systems promoted by Google as a mass user geographical visualisation system that can be browsed and customised by users in order to link information to certain geographical locations. The novelty is not only in its mass usage but also in its cross channel and cross platform nature; it can be accessed and used not only via Internet on an home desktop computer but also whilst away from the desk through a mobile phone, a PDA, or some other device that is connected to Internet.
3. A synthesis of the report can be downloaded for free at the ITU website, http://www.itu.int
4. http://www.field-works.net/
5. http://edinburgh.murmur.info
6. http://www.gpsdrawing.com
7. http://www.terirueb.net/choregraph/index.html
8. http://cabspotting.org/projects/intransit
9. http://www.pigeonblog.mapyourcity.net/index.php
10. http://www.beatrizdacosta.net/air.php
11. http://web.media.mit.edu/~tad/htm/tripwire.html

Bibliography

Bleecker, J. (2006), *A Manifesto for Networked Objects — Cohabiting with Pigeons Arphids and Aibos in the Internet of Things.*

Fraser N. (1993), *Rethinking the Public Sphere: a Contribution to the Critique Actually Existing Democracy.* In: C. Calhoun (ed.), *Habermas and the Public Sphere* Cambridge, USA: The Mit Press.

Graham S. (2004), *The cybercities reader.* London, Routledge.

Habermas J. (1964), *The Public Sphere: An Encyclopedia Article.* New German Critique (Autumn, 3), Pg. 49–55.

Habermas J. (1989), *The structural transformation of the public sphere: an inquiry into a category of bourgeois society.* Cambridge: Polity Press.

Hemment D. (2004), Locative Arts.

Nova N. (2004), Locative Media: a literature review.

Rheingold H. (1994), *The Virtual Community - Surfing the Internet.* London: Martin Secker & Warburg.

Rheingold H. (2002), *Smart Mobs: The Next Social Revolution*: Perseus Publishing.

Sterling B. (2005), *Shaping Things.* Cambridge, MA: The MIT Press.

Wilson, S. (2002), *Information Arts. Intersections of art, science and technology.* Cambridge (USA): The MIT Press.

Second Part

Technology, Physiology and Memory

Chapter 7

Neurosociological foundations of social networks: Distress-dependent synaptic architecture in social systems[1]

Thomas S. Smith, David Silon

Abstract

Social networks develop architectures akin to synaptic networks found in he human brain. This conclusion follows from studies of hyperstructures—distress-dependent neurosociological mechanisms active in attachment behavior and social bonding. Because hyperstructures recruit socially analgesic behaviors to sites of distress and pain, they give rise to a network-based immune response in social life. A prototype of this appears first in infant-caregiver attachment, and thereafter generalizes to much social interaction. Around sites where distress appears, networks grow and expand, constituting "network depots" that support a stress-buffering immune function. The expansion of these networks occurs as distress spreads outside of local neighborhoods—a process controlled by counter-current effects of depot exhaustion and replenishment. Viewed macroscopically and functionally, the neural architecture of these synaptic neighborhoods also reveals that they exhibit learning and memory—that, on the one hand, they embody forms of structural memory that are innate while, on the other, they also evolve by incorporating novel structural learning. We discuss network learning and memory as outcomes influencing one another.

Introduction

Patterns of attachment and interaction between newborns and their mothers have been conserved in mammalian species because they raise the survival odds of offspring, and hence contribute to the fitness of the species (Taylor 2002; Uvnäs Moberg 2003) Considered outside of the framework of evolutionary argument, these patterns point to profoundly interesting conclusions about social life. Simply put, social life sometimes functions as an extension of human physiology. Minimally understood, this says nothing more than that behavior serves to regulate essential physiological functions. There is little new in such an observation. It takes interaction to make a baby—or at least it used to. But there are other, at times macroscopic, implications of this evolutionary wiring of social life. Least obvious, perhaps, is how physiology's grip extends beyond dyads to shape social networks.

To understand physiology's network implications requires rethinking a number of basic ideas in network theory. In what follows, therefore, we introduce a new approach to conceptualizing network dynamics—one that, we also believe, may be useful in intellectual settings outside of sociology. We shall start by building upon and revising the theory of the hyperstructure (Smith, 1994, 1996, 2004; Smith and Stevens 1996, 1999, 2002). The original model of the hyperstructure introduced a theory of infant-caregiver attachment that was based on recent discoveries in neuroscience, particularly those concerning the part attachment played in stimulating the release of endogenous opioid peptides—morphine-like neurochemicals that function analgesically to buffer stress and modulate pain responses. A revised approach to exploring this mechanism arose from asking a basic question: Does nature provide us with comparable models of how biological systems meet stressors? The immune system provided such a model, along with related descriptions of how the human immune response is organized—specific recent discoveries of functional links between immune cells and neural activity. Most significantly, immune cells produce neuropeptides such as β-endorphins and other neurotransmitters, and neurons make cytokines such as IL-1 that produce inflammatory responses (see Blaloch 2002). These cells are linked in an immune network that mirrors infant responses to maternal behavior–specifically in the opioid effects of maternal attachment behavior and, in a counter-current fashion, in the HPA effects of maternal separation.

The hypothesis that hyperstructures function as part of a wider socially-based immune system has far-reaching implications. From this conceptual vantage, for example, hyperstructures need to be understood as synaptic generators in social life—mechanisms for establishing interpersonal links that transport the immune resources embedded in social life to sites where distress has been detected. Like the immune system itself, this mechanism functions as a sensory organ: it has been conserved to recognize distress as it arises in social life. Likewise, it also depends on an immune network—or, better said, it gives rise to responses that establish a network, both benefiting from and encouraging a specific kind of biological synchronicity among persons brought into interaction. The analgesic function of these networks is evidence that social life is wired to produce an immune response to stressors; at times and in part, that is, social life performs as an extension of the human immune system. Social immune networks, moreover, have an architecture resembling synaptic networks found in the brain, a conclusion that has implications for understanding what we shall describe as structural memory and learning.

The architecture of immune networks is itself evidence of ancient learning—adaptive responses that emerged on distant evolutionary landscapes and that now exist as a kind of structural memory embodied neurosociologically. Structural memory appears first of all in conserved attachment patterns, including the synaptic links by which social networks grow. What we shall see below is that hyperstructures are mechanisms not only for the reconstruction of such conserved structural memory but also for forming new structural memory, such as the construction of new attachments and network patterns.

These are in many ways bold claims, requiring further elaboration. We begin by returning to the study of infant-caregiver interaction, a foundation for building a new way of thinking about social networks.

Hyperstructures

It once seemed obvious what mothers were doing when they moved to comfort a baby in distress. They would pick up the baby, gently rock it, and attempt to soothe it with comforting gestures and gentle talk. Yet only in the past thirty years have the hidden physiological forces behind such behavior been discovered. Attachment damps distress by triggering soothing neurochemical changes in the baby, such as the release of endogenous opioids and of oxytocin. These neuroactive substances actually explain why a mother's touch is analgesic—why it can soothe a scraped knee, reduce the hurt of a bee's sting, or mitigate a pinched finger.

The theory of the hyperstructure includes a model of such effects. It shows how two mutually exclusive forms of maternal behavior, attachment-to-baby or separation-from-baby, control activity in two functionally coupled brain systems—the opioid system and the arousal system. Attachment stimulates the opioid system; separation stimulates norepinephrine and the arousal system. Since babies are dependent on their mothers to produce these behaviors and since one behavior excludes the other, there is a reciprocal relationship between activity in the two systems—when one is active, the other is inactive. The transition from activity to inactivity occurs when one system is depleted below a threshold associated with comfort. In the case of the transition to opioid activity, this change occurs when opioid levels pass below a threshold at which the baby begins to experience the onset of opioid withdrawal. For infants, withdrawal manifests itself as a distress call—cries that reach the mother, who then is moved to re-attach herself to the baby, thereby stimulating another opioid cycle.

Consider these two systems again, this time from a conceptual angle more suitable to discussing network effects. Figures 7.1a)-b) are schematics of the opioid cycle and the arousal cycle respectively. For the special case of infant-caregiver interaction, these two cycles can be understood as competing with one another; each cycle depends on one of two mutually exclusive forms of maternal behavior, attachment or separation. A new way to recognize implications of these cycles is to reconsider them in relation to the concept of a depot—a reservoir of neurochemicals (opioids or, for the arousal system, norepinephrine) that can be depleted or can be replenished, depending on the phase of cycle.[2] These ideas are represented in Figures 7.1a)-b). The symbols O and N in these graphics refer to opioids and norepinephrine, respectively; A and S refer to attachment and separation; and the subscripts i and c to infant and caregiver. The opioid depot cycle, in Figure fig07.01a), shows that opioid depletion, O_i^-, gives rise to withdrawal pressure in the infant, $W(O)_i$ that signals a need for depot recovery, and that these signals stimulate maternal attachment, A_c, commencing at a depot thresh-

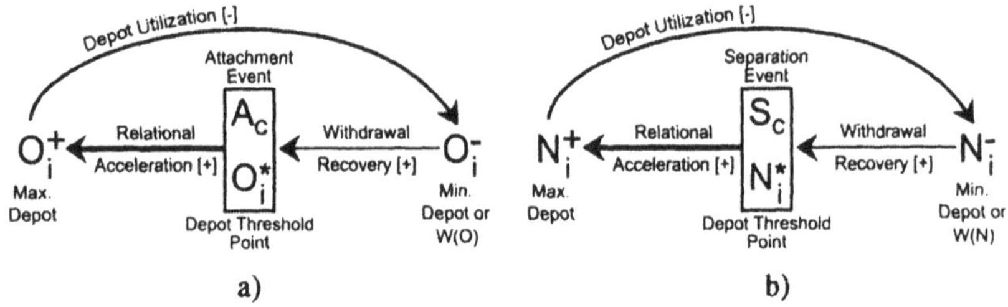

Figure 7.1 a) *Opiod Depot Cycle.* b) *Arousal Depot Cycle.*

old point we have called O_i^*. When this threshold is passed, maternal attachment accelerates depot recovery up to the depot maximum, O_i^+, where subsequent use of the depot again depletes it.

A comparable logic appears in the arousal depot cycle. Like an empty reading on your fuel gauge, N_i^-, the depot minimum, is associated with the gradual onset of pressures, $W(N)_i$, for depot recovery. When the N_i^* depot threshold is passed, maternal separation S_c begins to stimulate depot recovery up to N_i^+, where use of the depot again gradually depletes it.

When we couple these two cycles as they work in relation to one another, we get a system represented in Figure 7.2. Figure 7.2 helps make it clear that hyperstructures are really *biological oscillators*, with thermodynamic principles at work in the coupling of the two depots. We suggest this by placing, in the center of the graphic, a Depot Entropy sphere, indicating a withdrawal inflection point $W(O)_i/W(N)_i$. The coupling of the two cycles, that is, is always marked by simultaneous and competing pressures for depot replenishment and metabolism,

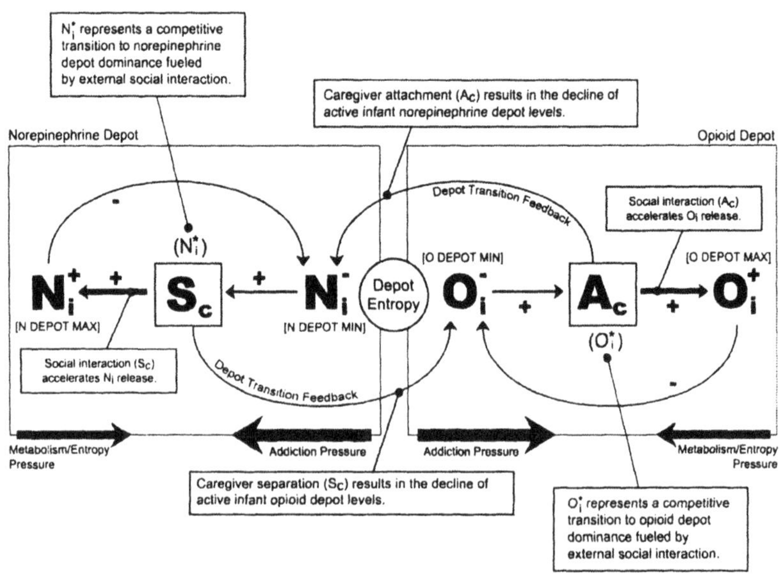

Figure 7.2 *Coupled Opioid and Arousal Depot Cycles.*

pulling in opposite directions. These appear as counter-current pressures—toward, on the one hand, the metabolism of the resources in the depot (the movement from depot-maximum to depot-minimum states), and, on the other hand, toward the replenishment of the depot (from depot-minimum to depot-maximum states). We label the latter "addiction pressure" because addiction is a potential outcome, as when changes in the threshold points, O_i^* or N_i^*, lead to ongoing and entrenched pressures for the replenishment of either depot. Here we also make explicit the notion that the behaviors associated with replenishment of each depot need to be understood functionally—that is, they denote classes of behaviors defined by their capacity to stimulate each system.

In these terms, the hyperstructure can be understood as a biological oscillator working to synchronize the behavioral and chemical depots of two (or more) interactants. In Figure 7.2, the infant's neurochemical states are shown linked to the caregiver's behavior. But in other social interaction, we can have behavior and chemical depots for two adult interactants considered simultaneously. Figure 7.3 graphs this coupling by bringing forward a 2-part notation for hybrid states indi-

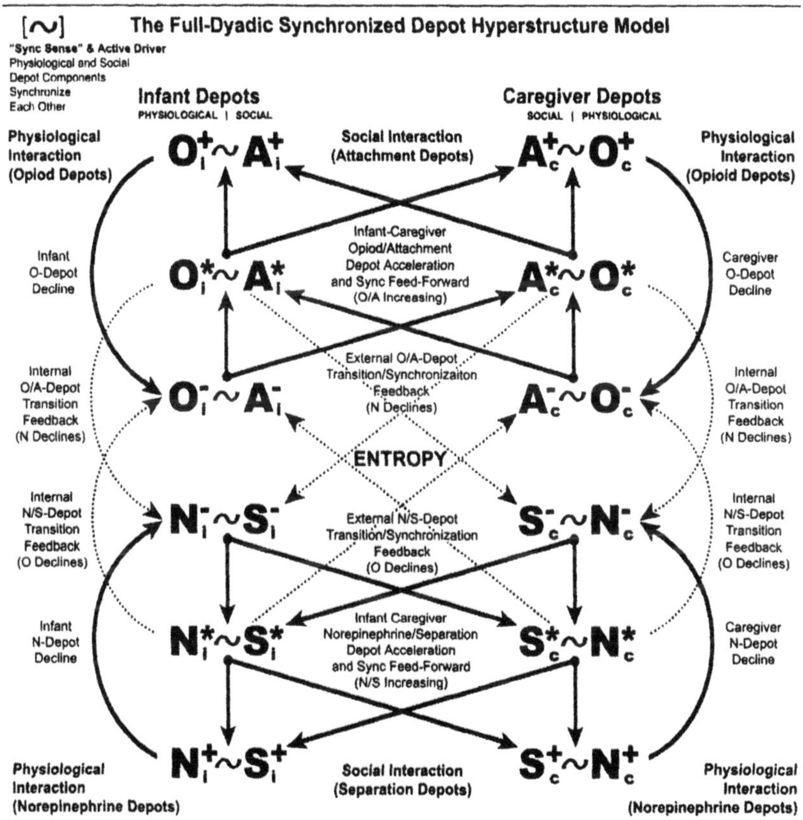

Figure 7.3 *Dyadic Synchronization of Behavioral and Chemical Depots*

cating the synchronizing ($\sim$) of chemical-and-behavioral depots—linkages such as those for $N_i^- \sim S_i^-$ or $O_i^- \sim A_i^-$. Implicitly here we are now relaxing the assumption that distress communicates itself only interpersonally, as it would if we were still talking only about infants. But as persons mature beyond the dependent condition of the infant, they develop capacities to modulate their own depot states, using personal depot resources, such as cognitive structures and behaviors. The effects of these personal depot resources appear in the two columns of the graphic—on the left-hand side (for person i) and on the right-hand side (for person c). For instance, $O_i^- \sim A_i^-$ represents the depleted opioid depot of interactant i,a condition paired with depleted personal attachment resources. This hybrid state gives rise, within person i, to withdrawal pressures. Some of these make their way into boot-strapping operations by which $i$may up-regulate his or her own opioid depot by using resources from a personal depot. Yet the signals of depot depletion are not normally left to personal resources alone, but are communicated interpersonally as withdrawal symptoms—as signals of discomfort such as anxiety.

These interpersonal communications of distress are shown by the crossbar arrows connecting the two columns. Figure 7.3 therefore shows both intrapersonal and interpersonal linkages. Though these dual linkages function side-by-side, the interpersonal links actually confer evolutionary advantages on the species, and work far more efficiently to modulate depot states. Evolution has wired us to fall back on social interaction as a means of managing distress.

Note that the empty depot state does not always have to produce interpersonal links. Links do not emerge if a person is able to self-synchronize and self-modulate depot states—or is for some reason isolated from social contact. But such links do emerge when distress spills into the interpersonal pathway. When this occurs, the hyperstructure begins to exhibit its conserved tendency to recruit behavioral resources to sites of distress—in effect, operating again as in Figure 7.2 but now supplementing the interpersonal channel by the personal depot strengths of individuals. This is a social immune response to stressors, and it creates synaptic architecture.

Synaptic networks

When distress spills into an interpersonal pathway, it recruits attachment responses from others, and when this recruitment begins, it sets up an imperative for interactants to synchronize their depot states. Without synchronization, depots remain unregulated, and chronic depot depletion (or saturation) can then lead to pathologies at both the individual and social level.[3] An imperative to synchronization arises initially through the discomforts that otherwise plague interactants—like the distress that accumulates in the neighborhood of a baby who is left unattended. The mechanism thus creates pressures toward interpersonal synchronization, and we observe emergent links that are in effect social synapses.

Such synapses are the emergent arteries of social immune networks—circuitry that comes into existence to convey analgesic resources to a site of social distress. The emphasis here is on emergent—such networks do not exist *a priori* as a structural framework for an immune response. They develop from the work of the hyperstructure, now understood as a synaptic pattern generator that reconstructs network memory, patterns that have been conserved evolutionarily. All infant-caregiver dyads are themselves reproductions of these conserved network patterns, reflections of structural memory generated and regenerated through time and independent of culture.

This same synaptic generator, moreover, potentially wires the rest of social life, recruiting attachment resources when it detects distress. This effect may well be scaled by kinship proximity, but it also exists incorrigibly as part of a wider neurosocial landscape. In effect, recruitment establishes an extended social network as a depot—an interpersonal depot that stands atop the chemical depots discussed above. Such extended social depots are really outgrowths of the network depot that exists in the interpersonal world of the infant. In that world, as we have seen, the organization of the network depot is innate—generated as a reconstruction of structural memory preserved in our neurosocial wiring. Parents "pool" their emotional resources in this family network depot, as part of an emotional division of labor that spreads the burden of childcare over several persons (Smith and Stevens 2002). The same pooling principle then generalizes to explain the formation of non-kinship distress-dependent synaptic structures. A network forms as persons are recruited into depots by the spreading of distress and the pooling of their analgesic resources. Recruitment can occur one-on-one, via the formation of dyadic links; it can arise when distress is broadcast into interaction neighborhoods, where persons are independently drawn, one-by-one or concurrently, into synaptic links; and it can occur sequentially, as when distress is handed off to persons recruited one after another. Synapses, once formed, are subsequently active as arteries that channel comfort-enhancing resources among the persons they link.

The concept of a synapse is important here because it suggests a wiring principle at work in networks. It is akin to the "fire together/wire-together" Hebbian principle governing the growth of neural networks in the brain, and it is coupled to a corollary rule—"out of synch/lose the link." In the brain, neural links appear when synchronous neural firing occurs, and they weaken or disappear when neural firing is asynchronous. Something like this also appears to be the case in social immune networks—that is, such networks grow and are sustained by synchronization and they weaken without synchronization. In families, for example, the immune network depends on attachment resources being recruited to the baby's distress. We know from decades of attachment research that pathologies and dysfunctions appear in families when attachment isn't synchronous. When mothers are unable to care for their babies, for example, babies develop insecure (weak) attachment styles. Families giving rise to such styles in children have networks marked by depleted immune depots, and accordingly generate "anxious/avoidant" or "anxious/ambivalent" styles of attachment (Ainsworth 1974). Children thereafter grow up reproducing these styles in their adult networks, indicating that their family experience has taught them that attachment has emotional costs (Hazan and Shaver 1987). By contrast, children from secure families reproduce secure attachment patterns—patterns based on the pooling of adequate emotional resources in a network depot and marked by synchronization of needs and responses. By reasoning with depot concepts, we can derive such attachment patterns from a network generator.

Structural memory

Important here is the notion of network learning and memory. A social synapse forms when a person is recruited to a site of distress and brings attachment resources to bear, synchronizing an immune response to the target. The synchronization is equivalent to the fire together/wire together rule. That is, it results in a synaptic link, comparable to the growth of a connection between two neurons. When such a link forms, the network can be said to have learned, and there is subsequently a tendency for the synapse to function as a pathway for immune and other activity. The tendency to reproduce the synaptic link is supported by such subsequent reinforcing synchronization. When a stressor activates a link, or propagates outward into adjacent links in a network, it will produce a reaction that is akin to inflammation in the immune system: comfort-enhancing resources, distributed among the links in the network, will be recruited to the site of the stress. Persons previously associated analgesically will again be drawn together. In their recurring interactions, synchronization will be at work, serving as a synaptic reinforcement. In proportion to the multiplication of links in the same network, we witness the growth of an analgesic network depot. Network memory is implicit in this depot of comfort-enhancing resources. As these links are reinforced, structural memory thereafter conserves and reconstructs them.

The simplest empirical illustration of this linking is provided by research on joint reconstruction of autobiographical memory, also called conversational remembering (Middleton and Edwards 1990). What we see in settings where people reminisce about photographs or discuss shared experiences from the past is "distributed cognitive activity." In their conversations, they jointly reconstruct a past, and their joint reconstruction in effect "pools" cognition. Thinking is not then something that occurs inside the head but is instead a public activity (Smith 2004)—it emerges in their conversation. Such conversations—linked utterances—are a collective product. And once the collective product is in place, it reinforces the tendency for the same persons subsequently to reconfigure themselves as a dyad—as a dyadic cognitive system. It is the collective working out of the memory itself—through the synapse—that is evidence of the attractor. It draws them together from time to time.

This is structural memory in the literal sense of memory: a jointly reconstructed, distributed cognitive structure. When such structural memories emerge and are learned by a network, the network has grown a new layer of controls. The memory itself then exerts influences on behavior, reshaping it into patterns the network has learned. Why do these synaptic links reproduce themselves? Clearly, there are forces behind the learned pattern, embedded in the physiology we have discussed. For one thing, shared thinking is profoundly intimate, a feature of adult attachment. It does not happen automatically but requires interaction that first establishes trust. Interestingly, this is an effect that, physiologically speaking, appears to be mediated by the release of oxytocin, one of the primary hormones at work in attachment. Experimental subjects given nasal doses of oxytocin, for example, begin to trust that others will behave honestly and fairly (Kosfeld et al. 2005).[4]

Oxytocin also changes the habituation threshold for endogenous opioids, prolonging their comforting and pleasurable effects (Panksepp 1998). Whereas oxytocin can yield its effects even among strangers, attachment behavior also triggers its release. It then becomes embedded in the process by which persons familiarize themselves with one another, and encourages them to grow together functionally. Crossing that threshold, persons begin to supplement each other as part of a distributed cognitive system, with the effect of drawing each other into a world over which the other has some control. It is this outcome that places persons on the same cultural landscape—requiring them to function together in a world constructed with others and not distorted by the error-propagating effects of deliberate solitude or forced isolation. The social synapse linking them is an efficient biologically mediated pathway for error correction and cultural integration, and all of the pressures behind cultural and normative integration in a community reinforce it.

Distributed cognitive activity entails one of the most important mechanisms tying persons into a social world. The array of such culturally and cognitively synchronized links becomes a structural magnet. Reconfiguring a network with the same persons is always an attractive neurosocial inducement, because such learned links function efficiently and without startup costs. Reactivating them is akin to a homecoming. Yet when links cannot be opened (as when particular persons aren't available), another effect commonly appears. Structural pressures begin to accumulate—pressures to recruit newcomers, to incorporate replacements into the network template. Absent network progenitors in this sense leave a structural vacuum that manifests itself in such recruitment pressures—in a "structure hunger" whose aim is to replace those lost to the network. Grief and mourning are classic illustrations of this hunger, equivalent to withdrawal symptoms when a depot is depleted.

Likewise, family patterns tend to reproduce themselves in interaction with non-family members, as clinical studies of "transference phenomena" have long made clear (Smith 1992). Depot depletion triggers a search whose aim is to regenerate a structural memory, and any person drawn into such a synapse is then a stand-in for an absent predecessor. When this occurs, persons recapitulate family patterns in non-family settings.

Autoimmune network patterns

There are other kinds of structural memory, and these also depend on something like the pooling that occurs when we witness socially distributed cognitive activity. We know that families have overlapping dyads and triads, and that pressures arising in one dyad can spill into others, bringing about various dynamic patterns. Family networks contain these patterns as structural memories, some with effects that survive intergenerationally. For example, clinicians have long recognized that certain addictive disorders can emerge as family patterns. A child can be transformed into a "family problem" by a "family projection process." Anxieties, signals of depot depletion, are dissipated by being focused on a child who is transformed into a delinquent or a substance abuser. Since anxiety in the "family system" otherwise results in conflict and high levels of discomfort, the child internalizes this projected role in response to the function of stabilizing the family. In effect, the child who becomes a "family problem" has all the family discomfort detoured through it, and thereby damps the anxiety that would otherwise burden other links in the family network.

One can think of such family patterns as examples of autoimmune network responses in social life. Normally the immune system's specialized cells–monocytes, lymphocytes, basophils, neutrophils, eosinophils—protect the body from harmful substances, called antigens. The immune system produces antibodies that destroy these substances. But in patients with an autoimmune disorder, the immune system can't discriminate between healthy body tissue and antigens. The result is an immune response that destroys normal body tissues, a process that is described as a hypersensitivity reaction, similar to allergies. In these cases, the immune system responds to substances it normally would ignore. In allergies, the immune system reacts to an external substance that would normally be harmless. With autoimmune disorders, the immune system reacts to normal body tissues.

When families project anxiety into a child and thereby create a "problem child," they are in effect producing an inflammatory response where there is no problem. This can injure a child, even though it might transiently stabilize the family system. Equivalent projective processes occur in many social networks, from communities to nation states. In a community setting, an equivalent to the "sick child" is the social deviant, and focusing on the deviant promotes the appearance of community pressures behind conformity behavior (Smith, Bernard 2005. Whereas many autoimmune disorders of the body are not well-understood, the origins of these social autoimmune disorders would seem to have an explanation. They are products of anxiety and distress—signals of depot depletion, on the one hand, and of asynchronous interaction, on the other. More common autoimmune effects in social networks include conflict among linked persons or "distancing" between them.

Of course, illness is more commonly real, and illustrates the way families can use an immune response to engineer depot recovery and transiently enhance comfort. A person exhibiting the symptoms of illness calls forth an immune response. "Sickness behavior," including a group of symptoms such as fever, is triggered physiologically when cytokines derived from immune cells produce inflammatory responses in the body. These are then met with various antinociceptive effects, including the recruitment of endorphins. When viewed interpersonally, "sickness behavior" extends beyond the sick person to include the adjacent social immune network. Sickness behavior always embodies a distress call, and this can give rise to the recruitment of social immune resources. We see this in diluted ways in every culture, as when persons greet one another with solicitations of distress—saying "How are you?" or "Come stai?" or "Wie gehts?" Solicited distress in this sense is the occasion for forming a synapse, recruiting persons into synchronizing interpersonal connections, inviting intimacy that opposite tendencies in social life ordinarily tend to inhibit. The same fact means that a family can use a sick child to improve its own comfort—to bring its members back into comfort-enhancing connections to one another.

Once established, a network preserves neurosociological interdependence, embedding persons in webs where behavioral and chemical resources are can be distributed across synaptic links. Such a *distributed* depot reinforces itself by transporting resources where they are needed, maintaining synchrony at the level of embedded individual links. Synchrony then is a distributed immune resource—a sensitively dependent harmonic property of networks. Some interaction networks—team sports, military formations, monastic orders, orchestral ensembles—make it plain how synchrony is a central ordering principle of social life. They are settings in which synchrony is trained into their members, piggybacking disciplines such as rowing or marching or chanting on top of the innate oscillator at work in attachment behavior. Yet in all social systems, the synchrony that potentially emerges in ordinary interaction is unifying and comforting.

Some conclusions

Hyperstructures are both neurosocial oscillators and synaptic network generators. In each of these respects, they introduce new ways of thinking about the neurosocial foundations of social networks. Not the least of these is that many networks exhibit patterns that can be understood as outgrowths of the human immune response. They show that persons naturally become part of linked depots of care, and they alert us thereby to the critical importance in social life of interpersonal and network synchronization. Several of these conclusions are new to the emerging theory of social networks, and the reasoning behind them has many other implications—many familiar, others novel.

At the very least, classic reasoning about the strength of ties in networks (Granovetter 1973) can be reconceptualized in view of Hebbian rules governing synapses—fire together, wire together. An important correlate of the strength of synapses in this sense is their information load, a property that depends on how often they convey signals or information. This is a variable that translates socially into the multiplexity of ties. The more heterogeneous the uses of social synapses, the stronger they are. Diffuse ties are strong ties, socially speaking. Yet because of the tendency for strong ties to form closed systems of links, strong ties alone can weaken a network's overall architecture, as Granovetter showed. Counteracting this fragmenting tendency is the part played by weaker ties—those that, in our terms, are only occasionally reinforced. These ties bridge among otherwise isolated clusters, and function to stabilize networks—to create the gridwork that brings all parts of a large network or social system into an approximation of a global oscillator.

Indeed, it was Durkheim's point that societies are societies in a network sense only when they do function as global oscillators. This in effect is what he reasoned in *The Elementary Forms of Religious Life*—that sacred time and profane time are alternating phases of social life. Where you have alternating phases, you find an underlying oscillator. Sacred time in effect replenished the depot associated with attachment, recruiting persons into a communal network that otherwise remained latent in the dispersed condition of everyday life.

Conclusions such as those in this paper point toward further refinements in our thinking about networks. One is suggested by the models we have introduced to represent the neurosociological foundations of networks. This work, which we have mainly sought to reconcile with ideas and research in neuroscience and physiology, would benefit by being brought into closer alignment with concepts that play a part in explaining synchronization in natural systems, particularly those that describe coupled oscillators.[5] Clearly, concepts such as coupling strength and frequency are ideas we have discussed above, developed in relation to biological oscillators that have been represented mostly in qualitative terms. Another direction follows a computational route, one we have used in earlier research (for example, Smith and Stevens 1999, 2002). Hyperstructures have implications that are difficult if not impossible to discover without bottom up computational experiments. Many of these remain to be disclosed in future research.

Notes

1. This paper was originally prepared for interim meetings of the ISA WG03, "Memories in Action: The Role of Memories in Social Change," Rome, 1995. It has been revised and extended for this publication.
2. What is a depot? In abstract terms, a depot is simply a pool of resources. In the context of studying complex systems, however, the concept of a depot is an important and useful. With depots, or pooled resources, systems have the ability to accomplish work, such as maintaining equilibrium, responding to various demands, or damping and distributing distress. Logically, depots also require associated depot structures that generate and store depot resources for later use and can detect when more resources are required. By identifying depot mechanisms and depot structures, and by analyzing their related macroscopic functions, it becomes much easier to understand and model the behavior of complex systems.

   For example, the human immune system consists of a depot of mulitple, specialized cell types (monocytes, lymphocytes, basophils, neutrophils, eosinophils). These cells arise from our primary blood-building depot structure, the bone marrow, a depot of stem cell-generating tissue. There are also associated central and peripheral immune structures, such as the thymus, spleen, and lymphatic system. Our immune systems respond to, and are recruited by, physiological distress. Also, since immune cells have a finite life, they must be constantly renewed, or the result would be eventual immune depot decay.

When individuals are immunocompromised from cytotoxic chemotherapy or infection with the HIV virus—or, in other words, have low immune depots as a result of depot decay—they are highly susceptible to infections and other disease. Chemotherapy, in particular, destroys immune depot progenitor cells, thereby impairing depot resource creation and renewal.

In chemotherapy patients who develop low immune depots, depot resource creation can be stimulated with medications such as Neupogen/filgrastim. Neupogen is a man-made mimic for G-CSF, or Granulocyte Colony Stimulating Factor, the compound our bodies use to maintain the immune depot by stimulating the production of new immune cells within the bone marrow. Other immune depot-mobilizing substances used to treat cancer and other diseases that mimic the body's own mechanisms of depot amplification include the interleukins, also called lymphokines.

Conversely, when individuals have any one of various forms of leukemia or lymphoma caused by cancer within the immune depot structures (bone marrow, lymphatic systems) and blood, they have very high immune depots. The depots increase out of control, locally or systemically, eventually impairing other important physiological functions, resulting in death. The treatment for these patients involves chemotherapy or radiation therapy, which destroys the rapidly proliferating immune depot cells and structures, or seeks to balance the immune amplification with radical depot attenuation.

In most people, however, the immune depot and depot structures operate correctly and we remain defended against disease. There is ample storage for immune depot resources; the depots are maintained at equilibrium or are increased as necessary (by the mechanisms of immune recruitment) to fight bacterial and viral invaders, as well as our own cellular decay; the immune system "teaches" its cells to respond to new infection by imprinting new patterns for recognition; and we live normal and mostly healthy lives.

Without an immune depot ready to perform the work of maintaining our bodies and responding to the first sign of infection, we would quickly die. So, the evolution of our immune depot function is a critical part of our survival.

3. Failed depot regulation is always associated with discomfort—with cumulating distress and anxiety. Chronic depletion of norepinephrine, for example, is behind the symptoms of PTSD (van der Kolk 1987); opioid depletion predisposes addictive patterns involving substance abuse. At a social level, depot depletion is likely associated with extreme states of communal disintegration or with extraordinary integrative pressures such as those behind moral panics.
4. Michael Kosfeld and colleagues in Zurich got students to play a game that involved exchanges of real money between two persons. One played the role of "investor," the other "trustee." The subjects had received either an inert placebo via a nasal spray or the neuropeptide oxytocin. Those who received the oxytocin subsequently displayed elevated levels of trust. In other experiments, oxytocin has also been shown to increase generosity (Zak, *et al.* 2007).
5. Splendid informal introductions to these ideas are presented by Strogatz (2003) and Watts (2003). More formal treatments appear in Strogatz (1994) and Pikovsky *et al.* (2001).

Bibliography

Barclay, C. and Smith T. (1992), *Autobiographical Remembering: Creating Personal Culture*. In: *Theoretical Perspectives in Autobiographical Memory*, Pg. 75–98, Edited By M. Conway, D. Rubin, H. Spinnler, and W. Wagenaar. Dordrecht: Kluwer.

Bildack J. (2000), *Detection and Function of Opioid Receptors on Cells from the Immune System*. Clinical and Diagnostic Laboratory Immunology, Vol. 7, Pg. 719–723.

Blalock J. E. (1984), *The Immune System as a Sensory Organ*. Journal of Immunology, Vol. 132, Pg. 1067–70.

Blalock J. E. (2005), *The Immune System as a Sixth Sense*. Journal of Internal Medicine, Vol. 257, Pg. 126–138.

Bowen M. (1978), Family Therapy in Clinical Practice. New York: Aronson.

Cacioppo J. and Berentson G. (2002), *Social Neuroscience*. In: *Foundations in Social Neuroscience* . Pg. 3–10, edited by J. Cacioppo, G. Berentson *et al.*, Cambridge: MIT Press.

Cacioppo J., Berntson G., Adolphs R., Carter C. S., Davidson R., McClintock M., McEwen B., Meany M., Schachter D., Sternberg E., Suomi S. and Taylor S. (eds.) (2002), *Foundations in Social Neuroscience*. Cambridge: MIT Press.

Csikszentmihalyi M. (1990), *Flow: The Psychology of Optimal Experience*. New York: Harper and Row.

Durkheim E. [1912] (1995), *The Elementary Forms of Religious Life*. Translated by Karen Fields. New York: The Free Press.

Edwards D. and Middleton D. (1988), *Conversational Remembering and Family Relationships: How Children Learn to Remember*. Journal of Social and Personal Relationships, Vol. 5, Pg. 404–416.

Eigen M. and Schuster P. (1979), *The Hypercycle: A Principle of Natural Self-Organization*. Part C. "The Realistic Hypercycle", Naturwissenschaften, Vol. 65, Pg. 341–369.

Field T. (1985), *Attachment as Psychobiological Attunement: Being on the Same Wavelength*. Pg. 415–454, In: *The Psychobiology of Attachment and Separation* edited by M. Reite and T. Field. New York: Academic Press.

Greenberg J. and Mitchell S. (1985), *Object Relations in Psychoanalytic Theory*. Cambridge: Harvard University Press.

Halbwachs M. [1950] (1980), *Collective Memory*. Translated by Mary Douglas. New York: Harper and Row.

Heinrich M., Neumann I. and Ehlert U. (2002), *Lactation and Stress: Protective Effects of Breast-feeding in Humans*. Stress, Vol. 5, Pg. 195–203.

Kerr M. and Bowen M. (1988), *Family Evaluation*. New York: Basic Books.

Kosfeld M. *et al.* (2005), *Oxytocin increases trust in humans*, Nature, Vol. 435, Pg. 673–676.

Newman M., Barabási A.-L. and Watts D. J. (eds.) (2006), *The Structure and Dynamics of Networks*. Princeton: Princeton University Press.

Panksepp J. (1998), Affective Neuroscience. New York: Oxford University Press.

Pikovsky A., Rosenblum M. and Kurths J. (2001), *Synchronization: A Universal Concept in Nonlinear Science*. New York: Cambridge University Press.

Smith T. S. (1991), *Strong Interaction*. Chicago: University of Chicago Press.

Smith T. S. (1994), *Catastrophies in Interaction*. Social Psychology Quarterly, Vol. 57 (3), Pg. 274–282.

Smith T. S. and Stevens G. T. (1996), *Emergence, Self-Organization, and Social Interaction: Arousal-Dependent Structure in Social Systems*, Sociological Theory 14 (July): Pg. 131–153. [Also: Santa Fe Institute Publication Series, Paper No. 94-08-046.]

Smith T. S. and Stevens G. T. (1999), *The Architecture of Small Networks: Strong Interaction and Dynamic Organization in Small Social Systems.* American Sociological Review, Vol. 64, Pg. 403–420.

Smith T. S. and Stevens G. T. (2002), *Hyperstructures and the Biology of Interpersonal Dependence.* Sociological Theory, Vol. 20 (1), Pg. 106–130.

Smith T. S., Stevens G. T. and Caldwell S. (1999), *The Familiar and the Strange: Hopfield Network Models for Prototype Entrained Attachment-Mediated Neurophysiology.* Pg. 213–245, In: *Mind, Brain, and Society* (Volume 5 of *Social Perspectives on the Emotions*) edited by David Franks and Thomas Smith. Stamford, CT: JAI Press.

Strogatz S. (1994), *Nonlinear Dynamics and Chaos.* Cambridge MA: Westview Press.

Strogatz S. (2003), *Synch: The Emerging Science of Spontaneous Order.* New York: Hyperion Books.

Zak P. J., Stanton A. A. and Ahmadi A. (2007), *Oxytocin increases generosity in humans.* PLoS ONE, Vol. 2 (11), e1128.

Chapter 8

Technology, memory and the (mind)body: towards the integration of the biological and the cultural in the analysis of human social behavior

Sidney M. Greenfield

Abstract

This paper presents a paradigmatic framework to integrate the analysis of biological, social and cultural data to more accurately explain human social behaviors. An information/communication model that focuses on the flow of messages between parts of what is conceptualized as an integrated whole replaces the conventional paradigm, Cartesian mind-body dualism, that heretofore has produced contrasting and, at times, competing views of genetic or cultural determinism. This imagery provides a new meaning to the concept of embodiment by conceptualizing it as the product of a culturalbiological process instead of the present metaphoric transference.

Introduction

Remnants of tools usually are found in early archeological sites intermixed with human skeletal remains. The cultures of primordial societies frequently are named after the tool types associated with physical remains. The presence of tools in assemblages of human body parts is not fortuitous. When the ancestors of Homo sapiens ventured out of the trees and began to walk upright on the African savannahs, to begin their eventual occupation of the rest of the planet, a series of changes in their physical structure resulted from their adaptation to bipedal locomotion in their new environment. One of these was the significant increase in the size and shape of their skulls to accommodate an expansion of their brains. Associated with this increase in size and complexity of their brains was the emergence of the ability to make and interact with each other in terms of symbols.[1]

Symbols[2] are a class of signs – at first verbal and later transmitted and exchanged by means of instruments of various kinds – in which the relationship between a sign and what it stands for (its meaning) is arbitrary and determined by those who use them. Members of a human group can bestow their own meanings on sounds, things and events and this is a unique ability of the species. Homo sapiens have the potential to participate in a symbolic world whose actualization is the result of their being enmeshed in the interpersonal dynamics of a particular social entity. Over time, meanings are attributed to a body of symbols facilitating continuing interactions among individuals and between them and their environment. Prior to the emergence of this unique capacity[3] evolutionary adaptation was specific, direct and almost exclusively biological. Genetic mutations would appear in some members of a particular population that enabled them to obtain greater amounts of food or otherwise be more successful than others in the environment in which they lived. As a result, more of the offspring that inherited the mutant genes were able to survive and reproduce than those without the new physical features. With the emergence of symbols a new form of evolution came to complement the biological: members of human groups could invent and use tools with which to adapt more successfully to their environments; they no longer were dependent exclusively on the randomness of biological happenstance.

While many animals make tools, only humans are able to conceptualize and modify them symbolically – in their heads – for use in specific circumstances.[4] Once conceptualized, production is taught and learned repeatedly often far away from the dangers of real-life situations. Simply stated, tool making and its uses are based not on the mutation of genes, but on changes in the culture of a group. These innovations may be transmitted to future generations and also adopted by other populations. Thus a second process of change combines with genetics to explain how our early ancestors adapted to and survived in a broad range of environments.

This brief introduction emphasizes the differences between the way anthropologists and other social scientists see human behavior and the picture presented by geneticists, sociobiologists and evolutionary psychologists. The latter postulate a genetic embedding of the behavioral patterns of our human ancestors at the time they roamed the African savannahs. Following Francis Crick's formulation of the "central dogma" of DNA —-> RNA —-> protein (Crick, 1958; see also the critique by Lewontin *et al.*, 1984, Pg. 58), ontological priority is ascribed to genes, negating fundamental changes that took place in the evolutionary process with the emergence of the human ability to think and behave symbolically. To be able to strip away the diversity and complexity associated with human behavior and reduce it to a simplified and abstracted basic nature, these scholars, in the tradition of European Enlightenment's desire to find similarities across societies, jumped from archeological and paleontological descriptions of our ancestors millennia ago directly to the contemporary Western world. They have bypassed the voluminous body of data on the many other peoples and social groups who have inhabited the planet and the variability of their cultural beliefs and practices.

Some writers in this tradition have attributed motivation and agency to genes. Not surprisingly, genes are depicted by them as similar to Thomas Hobbes' (1963) pre-social individuals whose competition with each other is so threatening that it leads them to exchange what is stated symbolically as their inherent rights as individuals for protection to Leviathan (the state) and Adam Smith's (1937) pre-social higglers and hagglers who strive to maximize their individual advantage or gain in the market. If competition and maximization are encoded in our genes, as they maintain, it follows that the contemporary institutions of the Western world, since they are based on the same set of assumptions, are also the products of our genetic inheritance. Human beings, however, never have been found living as isolated individuals in the state of nature imagined in the writings of Western intelectuals, separate from and independent of others. They do not and never have existed apart from groups and the symbolically based institutionalized patterns of a specific culture. It is a fantasy for scholars to presume humans ever lived in an abstracted state of nature. They do not compete – or cooperate – simply because of their genes. Advocates of this position negate the contributions of their anthropological and other social science colleagues by failing to acknowledge the ingenious ways humans continuously invent, adapt to and live in the world.

At the opposite extreme there are social scientists that, in their desire to portray the rich diversity of cultures, have given short shrift to the genetic underpinnings of human behavior often attributing what appears as ontological primacy to socially – and symbolically – constructed patterns of culture. It may seem at times as if there are no people involved in their presentations. At first these were of little more than abstracted cultural beliefs and practices, but after a (post modernist) paradigm shift they now are of narratives and discourses to which the metaphors of agency and embodiment are applied resulting in what emerges as an implicit assumption of explanatory directionality. Narratives and discourses that express beliefs and values (all symbolic) are placed metaphorically within the (physical bodies of) individuals that constitute social groups, often using terms from biology and physiology. Once "embodied" in this way, they further assume that the narratives and discourses become part of the social memory of the group and are transmitted from one generation to the next.

This results in two polarized groups of scholars who study and write about humans and their behavior based on contrasting views of the world. Though established on different sets of assumptions as to what exists and what has (primary) causal efficacy they often encompass the terms and concepts of the other. Genetic determinists may speak of culture and values, but see them as the product of the actions of genes. Social scientists write of narratives and discourses conceptualizing them to be embodied as if they were part of the same biological processes genetic determinists examine. The result is a contradictory and confusing literature that though it may contain the same words and concepts, uses them to mean something different, leading their respective proponents to speak and argue past one another.

This oppositional dualism, I suggest, has its roots in the way science developed as it split from philosophy in the 17th and 18th centuries. When Copernicus, Kepler, Galileo, Newton and others proposed an alternative explanation for the movement of the planets, they refuted not religion per se, but rather a given set of beliefs about how the world worked championed by the then hegemonic Roman Catholic Church. Fledgling scientists questioned neither the belief in God's existence or in His having created the universe. Instead they proposed that the epistemological view that assumed God's active intervention in the world as the explanation for everything be partitioned into two distinct and separate realms: the natural, claimed by science, and the supernatural left for God. Scientists then re-conceptualized God to be a master architect and mathematician who designed the material world in the image of a complex machine from which He stepped back, leaving it to run according to His design.

René Descartes (1980) provided the philosophical foundation and guiding imagery by which science, understood as the project to discover the assumed order (or laws) in the (natural) universe was to be carried out. Skeptical of all forms of knowledge, Descartes argued that *res cognitas* – "thinking substance, subjective experience, spirit, consciousness, that which man perceives as within – was understood as fundamentally different and separate from *res extensa* – extended substance, the objective world, matter, the physical body, plants and animals, stones and stars, the entire physical universe..." (Tarnas, 1991:278). The sciences that studied the *res extensa* phenomena of the natural world, such as physics and chemistry, used the image of a machine to orient them. This worked well until scientific thinking was applied to humans. Only in this species, as Descartes observed, "did the two realities come together as mind and body" (Tarnas, 1991:278).

Medical science, the first discipline to apply the scientific approach to the study of human beings, placed a proverbial "black box" over all that was mind. The body, by then conceptualized and treated *as if* it were a machine, was believed uninfluenced by such aspects of *res cognitas* as feelings and emotions. Finally, when the box was opened, the study of mind, first thought of as that which ran the machine, was taken on by the new science of psychology whose practitioners, like geneticists and biologists, conceptualized the isolated individual and his or her behavior as the largest unit of investigation. The social and cultural dimensions of human life, that became part of science with the development of sociology and anthropology, have extended the mind side of the dualism taking as their subject matter much of what Descartes and his followers had placed in abeyance. To complicate matters, social scientists mostly followed Durkheim and his "relentless insistence upon sociological explanations in terms of superindividual – that is, social and cultural – processes" (Nisbet, 1965:4). They resisted efforts to reduce their subject matter to individual psychology – let alone to genes. Anthropologists and sociologists developed images and paradigms that were independent of and paid little attention to the Enlightenment vision and consequently, vigorously opposed all attempts by scholars, be they geneticists, biologists or psychologists, to explain human behavior in what for them were reductionist terms. Furthermore, when they analyze and explain human behavior they are not wedded to the image of the machine and the Cartesian dualism. For science to bridge this chasm that is the product of the mind and body separation and opposition, the Cartesian view and its dualism must be replaced. Only with the acceptance of a new vision of what the world must be like for us to know it – to paraphrase

Thomas Kuhn (1970) – that is inclusive of both sides of the dualism will a realistic unification of the sciences be possible.

A paradigm to integrate mind and body

In the following pages one such imagery, proposed by a number of theorists and clinicians at the forefront of recently created disciplines such as neurobiology, psychoneuorimunology and psychobiology whose works focus primarily on health and healing, is presented. Mind and body, and their respective analytic components, are treated as part of a more general information/communication system that seeks integration rather than separation and opposition. In this representation information is pictured as flowing in several directions within the individual organism:

1. from genes to molecules to cells and to organs;
2. simultaneously from the social and cultural beliefs shared by the members of a (specific) society thru individuals to their genes and back to their respective organs and other sub-subsystems.

Along the way, to provide an integrated picture of human beings as biological organisms who live in societies that are organized in terms of culture – with its diverse symbolic arrangements – the information that is flowing is mixed and exchanged. Neither genes nor culture are attributed ontological priority. They are simply polar ends of an interrelated and integrated system in which the interaction of their respective information content contributes to understanding and explaining the subject matter under investigation. Each system, be it genetic, hormonal, immunological, psychological or cultural, has its own code by means of which its messages are transmitted. The key to one comprehending the other, enabling it to react and respond to the message, is a form of translation called transduction.[5]

The new paradigm is the product of contributions from scholars on both sides of the divide. Geneticists made it clear that RNA and DNA could best be understood as transmitting information that leads to the making of proteins that form enzymes and eventually organs and other body parts. In a series of groundbreaking experiments, neurobiologist Candace Pert and her colleagues (Pert, 1997; Pert *et al.*, 1985), and others, demonstrated how information flows through the body carried by chemical molecules that lock on to receptors on the cells. They showed that brain function is modulated by the flow of chemicals.[6] Mind, conceptualized as brain, including human emotions, and body are integrated with the functioning of their component parts and the working of total organism being the result of information exchanges within a single system.

While this might have been sufficient for scientists wedded to the *res extensa* side of the dualism, psychologists clinically involved in using hypnotically facilitated therapy extended the thinking to come closer to bridging the divide. When they placed patients with specific physical symptoms in this altered state of consciousness (hypnosis) and made suggestions directed at amelioration, and it worked, they were forced to ask why and how? Barber (1984), for example, summarized a series of studies showing that changes in (assumedly unchangeable) bodily processes in response to hypnotic interventions resulted in a reduction in or elimination of a number of illnesses. His conclusions have been corroborated by more recent research. How do words, expressed symbolically as suggestions to patients in an altered state of consciousness cross the mind/body divide and influence physiological systems so as to aid the patient?

Bowers (1977), following Black (1969), provided the paradigmatic imagery in terms of which the question could be reformulated and answered. "The entire human body," he wrote

> can be viewed as an interlocking network of informational systems – genetic, immunological, hormonal, and so on. These systems each have their own codes, and transmission of information between systems requires some sort of transducer that allows the code of one system, genetic, say, to be translated into the code of another system – for example, immunological.

Mind, he added,

> with its capacity for symbolizing in linguistic and extra-linguistic forms, can also be regarded as a means for coding, processing and transmitting information both intra- and inter-personality. If information processing and transmission is common to both psyche and soma, the mind-body problem might be reformulated as follows: How is information, received and processed at the semantic level, transduced into information that can be received and processed at the somatic level, and vice-versa? That sounds like a question that can be more sensibly addressed than the one it is meant to replace... (Bowers, 1977:231).

Ernest L. Rossi (1993; see also Rossi & Rossi, 1996), a psychologist who studied and practiced hypnotically facilitated therapy, then proposed an answer to Bowers question. "The major locus of brain-body communication," he writes (1993, 140 italics in original),

> is the limbic-hypothalamic-pituitary system that is like a great funnel drawing in information from all the neural networks from the higher brain of the cortex and transducing this neural encoding of information into the messenger molecules of the body.... *There are cells in the hypothalamus that can transduce information encoded in the form of the electrical impulses carried by the nerves of the higher cortex of the brain into the messenger molecules of the body....*

Rossi (1993:147) begins by discerning three major loops of communication within each of the body's cells. The first are the receptor, or "receiving stations" that pick up "primary messages," such as those of the hormones from the hypothalamic-pituitary system. Second is the energy and metabolism loop that involves enzymes and the mitochondria "energy generators" that power all the basic life functions of metabolism within the cell. Third are the protein factories that make more receptor and messenger molecules that will communicate to the brain and other cells.

Clarifying the difference between the mutation of genes and their expression,[7] he observes that some of the proteins produced, called "transcription regulators," modulate the expression of other genes. That is, they activate, turn on and off a specific category of genes (Rossi, 1993:151).

There are a core group of genes that are in constant activity in most cells in the body. They maintain what is thought of as the basic housekeeping functions of life, such as metabolism, energy production, respiration, growth, or the removal of waste. A second set of genes, the so-called "clock genes," set time cycles such as sleeping, waking and dreaming. Most other genes are turned on or off to perform their specific functions in response to signals (information) coming from outside the cell, usually referred to as the environment by geneticists and others still wedded to the old paradigm. In response to these messages the activated immediate early genes, as they are called, begin, within a range of time delays that may be as short as a few seconds, to make (messenger) mRNA, that initiate the production of proteins and enzymes that flow throughout the body resulting in a range of new or modified activities and behaviors. Some of the information that turns on immediate early genes comes in the form of electrochemical impulses of neurons in the cerebral cortex that encode the phenomenological impulses of the "mind" (see Rossi, 2000:182). That is, information contained in ideas and beliefs carried in the neurons of the brain can be transduced from the psyche of an individual to initiate gene activity that leads to subsequent changes in bodily function that in turn effects the body and brain in the form of feelings and emotions.

Do ideas from the brain activate genes and bodily functions only when an individual is in a hypnotically induced state, as much of the research seems to indicate? Do ideas and thoughts expressed symbolically only influence the soma when one is in an altered state of consciousness?

Milton Erickson, who reintroduced the use of hypnosis in therapy in the United States, is reported to have observed, "that people in everyday life ... naturally drifted between subtle but distinct mindbody states" (Rossi and Rossi, 1996, 129). When he worked with patients for at least an hour and a half or two —in contrast with most therapists who saw patients for only 50 minutes— he found they would go through distinct changes in their consciousness and states of being.

> During these lengthier sessions, for no apparent reason, the patient's head might start to nod rhythmically; eyelids would blink slowly, and then close over faraway-looking eyes. The body might go perfectly still, with fingers, hands, or legs apparently frozen in an awkward position. Sometimes there was a subtle smile on the person's face or, more often, the features were passive and slack.... On some occasions, Erickson did nothing to direct people to go into trance; it just seemed to happen, sooner or later, all by itself (Rossi and Rossi, 1996:129).

Erickson called these natural periods "The Common Everyday Trance." They are times of "openness and vulnerability to outside influences; suggestions made during this time are sometimes more easily accepted" (Rossi and Rossi, 1996:130).

Half a century ago researchers observed that every 90 minutes or so throughout the night, sleep became a very "active process" for about 10 to 30 minutes during which time oxygen consumption increased and more blood flowed to the brain. Breathing, heart rate, blood pressure and gastrointestinal movements became more variable than during wakefulness. During these periods of Rapid Eye

Movement (REM) sleep brain wave patterns, as measured by electroencephalograph (EEG), became similar to the active pattern when awake. Researchers later confirmed that the 90 to 120 minute dream rhythm apparently continues during the day. Kleitman (1982, 1969, 1963) one of the first to observe the REM activity pattern, speaks of a Basic Rest-Activity Cycle (BRAC) in sleep and wakefulness. If all human beings pass through these cycles controlled by their clock genes at specific points of which they are more open to outside influences, could these be the times at which ideas and aspects of cultural belief are best learned and transduced from the psyche to activate genes and bodily processes?

Recent neurobiological studies of

> associative learning have shown that dopamine neurons respond physically to rewards in a manner compatible with the coding of prediction errors, whereas slower dopamine changes are involved in a larger spectrum of motivating events.... During initial learning, when rewards occur unpredictably, dopamine neurons are activated by rewards. They gradually lose the response as the reward becomes increasingly predicted.... Dopamine neurons also respond to novel, attention-generating, and motivating stimuli, indicating that attentional mechanisms also contribute. Results from neuronal modeling suggest that the responses of dopamine neurons to primary rewards and conditioning stimuli may constitute particularly effective teaching signals for associative learning and embody the properties of the teaching signal of the temporal difference reinforcement model (Waelti *et al.*, 2001:43 quoted in Rossi, 2002:137).

New and stimulating ideas expressed to an individual during specified periods in a BRAC cycle, an alternative way of conceptualizing a state of consciousness the equivalent of what in Western scholarship is referred to as hypnosis, in effect may not just be learned better, it also may be transduced via the psyche and be embodied in the soma.

The content of most of what is learned and remembered by an individual human being are aspects of the symbolic meaning system shared by members of the specific social group in which he grew up, was socialized and lives: his culture. That is, segments of the culture of a group, expressed symbolically – and conceptualized as information – as they are learned most effectively by individuals during critical periods of cycles normal to them and remembered through replays in later cycles enters the communication loop, mediated by the brains of individuals through which it is transduced via their psyches to their genes, turning them on or off in a regular interplay.[8] The mind influences the genes, as do the beliefs and rules for appropriate conduct shared by the members of a socio-cultural group.

Those at the forefront of the new paradigm have not fully integrated this, and its implications, primarily because they are products of Western culture and its worldview. They assume a universe of autonomous individuals each in control of their own actions. Behavior over which the individual appears not to have control, or that differs from their "ordinary" behavior, often is said to be the result of the person being in an "altered state of consciousness" (ASC). "The traditions of European peoples and their New World descendants," as anthropologist Michael

Winkelman (2000:3) writes, "have tended to deprecate ASC in their emphasis on rational thought."

The assumption of altered states of consciousness presupposes the existence of a regular or customary mode. Western scientists refer to this as "the normal waking state" during which the individual is assumed to act "rationally" and is in control. A consequence of this assumption is that those who do not behave as expected by others with whom they regulary interact, such as the patients who are treated by therapists, or who study them, are said to be in an ASC. As Goodman reminds us, people who enter such states, as for example during the performance of rituals, until recently "had the dubious distinction of being... singled out by Western psychiatry as being abnormal and hence insane" (Goodman, 1988:36).

Anthropologists, who study other, non-Western peoples and their ways of life, generally have accepted this formulation of a normal state vs. alternate ones. Consequently, they write about people, especially those participating in religious and other rituals, as being in altered states of consciousness.

Erika Bourguignon and her students demonstrated many years ago that most human beings enter into altered states regularly in the performance of religious rituals. In a sample of 488 small-scale societies, she found that almost 90 percent showed evidence of religious trance behavior (Bourguignon, 1968, 1973; Bourguignon and Evascu, 1977). "Unless one wanted to maintain that the overwhelming majority of humanity was insane... the conclusion," according to one of her students, "was inescapable that religious trance was a perfectly normal human experience" (Goodman, 1988:36).

Erickson, Rossi and others maintain hypnotic or other trance states are "normal." All human beings go through regular periodic alterations in their bodily states (of consciousness and susceptibility) in reaction to genetically based cycles. Furthermore, we learn new information best when exposed to it during these periods. Accepting this paradigm permits us to have a way to conceptualize the relationship between culture, learning and memory and its interaction and interdependence with the body.

Claude Lévi-Strauss (1963, 1966), Victor Turner (1979 [1964], 1967, 1969), and others may have had this in mind in their analyses of rituals. In the course of the unfolding of rituals, information about what Goodman (1988) calls the "other reality," with its powerful forces and beings, is communicated to the participants. Rituals, as psychologist Stanley Krippner (2005:97) notes, "are ways in which mythological themes are performed." These cultural systems contain "storehouses of meaningful symbols by which information is revealed and regarded as authoritative, ... (Turner, 1968:2). This stating "of *sacra*," as Turner (1979:242) adds

> teaches the neophytes how to think with some degree of abstraction about their cultural milieu and gives them ultimate standards of reference. At the same time, it is believed to change their nature, transform them from one kind of a human being into another.

Most rituals last long enough for the participants to experience at least one BRAC cycle (of 90 to 120 minutes) during which they are exposed, either anew or in repeated and more dramatic and exciting ways, to knowledge, information deemed of great importance to the group. As each individual passes through his or her own cycle, while being subjected to repetitions of the valued cultural symbols, each transduces it via its psyche into its memory while it literally is being embodied into their physiology as (immediate early) genes are turned on and activated. In this manner the new paradigm enables us to conceptualize an integrated culturalbiological process in which knowledge is embodied in each participant in a ritualized teaching-learning activity while becoming part of the memory of the social group that together shares in the ceremonial experience.

For tens of thousands of years Homo sapiens have been repeating the process described above as members of specific, but diverse social groups, with each using its own, distinct symbolic content while adapting to a given micro environmental niche. The ethnographic record is filled with descriptions and analyses that profitably may be reexamined in terms of this framework to see, at least hypothetically, how the culturalbiological process might have worked in specific instances. More importantly, the paradigm may be implemented in the design of future research making possible the collaboration of future generations of biological and social scientists.

Notes

1. From the time of Darwin (1871), as Stout (2008) recently writes, "Tool-use and toolmaking have been seen as 'prime movers' in a uniquely human evolutionary package linking bipedalism, dexterous hand, articulate language, culture, and a large brain" (see also Holloway, 1969; Washburn, 1960). Unfortunately, I believe, interest has moved from the package of intermixed traits that led to the emergence of humanity to a focus on intelligence and the brain. Since intelligence, like most characteristics used to compare humans and non-humans can only be conceptualized symbolically, I take symbolic ability as the single most distinctive characteristic of the human species.
2. The power of symbols, as Victor Turner (1982:9) writes, "inheres not only in the shared lexicons and grammars of spoken and written languages, but also in the artful or poetic individual crafting of speech through persuasive tropes: metaphors, metonyms, oxymora, 'wise words' (a Western Apache speech-mode), and many more."
3. That most probably was encoded in the genetic materials of the species by the time its first remains were recorded.
4. After more than a century of studying the sign-making of, and trying to teach life forms as diverse as horses, pigs, dolphins, chimpanzees and others our symbols, researchers still are unable to find species other than humans able to establish arbitrary meanings for a collection of signs they themselves are able to make.
5. "Transduction" is the process by means of which information is transformed from one energy medium into another. A microphone, for example, transduces the sound waves we produce when we talk or sing into electrical impulses, and our music system's loudspeakers transduce electrical impulses into the sound waves we receive via our ears. The technology at the grocery store's checkout counter transforms bar codes on purchases into laser light reflections, then electrical impulses, and finally a mechanically imprinted itemized bill.
6. "Many of these informal substances are neuropeptides, originally studied in other contexts.... ... most, if not all, alter behavior and mood states. We now realize that their signal specificity resides in receptors (distinct classes of recognition

molecules).... Rather precise brain distribution patterns for many neuropeptide receptors have been determined. A number of brain loci, many with emotion-mediating brain areas, are enriched with many types of neuropeptide receptors suggesting a convergence of information at these "nodes." Additionally, neuropeptide receptors occur on middle cells of the immune systems, monocytes can chemotax to numerous neuropeptides via receptors indistinguishable from those found in the brain. Neuropeptides and their receptors thus join the brain, glands, and immune system in a network of communication between brain and body, probably representing the biochemical substrate of emotion" (Pert *et al.*, 1985:820s).

7. Mutations are chance errors that, if beneficial, change the phenotypes of future generations. Gene expression is the making of copies of certain genes or groups of genes that act as blueprints sent out in the form of messenger ribonucleic acid (mRNA) to the "protein factories" of the cell where they serve as templates for the manufacture or 'translation of the mRNA into new proteins.
8. This of course should not be taken to mean that cultural beliefs are transduced and embodied identically in each member of a social group. Humans are able to question and challenge the symbolically expressed beliefs to which they are exposed, even, or perhaps better, when in an altered state of consciousness.

Bibliography

Barber T. X. (1984), *Changing "unchangeable" bodily processes by (hypnotic) suggestions: A new look at hypnosis, cognitions, imagining, and the mind-body problem.* In: *Imagination and healing*, ed. A. A. Sheikh, 69–127. Farmingdale, NY: Baywood Publishing Company.

Black S. (1969), *Mind and body.* London: Kimber.

Bowers K. (1977), *Hypnosis: An informational approach.* Annals of the New York Academy of Sciences, Vol. 296, Pg. 222–237.

Bourguignon E. (1973), *Religion, Altered states of consciousness and social change.* Columbus, OH: Ohio State University Press.

Bourguignon E. (1968), *A Cross-cultural study of dissociational states: Final report.* Columbus OH: Research Foundation.

Bourguignon E. and Evascu T. (1977), *Altered states of consciousness within a general evolutionary perspective: A holocultural analysis.* Behavior Science Research, Vol. 12, 197–216.

Crick F. H. C. (1958), *On protein synthesis.* In: *Biological replication of macromolecules.* Symposium of the Society for Experimental Biology Vol. 12, Pg. 138–163.

Darwin C. (1871), *The descent of man and selection in relation to sex.* Vol. 1. London: John Muray.

Descartes R. (1980), *Discourse on method and meditations on first philosophy.* D. A. Cress, trans. Indianapolis: Hackett. Originally Published in 1637, 1641.

Goodman F. D. (1988), *Ecstasy, ritual and alternate reality: Religion in a pluralistic world.* Bloomington: Indiana University Press.

Hobbes T. (1963), *Leviathan.* Cleveland, OH: World Pub. Co.

Holloway R. (1969), *Culture: A human domain.* Current Anthropology, Vol. 10, Pg. 395–412.

Kleitman N. (1982), *The basic rest-activity cycle - 22 years later.* Sleep, Vol. 5, Pg. 311–315.

Kleitman N. (1969), *Basic rest-activity cycle in relation to sleep and wakefulness.* In: *Sleep, physiology and pathology*, Anthony Kales (ed.), Pg. 33-38, Philadelphia: Lippincott.

Kleitman N. (1963), *Sleep and wakefulness as alternating phases in the cycle of existence.* Chicago: University of Chicago Press.

Krippner S., *Trance and the trickster: Hypnosis as a liminal phenomenon.* Journal of Clinical and Experimental Hypnosis, Vol. 523 (2), Pg. 97–118.

Kuhn T. (1970), *The Structure of scientific revolutions.* 2nd ed., enlarged. International Encyclopedia of Unified Science. Chicago: University of Chicago Press.

Lévi-Strauss C. (1966), *The savage mind.* Chicago: University of Chicago Press.

Lévi-Strauss C. (1963), *The effectiveness of symbols.* In: *Structural anthropology*, Vol. I. Trans. Claire Jacobson and Brook Grundfest Schepf, Pg. 186–205. New York: Basic Books, Inc.

Lewontin R.C., Rose S. and Kamin L. K. (1984), *Not in our genes: Biology, ideology, and human nature.* New York: Pantheon Books.

Nisbet R. (1965), *Émile Durkheim.* Englewood Cliffs, NJ: Prentice Hall, Inc.

Pert C. (1997), *Molecules of emotion: Why you feel the way you feel.* NY: Scribner.

Pert C., Ruff M., Weber R. and Herkenham M. (1985), *Neuropeptides and their receptors: A psychosomatic network.* The Journal of Immunology, 135:2, Pg. 820s–826s.

Rossi E. L. (2002), *The psychobiology of gene expression: Neuroscience and neurogenesis in hypnosis and the healing arts.* New York: W. W. Norton.

Rossi E. L. (2000), *In search of a deep psychbiology of hypnosis: Visionary hypotheses for a new millennium.* American Journal of Clinical Hypnosis, 42:3/42:4, Pg. 178–207.

Rossi E. L. (1993), *The psychobiology of mind-body healing: New concepts of therapeutic hypnosis.* Revised Edition. New York: W.W. Norton. Original Published in 1986.

Rossi E. L. and Rossi K. L. (1996), *The symptom path to enlightenment: The new dynamics of self-organization in hypnotherapy: An advanced manual for beginners.* Pacific Palisades, CA: Palisades Gateway Publishing.

Smith A. (1937), *An inquiry into the nature and causes of the wealth of nations.* New York: The Modern Library.

Stout D. (2008), *Technology and human brain evolution.* General Anthropology. Bulletin of the General Anthropology Division, 15:2, Pg. 1–5.

Tarnas R. (1991), *The passion of the western mind: Understanding the ideas that have shaped our world view.* New York: Ballentine.

Turner V. (1982), *From ritual to theatre: The human seriousness of play.* New York: PAJ Publications.

Turner V. (1979), *Betwixt and between: The liminal period in rites de passage.* In: *Reader in comparative religion: An anthropological approach*, 4th Edition, edited by W. Lessa and E. Vogt, Pg. 234–243. New York: Harper & Row. Originally published in "Proceedings of the American Ethnological Society". Symposium on new approaches to the study of religion, Pg. 4–20, 1964.

Turner V. (1969), *The ritual process: Structure and anti-structure.* Chicago: Aldine.

Turner V. (1967), *The forest of symbols: Aspects of ndembu ritual.* Ithaca, NY: Cornell University Press.

Waelti P., Dickinsen A. and Schultz W. (2001), *Dopamine responses comply with basic assumptions of formal learning theory.* Nature, Vol. 412, Pg. 43–48.

Washburn S. L. (1960), *Tools and human evolution.* Scientific American, Vol. 203, Pg. 3–15.

Winkelman M. (2000), *Shamanism: The neural ecology of consciousness and healing.* Westport, CT: Bergin and Garvey.

Chapter 9
Technology and the body in aboriginal identity making

Diane Hafner, Deborah Brian

Abstract

The Lamalama of remote northeast Australia are avid users of electronic communications technology. Mobile phones have rapidly become a preferred means of contact within this intensely socially-defined, kin-based culture. Their enthusiastic adoption of electronic communications technologies of this kind defies simplistic notions of cultural authenticity or hybridity, and instead presents further insight into the nature of the body-technology nexus. In this paper, Wilk's (2001, 2002) notion of 'cultivation' is employed to comment on the cultural dynamics of uses of technology by the Lamalama. The relationship between embodied cultural practice and the capacities of electronic communications technologies is discussed, as is the significance of specific modes of engaging with technology in the production and reproduction of cultural identity. Lamalama use of communication technologies intersects with a variety of other forms of technology. Their use of these differing tools demonstrates that technological reach is revealed through bodily interactions with it, rather than as residing merely in its functional design and associated applications.

The envelope of technology, or 'curtain of objects' (Stiegler 1998:57), that surrounds the human body effectively mediates our interactions with the physical environment. In the case of modern wireless technologies, such as cellular or mobile phones and internet connections, it grants us the ability to be in many places at once. Using such technologies, we can be physically present in one location, and virtually present in another. We can even be virtually present in other times. Technologies of image and sound recording and reproduction permit a person's 'presence' to extend beyond even the tenure of their physical body.

In this way, deceased members of a family or a community can maintain an enduring presence through photography, film and voice recording. Their images and sounds can be captured, reproduced and distributed in a range of forms, something greatly facilitated by the virtualities of the World Wide Web. Using audio and visual recording technologies, we can manipulate the usual spatial and temporal constraints on human action and render the absent present in socially meaningful ways. Thus, living individuals retain the capacity to enter into a kind of synchronous relationship with ancestors, mediated by these technologies of memory construction.

In a current research project, we[1] are engaged in eliciting the meanings attached to photographic representations of the forebears of a contemporary group of Indigenous people, the Lamalama of remote Cape York Peninsula in northern Australia. The Lamalama count among their ancestors a number of individuals encountered and photographed by the anthropologist Donald Thomson in 1928 and 1929. Thomson spent many months in eastern Cape York Peninsula, and his photographs document something of the common ethnological concerns of the day, but also the close nature of his relationships with many of the people depicted (Rigsby and Peterson 2005).Thomson also commissioned or collected some of the quotidian objects depicted in his photographs, and in total, this collection of items ties his images, fieldnotes, and published writings more completely to the individuals he portrayed.

The Donald F. Thomson Ethnographic Collection has been preserved by his family and is currently housed in Museum Victoria which is located in Melbourne, Australia.[2] Our current multidisciplinary project is a collaboration with the Lamalama that aims to incorporate their perspectives into the ongoing management and understanding of the Collection. As we work with the Lamalama to develop an understanding of contemporary meanings and values they apply to the Collection, we draw on the capacities of a wide range of information and communication technologies. We take digitised copies of Thomson images and digital photographs of material culture items from the Collection to Cape York Peninsula, and bring Lamalama people to Museum Victoria to view the Collection and discuss its management. We film every stage of the work, both in Cape York Peninsula and in Melbourne.

The visual data that result from this process allow us to capture and analyse the discourse, gesture and social interaction that occurs in response to individual Lamalama engagements with the Collection, but they are also the raw material for future cultural production. The Lamalama are likely to use one or more pieces of the resulting visual material in community contexts, including a demountable exhibition and traditional cultural training programs. In addition, individual community members have proposed a range of technological initiatives for recording and disseminating information about their heritage via the Web, PC-based games, video and audio recordings, and print materials.

However, in thinking about how the Lamalama relate, or plan to relate, to their heritage via electronic technologies, it becomes apparent that in many ways use of these technologies represents a continuation of existing cultural modes, an expression of the Lamalama *habitus* (Bourdieu 1977). Just as the Lamalama have embraced the use of mobile phones, and in some cases, email, to maintain contact with kin beyond their remote region, so too they seek to use web-based technologies and audio and visual recording to make aspects of their heritage tangible. By employing digital copies of images and objects from a distant museum in local tourism and educational contexts, the Lamalama are able to make their own cultural meanings material for their children and for visitors. A similar process is already in train in relation to materials of this kind that individuals own themselves. By displaying images and recordings of their deceased ancestors the Lamalama can ensure their ongoing 'presence' in ways that contribute to the transmission of culture across time.

In this paper, I explore the nature of Lamalama engagement with technology, and particularly their use of technology in negotiating the boundaries between the individual and the community, and between people and their environment. In doing so, I draw on notions of individual use patterns, and of 'cultivation' (Wilk 2002), and habitus (Bourdieu 1977) to provide some frames for thinking about the importance of local patterns in conceptualising the relationship between people and the things they use. In a more general sense, electronic technologies offer the potential for creative, communicative, intellectual and affective expression that continues to be revealed as new forms of technology enter the market. Existing needs, feelings or habits may be challenged in the changing social and technological environment thus produced.

Wilk (2001; 2002) describes this situation in terms of a changing balance between two opposing general processes, 'naturalisation' and 'cultivation'. Originating in Bourdieu's (1977) understanding of habitus, these concepts seek to explain both the *naturalisation* of discretionary choices or 'wants' to become deeply felt 'needs', and on the other hand, the *cultivation* of more open discourse and discussion around those choices previously regarded as 'needs'. In essence, 'naturalisation' describes a shift from heterodoxy to habitus, and 'cultivation' signifies a shift from habitus to heterodoxy (Wilk 2002). Wilk (2002) describes the generation of a multitude of behavioural choices, embedded in complex social and moral discourses that emerge in relation to previously unquestioned needs and practices in response to new global environmental imperatives. The concepts outlined, however, have much broader potential application in the analysis of human cultural choices in relation to the adoption of technologies and related behaviours.

Lamalama responses to technology lean more towards the process of cultivation, in the sense that they typically consist of pragmatic, everyday problem-solving within their changing technological environment. This has been directly and indirectly evident during the current project, in which both research aims and Lamalama wants and needs have coalesced around their engagements with the functions and uses of technology. While the current project focuses on the relationship between museums and their source communities, it is technologically mediated at all points: among the researchers, between the researchers and the community, among community members; between the community and their past; and between the community and their future, as expressed in collective identity and group endeavours. A brief overview of the current situation of the Lamalama and this museum research project provides some context for the interpretations that follow.

Research with the Lamalama

The Lamalama are a group of around 200 Aboriginal people whose traditional estates in land, known as their *country*, occupies the hinterland of Princess Charlotte Bay on the east coast of Cape York Peninsula. Although they live in a remote part of Australia, and continue to incorporate hunting and gathering among their economic strategies for survival, they are not removed from the broader social and economic systems of the nation. Many of the Lamalama live in the small town of Coen, where there is access to state-run schooling to secondary level, after which all children must leave town to attend boarding school elsewhere. On weekends and during school holidays, Lamalama people, including parents and carers of school-aged children, travel down to their country on the coast at Yintjingga (Port Stewart), a distance of about 95 kilometres. Here they fish and hunt, and relax in the company of their kin, away from the temptations and harassments of town, where misuse of alcohol and resulting social disruption sometimes prevails. A core family group of perhaps half a dozen people are permanent or semi-permanent Yintjingga residents.

Cape York Peninsula has a monsoonal climate, and during the 'Wet' season the Lamalama are frequently prevented from reaching their homelands by floods and the associated poor condition of the gravel road linking Coen and the coast. While a few individuals may remain behind at Yintjingga as caretakers, most people return to Coen to sit out the rains. Coen township consists of a hotel, two shops, a couple of buildings used intermittently as churches, a state-run Health Clinic and School, and community facilities, as well as residential areas roughly divided into 'whitefella' and 'blackfella'[3] sections.

Most of the Lamalama are unemployed, or underemployed at least. Younger women hold jobs as shop assistants, government employees in the Clinic and School (in teacher aide and clerical positions) and in welfare work (home care of the aged), while younger men generally obtain seasonal employment as stock workers in the local pastoral industry. A regional lack of industry and enterprise means there is broad-scale unemployment. Many non-Aboriginal people in town subsist on welfare payments, while Aboriginal people participate in the Community Development Employment Program (CDEP). This national program provides Indigenous people around Australia with payments equivalent to unemployment benefits for work within the community, such as rubbish removal, mowing, street-cleaning, and home duties.

In economic terms the Lamalama could thus be described as mixed-strategy hunter-gatherers, but it is more accurate to describe them as people who live within two separate social and economic systems. To some extent this depends on their location, bush or town, but as a kin-based society in which cultural patterns transcend such neat divisions, there is an overlapping pattern of behaviour that is reflected in the distribution and use of resources of all kinds, as well as of values, obligations, and expectations with regard to behaviour. This pattern is also evident in Lamalama approaches to technology.

Members of the current research team have worked with the Lamalama for twenty to thirty years in some cases, providing context and longevity to our observations of their changing circumstances. Aboriginal experience in Australia reflects a well-documented history of dispossession, abuse and neglect as a result of a colonial past (Reynolds 1981; Kidd 1997; Attwood 2005) which the postcolonial settler nation is still struggling to reconcile. This fact was recently recognised in the *National Apology to the Stolen Generations* offered by newly elected Labor Prime Minister Kevin Rudd on behalf of the Australian Parliament in February

2008 (Commonwealth of Australia 2008). While the long-term impacts of this new era in national governance and race relations are a continuing subject for speculation, there is significant scope yet for recognition of the diversity of historical experience among Indigenous communities.

The Lamalama experience of state-sanctioned dispossession and social disruption has continued well into the period of living memory. In 1961, a significant number of the Lamalama were forcibly removed from the Yintjingga region to a mission some hundreds of kilometres away, under the terms of prevailing legislation (Rigsby and Hafner 1993). Over time, and as political conditions relaxed, they drifted back to the Coen region and slowly began to re-establish their presence within their country. The term 'country' implies a holism that incorporates notions of a shared corporeality extended between the person and the landscape, its resources, and all that is known or remembered about human interactions within it.

Since 1991, the Lamalama have acquired formal title to parcels of land within their country as a result of successful engagements with provisions of the state through land claim and native title legislation and other processes.[4] These events have done little to change their economic status, partly because of the nature of the titles acquired[5], but for a range of other reasons as well, including the economic and social marginalisation of the region. One of the aims of the current research has been to work collaboratively with the Lamalama towards their goals of cultural revitalisation, and the development of digital and other products that would enable them to tap into the dormant tourist market. Each year, thousands of four-wheel-drive and adventure tourists pass through Coen on their way to 'The Tip' – the northernmost part of mainland Australia. Some stay overnight in Coen, or simply stop to eat and re-fuel their vehicles. Currently, however, the flow of tourist traffic through Coen represents an under-utilised opportunity for community economic development.

As part of a larger Aboriginal population centred in the town, the Lamalama have access to a community Cultural Centre, managed by the Coen Regional Aboriginal Corporation (CRAC), an obvious place to try to attract tourist spending through exhibitions and other activities. While the current research collaboration offers potential for development of this opportunity, this was only ever a secondary rationale for Lamalama participation in the research. Rather, it is the potential to engage with their own history and so forge alternative narratives of identity through examination of the Thomson Collection materials that has been of primary importance. The community's remote location and economic and social marginalisation has previously limited their capacity for engaging with their past through the materials in the Collection, and it was specifically at their request that we sought funding for this project.

We have aimed to investigate the broadest range of their experience of the Collection during the first two years of this three-year project, through the methodological approach of combined field and museum visits. As indicated already, these activities have been filmed, and as such provide a second-level data source. We expect different kinds or levels of meaning to emerge in the differing locations of the study, and to date, the potential for performativity that the camera invokes has been especially interesting. While that is the subject of other papers (Hafner, Rigsby and Allen 2007; Wilmot 2008; Brian & Hafner 2008), the nature of Lamalama engagements with technology are also a subject of interest and observation for us. Our view of the way in which the Lamalama engage with technology reflects some recent approaches that stress the need to examine accepted hegemonies with regard to interpretations of the relationship between people and the things they use (Shove and Southerton; Horst and Miller; Wilk 2001).

Technology and tradition among the Lamalama

The Lamalama have been enthusiastic in their uptake and use of technologies of modernity. Elsewhere (Hafner 2008) I have described their interest in such technology, and while they use devices such as motor vehicles, televisions, VCRs, and to a lesser extent, personal computers, communications technologies are especially popular. Telecommunications in Cape York Peninsula are provided by Telstra, Australia's majority telecommunications provider of such services. In its various iterations, the government owned and operated Telstra historically maintained the monopoly position in the field, but is now only one of several providers of telecommunications in the country. Although now partly privatised, Telstra retains its majority position in the field, particularly with regard to the provision of services outside urban areas. Until April 2008, people in remote regions such as Cape York Peninsula were largely dependent on Telstra's analogue-based CDMA service, unlike their fellow citizens in urban areas able to access a broader range of digital networks. The CDMA service has been phased out in favour of third generation technology, which promises to provide more services, better coverage, and faster internet access speeds (Telstra 2008) in accordance with international standards for such technologies.

Mobile or cellular phones are the most popular communications technology among the Lamalama, reaching close to total saturation within the community. Adults and children almost without exception own and use mobile phones, which raises questions about the cultural purposes they serve. Certainly it is the case that people use mobiles to call relatives and friends in distant locations, or children away at boarding school, or for similar practical purposes. Some people also maintain standard landline phone services, but these have given way in favour of mobile phones in recent years. As a kin-based society, the Lamalama live in very close proximity to each other, certainly in Coen (to date, third generation technology has not provided coverage to the Port Stewart region), where it would be as easy, and certainly less expensive, to walk down the road to speak to people, rather than call them on a mobile phone. Young people especially like to call each other on their mobiles.

One way to explain this situation is to place Lamalama use of technology within a framework of cultural adaptation. This approach potentially has both positive and negative facets. In Australia, it has been popularly used to argue that Indigenous uptake of technology reflects a loss of tradition rather than a process of adaptation (Ponte, Marsh and Jackson 1994). The normative use of four-wheel-drive vehicles and guns to replace the practice of hunters pursuing their prey on foot using spears has been interpreted as evidence of a loss of tradition. This view suggests that tradition resides in the materials by which it is produced, and ignores the fact that it is better understood as a set of practices through which people reproduce culture over time.[6] It is also not an argument that effectively explains why, among the Lamalama, the technologies that extend the reach of the body through space are so enthusiastically received over those that offer more passive recreational value, such as television, personal computers and the like. A view that takes account of specific needs and use patterns has greater explanatory power here.

The earlier approaches of the French scholars Leroi-Gourhan (1993) and Lemonnier (1993) have been influential in this arena. Lemonnier (1993), for example, suggested that technology should be understood as in reciprocal relationship to the social systems in which it is found. To understand the role of technology in evolutionary development, it is therefore necessary to account for human actions upon it, but also its impact on its material environment, and the knowledge that produces and uses it. That is, when considered together, sociality and technology produce technological systems (Lemonnier 1989:156). Both Lemonnier and Leroi-Gourhan were concerned to explain the role of technology within human evolution, and at least in Leroi-Gourhan's case, possible teleological implications need to be separated from his understanding of technology as socially embedded in systems of practice (Audouze 2002:286).

Also influential has been Bourdieu's (1977) application of the notion of habitus – "the whole realm of the unconscious, patterned rules and feelings that guide behaviour" (Wilk 2002:10) – which indicates that specific needs and patterns of use require investigation and explanation. Individuals experience needs as part of the habitus, in the more heterodox world of the present, but "the rate of expansion of existing needs, the generation of new ones, as a product of the changing balance between two general processes" (Wilk 2001:115) is increased in times of rapid social or cultural change. Wilk sees this as particularly so in modern times, when it is possible to challenge the habitus of both producers and consumers through the development of new technology and its associated functions. In Wilk's view, a cultural model of needs is implicated in the production of any artefacts, but is constrained by the habitus of the producer, "their bodily practices and taken-for-granted common sense notions of the world" (Wilk 2001:117).

New products therefore succeed in the marketplace less through specific design than because they cultivate new needs in consumers. In this sense "people are always discovering new ways to use things: in other words they have a dynamic relationship with objects which allows them to develop new needs" (Wilk 2001:117). For Wilk (2001:117), needs "lie on the boundary between the social and the material, but cannot be reduced to one or the other". This can be seen in Lamalama use of mobile phones, and, to a lesser extent, email. Their acceptance of these technologies reflects patterns of intentional use that were perhaps not envisaged by their developers. Instead, people invest them with the functionalities that embody their own heterodox purposes. In other words, people bend technology to their own purposes.

The Old People, the story, and the embodiment of culture

For the Lamalama, the field of social action is not bounded by the presence of living, embodied social actors alone, and this has implications for their understandings of animation and agency in both people and things. The Lamalama habitus associated with the use of technology is not constrained by distinctions between traditional and non-traditional items and their uses, or in some senses, by distinctions between body and object. All things have an ephemeral or durable pragmatic value, and are invested with specific cultural meaning, either quotidian or cosmological, and sometimes both simultaneously. Thus, TVs and VCRs provide relief from the boredom of long days spent indoors during the peak of the monsoon season, but may be discarded when broken down, while fishing spears hand-crafted on country using its resources are the enduring and fiercely protected property of their makers, despite the fact that the materials to make them anew are in abundance at Yintjingga. Animals they hunt and consume are understood as invested with an agency within the realm of living things akin to their own, which generates the need to treat them with respect. The tools with which they are caught, for example, are thus subject to ritual investments that ensure the protection of both hunter and hunted. The Lamalama technological habitus can therefore be understood as describing a cultural arc between cosmological rules and values, and pragmatic evaluation of a variety of objects and their functions.

Clearly, people such as the Lamalama live in mixed cultural and social systems. When Thomson visited Port Stewart in the late 1920s, he found an autonomous people living in kin-based clan groups, speaking several local indigenous languages, and subsisting largely by hunting and gathering around the Stewart River estuary. By 1961 they had been removed from the region by the state, and it is really only in the last 15 years their descendants have been able to formally acquire land there and re-establish their presence. The appropriation of their traditional estates through pastoralism and other purposes resulted in the loss of specific ritual practices, such as the *bora* (male initiation), but the system of belief and associated rules on which such practices depended remains vital in Lamalama senses of the self. To be Lamalama entails understanding of the individual's integral connection to kin and country, and this knowledge is generationally variable. Younger people depend on their elders for authoritative interpretations of the complex linkages between the now moribund clan structure and the rules for behaviour it generated, includes those governing marriage, descent and inheritance of cultural property. Although Lamalama society is now organised in terms of 'families', the clan system remains evident in expressions of both social structure and behaviour in the present. Families whose forebears were members of clans owning land south of the Stewart River mouth, for example, are still known as the traditional owners of that country, and it is they who have primacy in any decisions made about the area.

As a kin-based society, personal relationships among the Lamalama remain subject to the cultural rules that determined behaviour in the past. Indigenous kin terms and the behaviour associated with them are used to demonstrate respect for appropriate categories of relatives, including the rules associated with sexuality and marriage, commensality, and mortuary practices and bereavement. On death, the disembodied social actor returns to his or her country, in Lamalama belief, and is then referred to as one of the Old People – now-deceased relatives who remain in the landscape as dynamic social actors. They may transubstantiate, as in one such recent case where the individual in question returned to certain of his kin as a green frog,[7] or else appear as one of a variety of spirit beings including ghosts, or more commonly as an energetic presence able to effect action in the material world. The Old People remain within the social field for an undisclosed period of time, and become guardians of both the natural landscape and the actions of its living, embodied owners – it is because the Old People are also owners of country that they are understood to have that responsibility despite their disembodied or transubstantiated state.

Another category of important supernatural social actors are the Story, elsewhere in Australia referred to as the Dreaming. These are the ancestral beings who created the world and its natural features and placed people and languages within the landscape. For the Lamalama, the most important of these is Marpa Hamanhu, whose home sits off the coast of the Stewart River estuary in the group of small islands that make up the Cliff Islands National Park. Marpa and other such beings are located in numerous sites within their country, and behaviour determined to be appropriate in relation to them is inscribed in Lamalama cosmology. This includes rules governing behaviour associated with pollution – for example, the proscription against urination in saltpan areas – but also includes demonstrations of respect, as in the proscription on shouting and other boisterous behaviour at some sites. Such behaviour is determined to be *kintya*, or 'against the Law', and failure to observe *kintya* rules has consequences ranging from bad luck to illness, disability and death.

Thus, inappropriate actions within the landscape may be punished by either the Old People or the Story – failure to 'warm' lines and spears over a fire of ironwood leaves before going fishing will probably result in a bad catch, while failing to ritually introduce strangers by giving 'smell'[8] and 'calling out' their names to the Old People and the Story while on country can result in more serious consequences. People causing damage to Story places, or sacred sites, are likely to suffer whether or not they are Lamalama. This was the case many years ago when a pastoralist dynamited one such site to provide a waterhole for his cattle. The crippling disability he began to suffer soon after was interpreted by the Lamalama as punishment by the Story for his actions, and at least one senior person related this disability of the lower limbs to the specific powers of the associated Story.[9]

Relationships among living kin are equally as closely connected as those between the Lamalama and their deceased kin. There are several other Indigenous groups resident in Coen, and although they mix with them in various ways, the Lamalama demonstrate a preference for each other's company. While they also have social relationships with local white people, these are for the most part much more circumstantial, even opportunistic, and decidedly more tenuous in terms of shared affect. Most of the Lamalama live in family-based households, although in recent years those people of approximately 40 years of age or under have taken advantage of newly-available public housing in Coen, moving out of the family home to establish their own households. These new household units are variously successful, and a tendency for sets of siblings to live together prevails, ensuring greater financial viability to the household. Thus the Lamalama live in a remote part of the country in a world of intense social relations in which communication assumes special importance. Coen is accessible by road and serviced by a regional airline that offers flights in and out several days a week, but both geographic and social location can be seen to reinforce their existing habitus when their patterns of communication are examined.

In the cultural senses implied by their cosmology, the Lamalama sense of the self is not entirely contained within the boundaries of the individual body. This implies more than that the Lamalama regard themselves as members of a collective group with shared identity, although that is certainly true. It is rather that the relationship between themselves and their land is one of shared substance. People live on the land and by virtue of the exchange of bodily matter that occurs as a result of their passage over it, they understand themselves to share in its substance.

Several years ago, when the Lamalama were in the process of establishing a market garden in Coen, a woman who had devoted considerable effort to it through planting and tending of vegetables suddenly died, and the garden was immediately closed to the Lamalama. In common with other Aboriginal people in Australia, the Lamalama are careful to observe bereavement conditions, which include prohibition on naming of the deceased person for a lengthy period of time after death.

These practices were observed in this case, yet despite the abundance of produce available in the garden, none of it could be eaten by the Lamalama. To enter the garden would have meant they were trampling on 'her' in the sense of walking on ground into which she had dropped her sweat, where it mingled with the earth in which the vegetables grew. By eating the vegetables they would also have been consuming the essence of their relative.[10] A relationship of shared substance between individuals and the country to which they belong is thus indicated, and demonstrates the way in which their habitus includes a sense of embodiment and substantiation, through which the person expresses their identity as Lamalama.

Of phones and photos: patterns of communication and the functions of technology

By comparison, Horst and Miller (2005) discuss the use of third-generation cellular or mobile phones in a kin-based society in Jamaica. There, consumers used these new technologies to meet their own specific needs, rather than for their designed or marketed functions. Horst and Miller suggest it is the inexpensive voice facility of third generation phones rather than the new features they provide, such as video phoning, that accounts for their success in the marketplace.[11] Their observations on the 'linkup' functions of mobile phones has some resonance with Lamalama usage, although it has to be said that Jamaican use patterns are in some senses quite different. Jamaicans use mobile phones for a variety of networking purposes, including the organising of activities around church, commerce and child-rearing. However Horst and Miller (2005: 758) indicate that, for men, "the most extensive form of phone-based networking revolves around the potential of sexual liaisons", with the added objective, for women, of "finding a man sufficiently solvent and reliable to support them".

Lamalama uses of mobiles include some similar networking and flirtatious purposes but their use of them is also generated by their remote geographic circumstances. There is no doubt that the technology's primary purpose as a practical communication tool is important, but this does not provide a complete picture of its cultural importance. For instance, although there is a government-run clinic in Coen, there is no hospital. Regular visits are made by the Royal Flying Doctor Service's medical teams, but people in Coen need to travel to the regional city of Cairns if they are faced with a stay in hospital. Likewise, while the two shops in Coen offer a variety of general goods, the purchase of motor vehicles or large white goods, for example, also requires that people travel to Cairns or other regional centres such as Weipa or Cooktown.

Local people measure their remoteness in terms of driving time rather than distance in this part of Australia. While Cairns is approximately a six to eight hour drive away, the nearest towns are respectively three to four hours' drive to the north (Weipa), or four hours' drive to the south (Laura). Although the Peninsula Development Road, the public highway that connects Coen to the outside world, is slowly being sealed, large tracts of it are still gravel surfaces in constant need of maintenance. During and after the Wet Season, times taken to reach other centres are increased as a result; flooded causeways and the impact of vehicle passage over wet roads makes them impassable in places. When out on country at Yintjingga, people have some limited contact with Coen and the outside world via satellite phones, available mostly in the outstation settlement areas, but these are not the only areas they access. Fishing, hunting and other activities can take people substantial distances away and out of use range, sometimes for protracted periods. These circumstances have been alleviated to some degree by the advent of mobile phones, which provide the security of potential contact in situations of difficulty, just as they do in urban situations.

Yet, apart from these pragmatic needs, the Lamalama have taken enthusiastically to mobile phones for link-up purposes of their own. For them, the majority of these purposes reside in maintaining kinship networks. Unlike the situation of weak kin networking through mobile phone use in Jamaica described by Horst and Miller (2005), Lamalama senses of the self and collective identity are strongly embedded in and bounded by kin relations and the behavioural rules they dictate. Even among children and young adults, these principles are maintained as a matter of course, seen as the guiding principles for life, and expressed in the moral and jural code described simply as "Law". Mobile phones provide the Lamalama with a way to maintain the close social contact with kin that has always been the basis of their experience. When relatives are away in Cairns, for example, they may be called in order to check on the progress of events there (admission to hospital, time of flight departures, and so on), to pass on news from home, or for seemingly less consequential reasons. This is particularly the case among young people, who will briefly call each other simply to share jokes.

In Coen, much the same patterns of phone use pertain. The township of Coen is divided into roughly three separate areas: "town" (the central area with shops, hotel, School and Clinic); "bottom end", on the southern edge of town; and "nuther side", the residential area situated on the other side of Lankelly Creek, and given over mostly to Aboriginal housing. Although some people walk between these areas as need or inclination dictate, most prefer that kin take them where they want to go if a vehicle available. This is despite the fact it takes approximately 5-10 minutes at most to walk between each of these areas. While vehicle ownership is something of a moving feast – few people can afford to purchase and maintain the four-wheel-drive vehicles required to traverse the region – everyone seems able to maintain a mobile phone and even upgrade to third generation technology. The current phasing out of CDMA made this upgrade inevitable, but in Coen many people moved to the new technology well before it was necessary to do so.

Mobile phones assist the Lamalama in maintaining both immediate and long-distance relationships. This is done in ways that reflect cultural practice and so maintain Lamalama habitus. For instance, specific rules of etiquette apply here in relation to private spaces. At Yintjingga, people used tents and open hearth areas before houses were built. It was common practice then that any kin approaching another family's living area would 'announce' their presence by coughing, clearing their throat, or calling out, rather than simply walking in, out of respect for privacy. Similar rules apply to entering houses in Coen. in a continuation of this practice today, Lamalama people use mobile phones to call each other while in neighbouring houses. They use them to call each other in different parts of the street, or in different shops, thus maintaining contact with each other in a way that is reminiscent of Jamaican link-up purposes as described by Horst and Miller (2005). They maintain relationships with kin and friends in remote locations through brief calls of this kind, and some use email for similar link-up purposes.

One young woman who owns her own computer maintains contact with a string of between six and fifteen friends and relatives through the posting of chain letters, online jokes and aphorisms, as well as occasional news items of relevance. Another maintains contact with a string of between six and ten people, posting the same range of messages. By virtue of my relationship with these individuals, I am copied into both their lists, and in the last year, have received on average two messages a week from the first woman, and one every two months from the second. It is not usually necessary to reply to each email, or to enter into ongoing conversation, but the regular sending of emails serves to sustain links between people both locally and at a distance.

Similarly, people will call to remind drivers at the "bottom end" to collect them on the way to "nuther side", despite the fact the message could be as easily achieved by flagging down the vehicle on its way though town. Mobiles help to remind the Lamalama of their obligations towards each other – in this instance, even the few minutes of the drive home serves to reinforce the connection and associated companionship that exists between kin. The possibilities offered by new phone technology are not restricted to calls alone, allowing people to more easily send images of themselves and other relatives to kin and friends in other parts of Australia. Lamalama purposes in using communications technology are therefore based in the affective dimensions of their sociality. As an embodied practice, the use of mobile phones, in particular, extends their emotional reach beyond the physical boundaries of the body.

Yet, as with the Jamaican usage described by Horst and Miller (2005), calls are usually brief, not conversational, and in the ways described, simply serve the function of maintaining contact between family and friends. People do not generally engage in protracted conversations with kin and friends when they make mobile phone calls, although they may do so when using landlines. This is partly because of the associated cost, with landline charges being considerably cheaper, but it is also reminiscent of the link-up purposes described by Horst and Miller – what is said is less important than the act of saying it. As Fog Olwig (2005) notes, Jamaican cell phone use points to "the emergence of what one might call, using Appadurai's (1996) notion of the "-scape," a "chatscape" – a fluid, mobile, multipurpose ego-centered form of sociality based on fragmentary, informal chats that is negotiated by individuals in accordance with their particular life circumstances". In these ways, contact and coherence across group experience is maintained, and, in the case of the Lamalama, points to a particular style of sociality that they identify as "proper Lamalama way".

The fundamental principle at play in this notion of "proper Lamalama way" is the idea of *respect* – respect of the person, of the country, of strangers, of animals and the rest of the natural environment. The rituals of 'warming' and of 'calling out', described above, are part of a complex of activities within country that includes proper methods for hunting, killing, and butchering animals for human consumption, and for extracting plant resources, as well as observation of behavioural rules at Story places. The Lamalama also demonstrate respect for each other through the use of appropriate kin-terms, and show respect for country and the environment through practices such as harvesting wild resources and 'cleaning' the country (burning off debris). In behaving in this way they are honouring their ancestors and living in the Lamalama way. In terms of our current research, their visits to the Thomson Collection to advise on its management is seen by them also as behaving in the "proper Lamalama way", because they are applying the same principle of respect to their collective past.

As well as actively contributing to management of the Collection within the walls of the Museum, Lamalama participation in the project, and especially their engagement with the images in the Thomson Collection, allows them to familiarise themselves with ancestors generally known only by name. Through the Thomson images, the Lamalama are able to revisit their ancestors and through examination of the activities they are depicted as engaged in, they are able to revisit the culture of their past. Thus, for present generations, the images embody 'Lamalama-ness', although few of the people in the images would have thought of themselves as Lamalama, nor realised their future impact. Instead, they would have been more likely to identify themselves as members of particular clans rather than as members of an overarching tribal group – as Morrobalama, or Rimangud-dhinma, or any of the 35 or so other clans whose amalgamation over time has synchronically produced the Lamalama people of today. For their descendants, the images provide evidence of changes and continuities in Lamalama praxis through time, which in turn provides a sound basis for a kind of narrative development of identity in the present, to which technology, again, contributes.

We have noticed that potential constraints on naming or viewing images of deceased relatives have not generally applied in this work, and we extrapolate this is because the Thomson images depict people mostly unknown in life to the present Lamalama generations. In fact, the Lamalama are proud of the installation of images of their ancestors on show at Bunjilaka, the Indigenous exhibition space within Melbourne Museum. They have been able to observe visitors from around the world engaged in looking at the images, and have commented on feeling "proud" and "happy"[12] that it is their culture being viewed. In this sense they believe themselves to be honouring their ancestors in ways that are culturally appropriate. In discussion of this point, some Lamalama people have said that the body corporate, so to speak, would suffer if it were not appropriate for them to work on the images of their Old People with us. The Old People exist within the material plane to the extent that they can advise living relatives on strategies in local political struggles[13], engage young children in conversation, or effect physical punishment on individuals through illness when required. The Old People would act in some way to indicate their displeasure if engaging with them, and past practices, through the Thomson images was deemed as doing something that was not 'proper Lamalama way'.

Technologies of identity making

The technologies important to the Lamalama are thus all ones that move them through time and space in culturally appropriate ways. Both phones and photographs can express elements of the Lamalama sense of self, but at different ends of a continuum of cultural practice that emerges from their habitus Lamalama uses of technology express purposes that seem to relate to both traditional and modern goals, but posing a boundary between purposes in this particular way is not useful. Clearly, Lamalama engagements with technology simultaneously serve both contemporary pragmatic purposes as well as cultural and cosmological imperatives associated with much deeper temporalities. Throughout, there is a concern for engaging with technology in the "proper Lamalama way".

Notably, too, the body is implicated in multiple ways in Lamalama uses of technology, whether in the form of spears, phones or photographs Technological forms are used to extend the reach of the body through space to 'link-up' with kin and friends, as well as through time in the case of museum-held objects and images, used to relate to ancestors no longer living.

As well as extending the reach of the physical body, the Lamalama use modernist technologies to maintain tradition and bring about cultural transformation. Through the use of older photographic technologies, and more modern digital imaging and audiovisual recording technologies, the Lamalama are able to elaborate their own shared memories of the past, and to incorporate technologically-mediated representations of their past into the field of their living community memory. Throughout the duration of this research project, they are continuously brought into contact with their past in an exercise in 'collective remembering' (Wertsch; Hafner 2008) that integrates individual recollection with information in the images to forge new senses of the self Interaction with the Thomson images provides them with a visual representation of historically located, embodied social actors previously available only via oral recollection and testimony, but whose influence on the present continues.

The cultural principle of respect permeates Lamalama actions in the present, and their approach to photographic images reflects the same elements of the habitus as does their use of customary technologies such as spears. Just as they use spears in culturally appropriate ways to hunt and to sustain their corporeal body, so they use phones and photographs to sustain the social and affective body. If they were to engage with the photos in a different way they would lose sight of their own purposes. Central to the Lamalama's reasons for participation in the research we are conducting is their intention to further develop materials arising from the research for their own purposes of cultural education and training. This will be possible because of the specific methodology of the project, which involves them interacting with the objects depicted in the Thomson images, and the images themselves, whether or not these objects and processes otherwise have relevance to them in the present.[14] Nonetheless, they demonstrate positive emotions in discussion of these images, which bring them into contact with arcane elements of their own past at times.

The power of photographic imagery to establish a kind of virtual presence is well-known (Barthes 1977), and particularly demonstrated by such experience. Faced with social and economic fragmentation as they are, the Lamalama seek to strengthen their identity at the same time as they engage with technologies which might be seen to bring about further group fragmentation because they are so removed from 'traditional' forms. I do not believe this to be the case. The cultural record that results from the project will include a number of different catalogues of their experience, including the filmed material, a database of objects and processes, a demountable exhibition, and at least one book.

All of these outcomes, but particularly any of the material already in digital form, can be immediately used by the Lamalama for their own cultural purposes. They are especially keen to develop PC-based games and digital material for mobiles that provide simple, direct and engaging ways of educating their children in the 'proper Lamalama way'. They also foresee ways to develop similar outcomes for the tourist market. The importance of this lies more in educating strangers about the right way to behave on country than in developing a tourist enterprise, in reflection of the values that inform the Lamalama habitus, but they do provide limited economic potential. It was on one of the first research visits to Museum Victoria that this possibility was elucidated. This seems to me to reflect the kind of pragmatic decision-making associated with everyday issues that Wilk (2001) describes as the process of cultivation – "the expansion of existing needs in new directions, bringing bodily experience into open discourse, display, debate and contention".

Reconstituting oral knowledge into some kind of digital format is what will be required to achieve this. Although some Aboriginal cultures are particularly given to visual representations of their cultures – evident, for example, in the Papunya Tula style of dot painting now famous internationally – the Lamalama do not have a strong tradition of visual mythology. Although there may be problems of conceptual translation between oral and visual forms that arise in this process, the core notion is inherently 'Lamalama' in being an expression of the responsibility they share with their Old People to protect and care for the country. Actions such as defending their claims to land in the formal legislative forums of native title processes are dramatic and protracted acts of appropriate behaviour, but so these methods of cultural transmission are also fitting. A conceptual boundary between modern and traditional technologies is not useful because it is an artificial distinction concerning Lamalama ways of being. The virtual presence of the ancestors is available through the Thomson images and can be used in a mnemonic fashion to maintain culture. The process of engaging with their past through such technologies acts to liberate the affect and memories embodied in their current knowledge, and allows them to reconstitute their identity through communication with the visual evidence of their past.

Conclusion

Some of the technologies discussed here are directly communicative, while others are important for their power to transcend time and space and relocate individual bodies within a larger narrative of cultural identity. In this context, it is less useful to interpret mobile phone use, for example, as a phenomenon of modernity than as the embodied practice of satisfying communication needs. The style of Lamalama bodily interactions with technology, both traditional and electronic, and their aspirations for future technological expressions of the group self are demonstrated in their wide use of technologies for communicative purposes. It is also demonstrated in their pragmatic understanding that through bodily engagement, tools of all kinds can offer a technological advantage.

It remains to be said that although as part of our research we have considered the potential role of immersive environments to meet these various communicative and identity needs of the Lamalama, the difficulties they seem to pose in an uncontrolled physical environment of extreme weather conditions, dust, and humidity, and a lack of technical support makes them impractical for placement in Coen or at Yintjingga. The question would also be what additional purposes they could serve as tools of communication. The Lamalama regard it as part of their cultural responsibility to educate all people who are likely to enter their country about the potential risks involved in doing so. Immersive technologies offer one interesting tool for doing that. Among immediate plans for fieldwork in the middle of this year is discussion of their feelings about locating such technology within the Museum, where it would be less available to them, but more publicly available as a teaching tool. This would be developed in full consultation with them if it were to proceed. An elaboration of the project such as this raises fresh questions about the affective relationship between the body, the habitus and technology, when the distance between the people and the country concerned, and such a public, but nonetheless intimate representation of it is considered.

Notes

1. A multi-disciplinary research team working on the project 'Oral Tradition, Memory and Social Change: Indigenous Participation in the Curation and Use of Museum Collections', funded by the Australian Research Council Linkage Project funding scheme (LP0667418).
2. The Collection is located at the Melbourne Museum campus of Museum Victoria.
3. These terms are most commonly used by the Lamalama and others in Coen to distinguish between Aboriginal and non-Aboriginal residents.
4. While most of the Lamalama still reside in the Coen-Port Stewart region, their kin also reside in places such as Cooktown, the Aboriginal township and former mission of Hopevale, and the distant Queensland capital of Brisbane.
5. Generally speaking, the special status of these titles means they do no allow re-sale of the land or other kinds of exchange of it, or other purposes such as mining of its resources.
6. This became important in the formal context of land claims and native title processes, where claimants are generally required to prove cultural continuity back to the moment of European contact. This has often proven difficult because of the dispossession associated with settlement and consequent historical upheaval experienced by Indigenous people in Australia.
7. E. Liddy, personal communication, telephone conversation, March 2008.

8. In the process of giving 'smell', owners of country apply underarm sweat to visiting strangers, who thus take on some of their substance and so become known and familiar, affording them protection from harm.
9. S. Bassani, personal communication, Running Creek, 1993.
10. This raises the question of how important these considerations are in relation to living, embodied people. Community development workers involved in the project were suggesting the use of eco-friendly, composting toilets at the time, which would produce fertiliser for the garden. For the same reasons – they would be eating their own bodies – the Lamalama rejected the suggestion.
11. Horst and Miller cite the figure of US$125 billion in licence sales worldwide reported by *The Economist* in 2004.
12. S. Peters, personal communication, Melbourne, February 2007.
13. E. Liddy, telephone conversation, March 2008.
14. For example, in Thomson's time, their forebears were known as rain-makers, and one sequence of images depicts a rain-maker at work. While people are fascinated by this series, they have been reticent to engage with it – sorcery is involved in the practice, and while they admire the evident power of the sorcerer, they are historically removed from his knowledge.

Bibliography

Attwood B. (2005), *Telling the Truth about Aboriginal History.* Crows Nest, NSW: Allen & Unwin.

Appadurai A. (1996), *Modernity at Large: Cultural Dimensions of Globalization.* Minneapolis, Minn.: University of Minnesota Press.

Barthes R. (1977), *The Photographic Message.* In: *Image-Music-Text*, translated by Stephen Heath, Glasgow: Fontana/Collins.

Audouze F. (2002), *Leroi-Gourhan, a Philosopher of Technique and Evolution.* Journal of Archaeological Research, Vol. 10 (4), Pg. 277–306.

Brian D. and Hafner D. (2008), *Aboriginal Identity Construction, Memory and Virtual Representation*, unpublished manuscript, in preparation.

Bourdieu P. (1977), *Outline of a Theory of Practice*, Cambridge: Cambridge University Press.

Commonwealth of Australia (2008), Parliamentary Debates. House of Representatives, Official Hansard, No. 1, 2008, Wednesday 13 February, 2008, Pg. 167–177. Accessed online at http://www.aph.gov.au/hansard/reps/dailys/dr130208.pdf, on 5 May 2008.

Fog Olwig K. (2005), *Comment on Horst and Miller.* Current Anthropology, Vol. 46, Pg. 769–770.

Hafner D. (2008), *The Past, Present: Lamalama Interactions with Memory and Technology.* In: B. M. Pirani and I. Varga (eds.), *The New Boundaries between Bodies and Technologies*, Newcastle: Cambridge Scholars Publishing, Pg. 251–268.

Hafner D., Rigsby B. and Allen L. (2007), *Museums and Memory as Agents of Social Change.* International Journal of the Humanities, Vol. 5 (6), Pg. 87–94.

Horst H. and Miller D. (2005), *From kinship to link-up: cell phones and social networking in Jamaica.* Current Anthropology, Vol. 46, Pg. 755–778.

Kidd R. (1997), *The Way We Civilise: Aboriginal Affairs – the Untold Story.* St Lucia, Qld: University of Queensland Press.

Lemonnier P. (1989), *Bark capes, arrowheads and Concorde: on social representations of technology*. In: I. Hodder (ed.), *The Meaning of Things: Material Culture and Symbolic Expression*, London; Boston; Sydney: Unwin Hyman, Pg. 156–171.

Lemonnier P. (ed.) (1993), *Technological Choices. Transformation in Material Cultures since the Neolithic*. London: Routledge.

Leroi-Gourhan A. (1993), *Gesture and Speech*. Translated from the French by Anna Bostock Berger, Cambridge Mass.: MIT Press.

Ponte F., Marsh H. and Jackson R. (1994), *Indigenous Hunting Rights – Ecological Sustainability and the Reconciliation Process in Queensland*. Search, Vol. 25 (9), Pg. 258–261.

Reynolds H. (1981), *The Other Side of the Frontier: Aboriginal resistance to the European invasion of Australia*. Ringwood, Vic: Penguin Books.

Rigsby B. and Hafner D. (1993), *Lakefield National Park Land Claim Book*. Cairns, Qld: Cape York Land Council.

Rigsby B. and Peterson N. (2005), *Donald Thomson, the Man and Scholar*. Canberra: The Academy of the Social Sciences in Australia.

Shove E. and Southerton D. (2000), *Defrosting the Freezer: From Novelty to Convenience*. Journal of Material Culture, Vol. 5 (3), Pg. 301–319.

Stiegler B. (1998) [1994], *Technics and Time: The Fault of Epimetheus*. Translated by R. Beardsworth and George Collins, Meridian – Crossing Aesthetics Series, Stanford, CA: Stanford University Press.

Telstra (2008), *Next G*. http://www.telstra.com.au/mobile/networks/nextg.html, accessed Sunday 27 April 2008.

Wertsch J. V. (2002), *Voices of Collective Remembering*. Cambridge: Cambridge University Press.

Wilk R. (2001), *Toward an Archaeology of Needs*. In: *Anthropological Perspectives on Technology*, M. B. Schiffer (ed.), Albuquerque: University of New Mexico Press, Pg. 107–122.

Wilk R. (2002), *Consumption, human needs, and global environmental change*. Global Environmental Change, Vol. 12 (1), Pg. 5–13.

Wilmot S. (2008), *Documentary Facts and Lies*. Double Dialogues, in press.

Chapter 10
Bodies in canyoneering: an essay on socio-geographic mapping

André Suchet

Abstract

This chapter presents a socio-cultural study carried out through questionnaires and an ethnology of participants in canyoneering. The results highlight three principal ideal types of users: the *tourist-customers*, the *hedonistic participants* and the *purists*. In the first section, this chapter questions economic considerations in the choice of a site for practice (*homo-oeconomicus* theory). In the second section, sociology attempts to understand the influence of layout in the choice of a site for practice. Two concepts are used: the mechanism of distinction and the principles of the relation to the body in sport. The results will show that these two concepts of the same sociological theory do not complete one another but that they oppose against each other. The text discusses the mutual weight of these two forms of explanation.

Canyoneering came into the limelight with Nicolas Hulot's pictures in *Ushuaia Nature* or by those of *Carnets de l'Aventure* on Channel 2 (French TV). Canyoneering, or canyoning, is a symbol of leisure, outdoor sport, and adventure. In France, the development of outdoors sports has been studied by Loret[1], Pociello[2], Lacroix & Bessy[3] and Jorand[4]. Hybrid activity of trekking, climbing, caving and hydrospeed, canyoneering consists of the descent by foot and by swimming in an enclosed river which is called a canyon. Canyons have comprising natural projections, cascades and toboggans. In relation with the heavy tendencies of the new sporting activities which take place in the mountains, this activity has developed. This is because of "it's very enjoyable character, which gives an immediate pleasure"[5]. In this sense, even if there are not national statistics concerning participants, the practice of the canyoneering can be seen as a significant component of commercial and cultural activities of mountain territories[6]. This paper presents the results of an inquiry into socio-demographic and socio-geographical aspects of the practice of canyoneering in the French department of Pyrénées-Orientales. This research analyzes the distribution of participants across the sites by statistic and cartographic tools.

From a methodological standpoint, this study is based on a survey (questionnaires, N = 237, Q = 44) involving the users of 14 canyoneering sites during the summer months of 2006. The sites were chosen because they were representative (long/short, difficult/easy, aquatic/vertical...). Several interviews and a collective discussion were held for a preliminary inquiry. A space sampling, on the basis of Deborde's work[7], made it possible to identify 14 natural sites representative of the

French canyons. The questionnaires were distributed from 11am up to 8pm. This corresponds to a departure time for the participants of 7am for the morning and 6pm in the evening. The period over which the results were collected range from the high season (tourist season) to the end of August (when the tourists decrease in numbers and when the rain begins which swells river flows).

The indicators related to economic conditions of the people (for the first paragraph) and to the social characteristics, cultural and sporting preferences, marital status, political orientation, religion, to the relationship participants have with nature and their technical level in canyoneering (for the second paragraph).

In the first section, this chapter questions economic considerations in the choice of a site for practice (*homo-œconomicus* theory). The purchasing power of the participant is correlated to the distance covered to reach a site for canyoneering. In the second section, sociology attempts to understand the influence of layout in the choice of a site for practice. Two concepts are used: the mechanism of distinction[8] and the principles of the relation to the body in sport[9]. The results will show that these two concepts of the same sociological theory do not complete one another but that they oppose against each other. The text discusses the mutual weight of these two forms of explanation.

Bodies are not well studied by geographers. From an international point of view, we consider that only the American feminist geographer has considered the subject. Thanks to the work of Longhurst[10] or Moss & Dyck[11], and recently, Nelson & Seager[12] affirm in introduction "The body is the touchstone of feminist theory". As a matter of fact, some French authors have also dealt with this issue, but there writings are not well known. They see the subject differently from the feminist geographers. They study landscape as though they were studying a body[13]. For example, Corneloup[14] compares rocky peaks to phallus; he explains "the geography of the sexual bodies could be traced on the natural geographical spaces". This work doesn't inscribe in any of these perspectives. It's about explaining the spatial (the repartition of participants) with the social (and particularly by the type of body relations). In theoretical frame, this research subscribes to Pierre Bourdieu's sociology and his concepts of habitus, exis corporel and distinction[15]. And the discussion used the complementary questions of Georges Balandier[16].

The elimination of the economic hypothesis

If we consider the economical explanation, we can conclude that the lower class tends to stay closer to home instead of travelling to further destinations. We could assume that with the increase of living expenses people decrease their travelling distance. The following maps show us a statistical analysis, and look at the correlation between their declared revenues and their accepted travelling distance for practical canyoneering.

The data (figure 10.1) invalidates the assumption of economic correlation in the choice of the destination for practice. The recruitment of different canyons

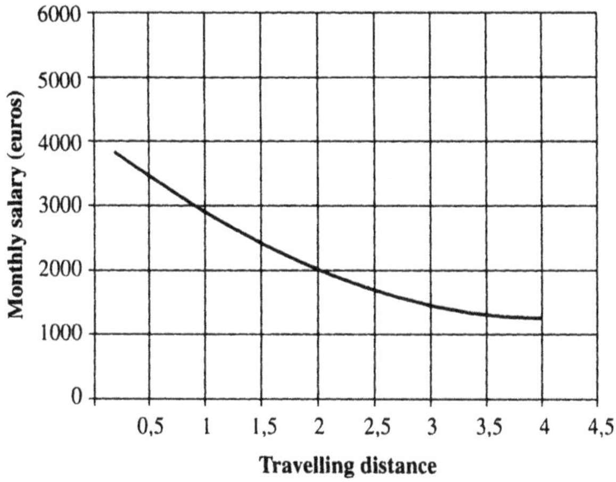

Figure 10.1 *Tendency line between declared revenues and travelling distance for canyoneering (N = 237).*

is not relative to the economic level of the participants. These results confirm the principal articles published on the subject[17]. As the analysis of the economic matters does give any light to the problem, social cultural diagnosis may give birth to a reasonable explain.

Cultural hypothesis

In this paragraph, we can interpret the data with socio-cultural theory[18]. We'll start by presenting a typology of people who practice canyoneering (fig. 10.2), and after we will expose interpretations of these results.

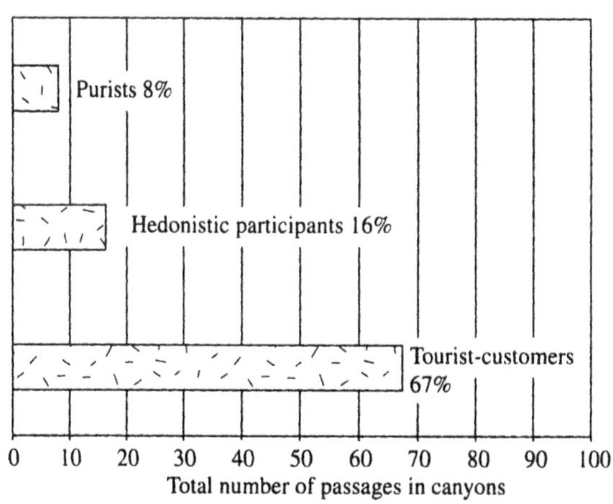

Figure 10.2 *Ratio of modalities (N = 237).*

Typology of canyoneering participants

The tourist-customers:

The *tourist-customers* are the beginners who are taught by a professional of the activity (canyoneering instructor). Most of them are non-residents (74%). Are the users of the sites which choose neither the course nor its characteristics and do not express a continuous investment in practice (hiring of the material, ignorance of the Internet forum and of the specific literature, non-adhesions in mountain activities clubs). For the majority of these people, it is the first descent (73%) and less than 13% of them have already descended less than five canyons. Without knowledge of canyon and technical skills, they are declared "beginners", "to discover"... Item "practise canyoneering to have a laugh with sporting buddies" collects 49% of answers. Those who practice with a canyoneering instructor and continue on their own is lower than 5%. For them, canyoneering is a "holiday activity" (39%) or a "natural sport" (40%) or even, a "practice of nature", but the ecological concern is not relatively present. *Tourist-customers* have a preference for the hedonist and enjoyable aspect of canyoneering "one came for the feelings, the jumps", "the shivers", "it is a sportsman", "it is amusing". For them, the canyoneering is "dangerous". People express interest in new forms of pleasure during their holiday. They respect the instructions of the instructor. They reject the pain, the difficulty and the danger.

Apart from a general taste for a walk, the canyoneering follows rafting (14%), canoe-kayaking (13%), or via-ferrata but always with an instructor and location of material. Only climbing, quoted with 34%, is practised with more time invested, but generally indoors. In addition one finds the mountain bike (14%), resort skiing (31%) as well as tennis, golf or scuba diving for the non-mountain activities, but the quantity of answers "no physical activity" is important (11%). If the context is family based, the practice itself concerns those who are 28-45 years, i.e. the parental couple (32%), or those who are 15-25 years (60%) the remainder of the family waits near the cars or is in an excursion near the site.

Characterized by a family, professional and school success[19], the *tourist-customers* declares a preference for, the theatre, reading, music but also bowling, cinema or "shopping". Lastly, they have a political tendency of being right winged (40%), and far-right (25%), a prevalence of the economic capital on the legitimate cultural capital and a certain culture of competitiveness (the more we do things and more the holidays are successful). Many of them being executives of small companies, in other words, we found here the traces of a "culture of private sector"[20].

The hedonistic participants:

Between the two categories described by Corneloup[21] (*hedo-mountainer* and *sporting-performer*), this second category is for those who practice canyoneering without an instructor and do 4 to 40 canyons per season in optimal conditions. They are young men (24-30 years) looking for excitement, sometimes with their partner "for a day of sport" or if not, with male friends (30-45 years) to see the view. But there are also those who come as part of associations (the French Alpine Club CAF, or the French Federation of Mountain and Climbing FFME). There are also the regulars who offer the children, the family or the friends the opportunity to start their initiation in canyoneering as well as to take part in non-sporting activities such as "aperitif", "convivial meal".

Technically, when a jump is delicate or a too vertical toboggan, they do not hesitate to attach a rope (instrumentation). For them, "the main thing is to give pleasure". The *hedonistic participant* often lives close to the sites, and won't travel more than two or three hours to practice their outdoor activity. They consider canyoneering "a real sport". For them, the canyoneering is "moderately dangerous" and constitutes a "nature activity", 22% indicating that they "practice canyoneering to have a laugh with sporting buddies". In all the case, they practices activities of mountain (trekking, climbing, alpinism or caving and canoe-kayak), or practice sea-activities (sailing, scuba-diving...) or other non motorized activities. Many of them are teachers (24%), technicians (approximately 20%), or employees (often in the main French companies: French group of electricity EDF, the Post office...), this second category comes mostly from the cultured middle class. In addition, their political view is left winged (53%), which confirms the presence of a "culture of public sector"[22].

The purists:

Men aged from 35 to 65, who reside in rural areas, *purists*, or "impassioned" as they are defined, practise all year round. They add up the greatest number of outings (from 40 to 100 canyons by year), and sometimes going at odd times (the evening "after work", the night...), the reasons given are "to avoid busy periods" or "the sunset". Beginning with other mountain activities (climbing, alpinism) or even with scuba-diving and caving, these *purists* consider canyoneering as a real physical activity, but dispute the word "sport". For 76% of them, it's not a risky activity. Interested by the "imposing" natural environment, they strongly reject the phrase "practice canyoneering to have a laugh with sporting buddies", for them: "in canyon, we take it seriously". They are often the first to venture in the canyons (discoverers of the sites) and who travel just to take part in this activity (Sierra of Guara and Andalusia in Spain, Italy...). They have a significant knowledge of canyons. Technically, they prefer to descend canyons using ropes as little as possible (on jumps of more than 20 meters, difficult toboggans, dangerous descents...).

When there is an absence of family structure (divorce, separation...), they are invested largely in the activity: "since my divorce, I do nothing but canyon(ing)". Retired, unemployed, technicians or employed, they usually come from the working classes. Sometimes in club (mainly the French Alpine club CAF), they refuse administrative engagements, criticize "excesses of regulation" and seek an "escape" elsewhere. Lastly, they have strong opinions on ecological matters (opposing hunting and fishing...), they feel strongly against other groups in canyons and tend to vote for the far-right, which seems to indicate a culture of "dramatization"[23].

They are however mainly former sympathizers of the left, even of extreme left who gave fun character to the canyoneering discussed by Loret[24] in the years 1970-1980. Traces of a hippie culture which is prevalent in their life style when they travel between impassioned friends to foreign locations in their motor homes. Utopian ideologies which are now disappointed have left a soured characteristic within the fold and a vote protester well studied by Duret[25].

From this socio-demographic data and regrouping by standard ideals of practice, it is possible to outline a general analysis of the results by replacing contours of this microanalysis in the structure of the Space of Social Positions (SSP, EPS en French) established by Bourdieu[26] and declined in the field of the physical practices by Pociello[27] at the University of Paris 11 - Orsey. We found different modalities of practice, and theses modalities correspond to different social positions.

Mechanism of distinction *versus* corporal dispositions

From the vertical axis of the figure which spreads out the volume of capital, one can question the place of the social process of distinction within the activity. As shown by Lefèvre[28] for the high mountain territories, distinctive logic[29] generally pushes those who have the most capital to strongly invest themselves in practice or to seek the prestigious and difficult sites. For Lefèvre, in terms of altitude, the mid-altitude/high altitude boundary, favourite area of the *female hiker* (bottom of social hierarchy), contrasts with the Himalayan peaks so dear to the *upmarket mountaineer* (top of social hierarchy). Lefèvre's results[30] highlight the existence of "a geo-alpinist structure with territorial marking of high-altitude mountain activities", already observed in other natural spaces such as the sea[31], and forest areas[32], where the spatial distribution is often synonymous with social distribution[33]. However, in spite of these present results in the literature, the data analysis reveals an investment in the activity inversely proportional to the volume of economic and cultural capital of the participant. For example, the *purists* represent the top of the hierarchy in the activity (greater number of sites per year, travel in foreign spots, better technical level...), but they are usually members of the working class.

Questioning the new dispositionnaliste approach, i.e. Lahire's work, it seems initially possible to attribute these results to a "dissonance of cultural consumption"[34]. The assumption is reinforced by a significant rate of social change of the *purists*. They have made many voyages and have often increased their social relations and thus have a lot of sources of influence. For example they changed many times their professional situations and thus their social experiences.

However, a specific knowledge[35] of canyoneering validates the assumption of a correspondence between the three types of practices and positions of class by mobilizing the principles in relation to the body which is socially acquired by the individual[36]. Indeed, independently to the prestige (the distinction) that a high degree of practice brings, the canyoneering (as certain other practices) imposes difficulties on the body: endurance, cold water, painful descent into the mud, going up into the scree to arrive at new spots, the weight of the bags containing the equipment... as many elements which correspond finally more to a habitus of working class than to that of the socially dominant fraction[37]. On one side we find technological bodies (descent by rope, Zip Line, ascending ropes and hauling systems, complete equipment, life line, frame-loaded rope clamps...) for the upper classes, and on the other side bodies as a projectile (jump, toboggan, nudity, practicing without the helmet...) for the lower class[38]. In other words, it is possible to explain the foundations of this paradox while basing it on the principles of the relation to the bodies defined by genetic structuralism which comes here to oppose the mechanism of distinction. To conclude this paragraph the sociology of the body constitutes the key to understanding this activity.

Conclusion

In conclusion this chapter used the data of a socio-demographic and socio-geographical study concerning the participants of canyoneering in the French department of Pyrénées-Orientales to understand the choice of a site in an activity of sporting leisure. After having rejected economic explanations, this study highlights a rivalry between the mechanism of distinction and the principle of relation to the body imposed by a high level in canyoneering. On one side, distinctive logic[39] generally pushes those who have the most capital to strongly invest themselves in practice or to seek the prestigious and difficult sites, but on the other side, intensive activity corresponds rather to a type of relation to the body accepted by the working class. With respect to the opposition between distinctive logic and the relation with the body; it is ultimately the physical side which comes out on top. All in all, to fully understand this activity, one must have a strong relation to the body. So, remarking that the study was carried out on a representative sample of the canyoneering sites in France, but in only one department: This constitutes a limit and gives the possibilities for continuation.

Notes

1. Alain Loret, *Génération glisse* (Paris: Autrement, 1995).
2. Christian Pociello, *Les cultures sportives* (Paris: Presses universitaires de France, 1995). Christian Pociello, "Le corps contre le sport? (Paragraphe introductif)", *Esprit* 62 (1982): 7–10.
3. Gisèle Lacroix and Olivier Bessy, "Glisse d'hier et surf d'aujourd'hui", in *Surf Atlantique. Les territoires de l'éphémère*, ed. Jean-Pierre Augustin, 25-48 (Talence: Maison des Sciences de l'Homme d'Aquitaine, 1994).
4. Dominique Jorand, "Du hang gliding au vol libre: l'émergence d'un 'sport californien' en France dans les années 70", *Stadion* 28, no. 2 (2002): 275–292.
5. Philippe Bourdeau, "D'eau et de rocher: le canyoning", *Cahier Espaces* 35 (1994): 50.

6. Clémence Perrin and Jean-Pierre Mounet, "L'insertion de l'offre de canyoning dans l'industrie touristique: une étude comparative entre le massif des Bauges et le Vercors", *Society and Leisure* 27, no. 1 (2004): 143–72. Arnaud Pinguet, "Le cadre socio-économique du canyoning: points de repères", *Cahiers du CSSM* 6 (1996): 20–25. Altimax and Versant sud, *Étude sur les activités de pleine nature dans les Alpes françaises: Via Ferrata, Canyoning, Parcours Acrobatique Forestier, Escalade* (Research Report for AFIT and DATAR, 2003). André Suchet, Dominique Jorand and Pascal Mao, *Les usagers des parcours naturels de canyoning en France* (Research Report for FFS and FFME, 2008).
7. Christophe Deborde, "L'offre française des parcours de canyoning" (Maîtrise dissertation, University of Grenoble 1, 1996).
8. Pierre Bourdieu, *Distinction: A Social Critique of the Judgement of Taste* (Cambridge: Harvard University Press, 1984).
9. Pierre Bourdieu, "Sport and Social Class", *Social Science Information* 17, no. 6 (1978): 819–40. Christian Pociello, "'La force, l'énergie, la grâce et les réflexes'. Le jeu complexe des dispositions culturelles et sportives", in *Sports et société. Approche socio-culturelle des pratiques* , ed. Christian Pociello, 171–237 (Paris: Vigot, 1981).
10. Robyn Longhurst, *Bodies: Exploring Fluid Boundaries* (New York: Routledge, 2000).
11. Pamela Moss and Isabel Dyck, *Women, Body, Illness: Space and Identity in the Everyday Lives of Women with Chronic Illness* (New York: Rowman & Littlefield, 2002).
12. Lise Nelson and Joni Seager, eds. *A Companion to Feminist Geography* (Oxford: Blackwell, 2004), 2.
13. Marc-André de Figueres, *Erotic Theory of the Collioure Bell Tower* (Perpignan: CEDACC, 2002). Jean Corneloup, "Le masculin et le féminin en escalade", in *Sport, culture et tradition Conference Proceedings*, ed. Charles Pigeassou, 321–330 (1995).
14. Corneloup, "Le masculin et le féminin".
15. Bourdieu, *Distinction*. Pierre Bourdieu, "Program for a Sociology of Sport", *Sociology of Sport Journal* 5, no. 2 (1988): 153–61. And Luc Boltanski, "Les usages sociaux du corps", *Annales* 26, no. 1 (1971): 205–33.
16. Georges Balandier, "Ce que 'disent' le corps et le sport", *Corps et Culture* 6-7 (2004): 181-84.
17. Monique Pinçon-Charlot and Paul Rendu, "Distance spatiale, distance sociale aux équipements collectifs en Île-de-France: des conditions de la pratique aux pratiques", *Revue française de sociologie* 23, no. 4 (1982): 667–96. Marijke Taks, Roland Renson and Bart Vanreusel, "Of Sport, Time and Money: An Economic Approach to Sport Participation", *International Review for the Sociology of Sport* 29, no. 4 (1994): 381–95.
18. Bourdieu, *Distinction* . Pociello, "'La force, l'énergie, la grâce et les réflexes'".
19. 25% married people, often with children, 48% of young unmarried couples. Average monthly income of the households is between 2 600 and 3 800 Euros (62%) or between 38 00 and 6 000 Euros (27%).
20. François de Singly and Claude Thélot, *Gens du privé, gens du public* (Paris: Dunod, 1989).
21. Jean Corneloup, "Les cultures sportives de la montagne d'aujourd'hui et de demain", in *Les Pyrénées entre deux mondes*, ed. Frédéric Bartczak and Johann Rage, 133–58. (Perpignan: Presses universitaires de Perpignan, 2003).
22. de Singly and Thélot, *Gens du privé*.
23. Jean Corneloup, "l'enquête d'opinion dans l'étude des pratiques sportives de montagne", *Bulletin de méthodologie sociologique* 83 (2004): 19–42.
24. Loret, *Génération*.
25. Pascal Duret, *Les larmes de Marianne: Comment devient-on électeur du FN?* (Paris:

Armand Colin, 2004).
26. Bourdieu, *Distinction.*
27. Christian Pociello, "Pratiques sportives et demandes sociales", *Travaux et Recherches en EPS* 5 (1979): 31–48. Pociello, *Les cultures sportives.*
28. Brice Lefèvre, "Contribution to the Study of the Social Structure of High-Altitude Mountain Activities: Example of Users in the Mont-Blanc Massif", *Journal of alpine research* 92, no. 4 (2004): 76–85.
29. Bourdieu, *Distinction.*
30. Lefèvre, "Contribution", 80.
31. Jean-Pierre Augustin, "Pratiques de la mer et territoires urbains: de nouveaux espaces de loisirs sportifs pour l'agglomération de Bordeaux", *Revue Géographique des Pyrénées et du Sud-Ouest* 57, no. 4 (1986): 589–609.
32. Bernard Kalaora, *Le musée vert: radiographie du loisir en forêt* (Paris: L'Harmattan, 1993).
33. Bourdieu, *Distinction.*
34. Bernard Lahire, *La culture des individus. Dissonances culturelles et distinction de soi* (Paris: La Découverte, 2004).
35. Balandier, "Ce que 'disent' le corps et le sport".
36. Bourdieu, *Distinction.* Bruno Frère, "Incertitudes sur l'habitus. Paradigme génétique et phénoménologie: vers un approfondissement de la notion d'habitus", *European Journal of Sociology* 46, no. 3 (2005): 469–94.
37. Bourdieu, *Distinction* . Bourdieu, "Sport and Social Class". Boltanski, "Les usages sociaux". Pociello, "'La force, l'énergie, la grâce et les réflexes'".
38. Pociello, *Les cultures sportives.*
39. Bourdieu, *Distinction.*

Bibliography

Altimax and Versant Sud (2003), *Étude sur les activités de pleine nature dans les Alpes françaises: Via Ferrata, Canyoning, Parcours Acrobatique Forestier, Escalade.* Research Report for AFIT and DATAR, Paris.

Augustin J.-P. (1986), *Pratiques de la mer et territoires urbains: de nouveaux espaces de loisirs sportifs pour l'agglomération de Bordeaux.* Revue Géographique des Pyrénées et du Sud-Ouest, Vol. 57, no. 4, 589–609.

Balandier G. (2004), *Ce que 'disent' le corps et le sport.* Corps et Culture, Vol. 6-7, 181–84.

Bourdeau P. (1994), *D'eau et de rocher: le canyoning.* Cahier Espaces, Vol. 35, 49–54.

Boltanski L. (1971), *Les usages sociaux du corps.* Annales, Vol. 26, no. 1, 205–33.

Bourdieu P. (1978), *Sport and Social Class.* Social Science Information, Vol. 17, no. 6, 819–40.

Bourdieu P. (1984), *Distinction: A Social Critique of the Judgement of Taste.* Cambridge: Harvard University Press.

Bourdieu P. (1988), *Program for a Sociology of Sport.* Sociology of Sport Journal, Vol. 5, no. 2, 153–61.

Corneloup J. (1995), *Le masculin et le féminin en escalade.* In: *Sport, culture et tradition. Conference Proceedings*, edited by Charles Pigeassou, 321–30.

Corneloup J. (2003), *Les cultures sportives de la montagne d'aujourd'hui et de demain.* In: *Les Pyrénées entre deux mondes*, edited by Frédéric Bartczak and Johann Rage, 133–58. Perpignan: Presses universitaires de Perpignan.

Corneloup J. (2004), *L'enquête d'opinion dans l'étude des pratiques sportives de montagne.* Bulletin de méthodologie sociologique, Vol. 83, 19–42.

Deborde C. (1996), *L'offre française des parcours de canyoning.* Maîtrise dissertation, University of Grenoble 1.

Duret P. (2004), *Les larmes de Marianne: Comment devient-on électeur du FN?.* Paris: Armand Colin.

Figueres de M.-A. (2002), *Erotic theory of the Collioure bell tower.* Perpignan: CEDACC.

Frère B. (2005), *Incertitudes sur l'habitus. paradigme génétique et phénoménologie: vers un approfondissement de la notion d'habitus.* European Journal of Sociology, Vol. 46, no. 3, 469–494.

Jorand D. (2002), *Du hang gliding au vol libre: l'émergence d'un 'sport californien' en France dans les années 1970.* Stadion, Vol. 28, no. 2, 275–292.

Kalaora B. (1993), *Le musée vert: radiographie du loisir en forêt.* Paris: L'Harmattan.

Lacroix G. and Bessy O. (1994), *Glisse d'hier et surf d'aujourd'hui.* In: *Surf Atlantique. Les territoires de l'éphémère*, edited by Jean-Pierre Augustin, 25–48. Talence: Maison des Sciences de l'Homme d'Aquitaine.

Lahire B. (2004), *La culture des individus. Dissonances culturelles et distinction de soi.* Paris: La Découverte.

Lefèvre B. (2004), *Contribution to the Study of the Social Structure of High-Altitude Mountain Activities: Example of Users in the Mont-Blanc Massif*, Journal of alpine research, Vol. 92, no. 4, 76–85.

Longhurst R. (2000), *Bodies: Exploring Fluid Boundaries.* New York: Routledge.

Loret A. (1995), *Génération Glisse.* Paris: Autrement.

Moss P. and Dyck I. (2002), *Women, Body, Illness: Space and Identity in the Everyday Lives of Women with Chronic Illness.* New York: Rowman & Littlefield.

Nelson L. and Seager J. (eds.) (2004), *A Companion to Feminist Geography.* Oxford: Blackwell.

Perrin C. and Mounet J.-P. (2004), *L'insertion de l'offre de canyoning dans l'industrie touristique : une étude comparative entre le massif des Bauges et le Vercors.* Society and Leisure, Vol. 27, no. 1, 143–172.

Pinçon-Charlot M. and Rendu P. (1982), *Distance spatiale, distance sociale aux équipements collectifs en Île-de-France : des conditions de la pratique aux pratiques.* Revue française de sociologie, Vol. 23, no. 4, 667–96.

Pinguet A. (1996), *Le cadre socio-économique du canyoning: points de repères.* Cahiers du CSSM, Vol. 6, 20–25.

Pociello C. (1979), *Pratiques sportives et demandes sociales.* Travaux et Recherches en EPS, Vol. 5, 31–48.

Pociello C. (1981), *La force, l'énergie, la grâce et les réflexes'. Le jeu complexe des dispositions culturelles et sportives.* In: *Sports et société. Approche socio-culturelle des pratiques*, edited by Christian Pociello, 171-237. Paris: Vigot.

Pociello C. (1982), *Le corps contre le sport ? (Paragraphe introductif).* Esprit, Vol. 62, 7–10.

Pociello C. (1995), *Les Cultures Sportives.* Paris: Presses universitaires de France.

Singly de F. and Thélot C. (1989), *Gens du privé, gens du public.* Paris: Dunod.

Suchet A., Jorand D. and Mao P. (2008), *Les usagers des parcours naturels de canyoning en France.* Research Report for FFS and FFME, Grenoble.

Taks M., Renson R. and Vanreusel B. (1994), *Of Sport, Time and Money: An Economic Approach to Sport Participation.* International Review for the Sociology of Sport, Vol. 29, no. 4, 381–395.

Chapter 11
Bodies in action: performing identity in dating sites

Begonya Enguix Grau, Elisenda Ardèvol

Abstract

This chapter presents an ongoing research about the presentation of the body on the public spaces of the Internet, paying special attention to the profiles of websites devoted to the establishment of emotional or sexual interactions. In particular, we will center on the analysis of a dating website devoted to the Bear community. On-line profiles are a good setting for the analysis of how people use the representation of the body to perform gender, sexuality and self identity. In online platforms centered on personal relationships, body and embodying practices are determinant for our understanding of the relations between humans, technologies and social life. Parting from the Goffmanian concept of "presentation of self", we will focus on how bodies are constituted as objects of consumption, interchange and power in sites where gender identity is performed through visual representations of the body and textual descriptions of one's self. We will focus on the discursive and representational strategies of participants when "filling up" the standard profiles that these sites offer and how they negotiate or "skip" their rigidity to present themselves to possible partners, the perception of these portraits as authentic or playful and how social and technological patterns shape users' practices. By doing so in a particular site, we will learn how human bodies and their digital representations construct a basis for emotional relationships. Summing up, this paper aims to contribute to the understanding of how embodied practices of gender identity are intertwined with digital technologies.

Bodies and identities in (and) cyberspace

The search for new relationships is one of the most popular and widespread uses of the Internet. Chats, social networks and online dating websites can be used as an opportunity to "flirt" or, simply, to look for companionship. Computer mediated interaction also offers a new form of personal relationship that emerges and can be maintained exclusively online, and that may even evolve into different forms of eroticism or virtual sex experiences.

The Internet social-technical architecture enables different interpersonal relationship strategies taking advantage of the media features, especially the chance to interact anonymously. However, apart from privacy, users consider the relationships over the Internet in terms of "ease" and "convenience", as they offer a reliable and easy way to establish contacts (Ardèvol, 2005).[1]

Anonymous online interaction, unlike other kinds of physical or online relationship with other people, allows users to interact with others and at the same time control the degree to which they are involved in the relationship. Theoretically, users can experiment with their otherness, playing with the heterogeneity of ethnic, social, geographical and gender background from an aseptic standpoint, without the risks and commitments these relationships entail when being held face to face. Interpersonal communication through this mediated system seems to offer a mode of relationship that fits into a lifestyle in which individuals consume technology and are curious to meet new people outside the closed circle of the family, and from educational and professional relationships. Sometimes, online dating may be considered as the beginning of an adventure without leaving home, or it may be used as a way to establish new relationships for those who have moved to a new country or city (Ardèvol, 2005).

According to Wakeford (2004:115), four themes define the existing research on gender, body an identity: a. identity and presentation of the online persona; b. the creation of queer virtual space (or gendered online social contexts); c. the electronic facilitation of social networks and virtual community; and d. the potential of new technology to transform erotic practice.

Presentation of bodies, strategies of manipulation and the issue of intertwining of bodies and identities become central in this type of platforms whose aim is the establishment of personal relationships. However, the question of disembodiment have been largely discussed:

> Cyberspace differs from an off-line community because it offers additional opportunities to perform *and* destabilize national, ethnic, class and sexual identities... The Internet often appears as a site of unlimited possibilities and identity games. In the early to mid 1990s, many scholars were fascinated with what the imagined as disembodied and free-floating identities in cyberspace. Later works challenged this post-modern hype of "be anyone you want to be in cyberspace". For example, feminist and critical race theorists further challenged the supposed freedom of identity performance on line by demonstrating that cyber-subjects are far from deracinated and degendered (Nakamura 2000, O'Brien and Shapiro forthcoming). Although cyberspace is often imagined as a borderless deterritorialized frontier, many studies have shown that if often becomes a battlefield of national (ist) conflicts and violence (Boym 2001, Mitra 1997, in Kuntsman 2004:6).

Idealized online disembodiment refers to independence, freedom, connection in perfect harmony without the limitations or needs of the physical body (Heim, 1993), desinhibition and multiple contacts (Le Breton, 1990). It also refers to desmaterialized identity play, simulation, anonymity, ubiquity and fragmentation of a unique self.

As Gómez et al.[2] have said, "classic works" like those of Reid, Stone, Turkle, Rheingold and all the contributions to the anthologies edited by Jones, were pretty much framed in this approach that stresses disembodiment of identity, a self that is constructed by a series of textual representations, a "free from the flesh" self in a playful way. In some of these canonical studies of the Internet, the capacity of the communication systems of the "cyberspace" to become "labs of identity", was emphasized as a way to perform the postmodern subject (Turkle, 1995) or as an "identity moratorium" similar to the adolescence crisis (Erikson, 1979) (in Gómez et al. 2007).

By separating the user's body as a "container" of identity from the communicative practices of identity as representation and as symbolic performance on line, people can create multiple and new identities and experiment with them (Kendall, 2000; Turkle, 1995 in Herrman, 2007). But they are materialized identities, as our online identities are often linked to a body that can or cannot be fictional and which we reinforce through our textual descriptions, such as nicks, icons, drawings, or photographs. Cyberspace offers the opportunity for identity play, for reinventing the self... but the meat is never left behind (Bell, 2001). In fact, current cyberspace interactions are *profoundly* embodied in the sense that they arise from the actions of individual bodies sitting at computers typing and reading (Bell, 2001), not to mention the case of online dating platforms, where body and its presentation is central to the extent that in some sites user's photographs are compulsory. When flesh is not present in and through images, it is put on the texts (Slater in Gómez et al, 2007). Cyberspace does not annihilate bodies but transforms and performs them through expressions of gender and identity. Paraphrasing Butler's well-known book *Bodies that Matter* (1993), we could ask how bodies matter in cyberspace and in dating sites and how "sexual identity affects and is produced in cyberbodies" (Zafra, 2005:25). Technology mediates presence in a social context, the cyberspace, which "initially affords a place where the body is "fully malleable"" (Whittle, 2001:156).

Furthermore, the possibilities of self-invention in these kind of platforms is limited by the ready-packaged profiles and genders that the different social networks, websites, chats or MUDs offers to participants. Or by actual interactions: in this sense, it is interesting to report Kaloski's study on gender-labelling on chats and how people self-labelled as "spivak" –a "neutral" gender- are recommended to change to "female" if they "want to have sex" (in Bell, 2001:175). In this sense, "what seemed to be the final rupture between hallucination and reality, the organic and the electronic, was not it because textual and visual representations of gendered bodies and erotic desire were less imaginative. It was a new technology but with the same stories" (Wajcman, 2006:110). In any case, the existing tension between disembodiment and re-embodiment can be considered as corporeal strategies that represent negotiations, uneasy articulations which speak, perhaps, of the durability of the flesh (Bell, 2001).

The phenomenological concepts of labelling and self-labelling are related to adscribed and inscribed identities and to Goffman's concept of "virtual social identity" (the stereotyped imputations of attributes and category we make of another met in the routines of social intercourse) as opposed to "actual social identity " (the category and attributes he actually may possess) (Goffman, 1968; Weinstein 1965). Identity adscription and inscription construct the self, but do not use it up.

Bodies can be seen as objects and subjects; eliminating corporal signs does not mean creating new identities (Stone in Wajcman 2006:108) nor issues of identity can be reduced to issues of self-presentation: 'the creation of an electronic persona online –whether through text or graphics- may permit misrepresentation as well as the opportunity for sustained fantasy or deception' (Wakeford,

2004:123–124). As Tucker emphasizes, bodies can be 'acted upon' or can be 'actors' (2006:433). Both processes can be identified in interactions in online dating sites.

> Social constructionism positions the body "as acted upon, be it by discursive formations, cultural norms or any other societally originating practice. Thus, the role of agency is, in essence, absent. Shilling accuses phenomenological approaches of the same failing: that of making prominent one 'side' of the body –society dualism. This time, though, he argues that the phenomenological emphasis on the subjective 'lived experience' of individuals, which is wholly 'housed' and emanates from our bodies –as actors- is overplayed. For Shilling, such approaches are correct to address issues of agency and the active (embodied) nature of human experience, but not to the almost complete occlusion of social processes. Structuration theory attempts to bridge the gap for Shilling, who singles out the work of Giddens, Bourdieu and Grosz as key authors. These theories, attempt to argue that the body is constructed by social processes whilst simultaneously actively working to reconstitute such process (Tucker 2006:434).

Bodies, identities and genders are fully significant when inscribed in the sex differentiation system, although, as Foucault and Butler explain, "establishing different categories of persons based on their bodies has been through time an strategy of control and domination" (Howson, 2005).

For Butler, agency is essential to the understanding of gender definitions as "gender is in no way a stable identity or locus of agency from which various acts proceed; rather, it is an identity tenuously constituted in time –an identity instituted through a stylized repetition of acts" (Butler, 1990:270). This conceptualization contrasts common social scientific models of gender, which portray persons as stable participants who express their identities (biologically given phenomena) during social interaction. Online dating sites are privileged contexts to analyse genders, identities, sex difference, sexual practice and desire which otherwise are difficult to copse. Although this is not the place to articulate discourses on the contingency of the system sex/gender/sexual practice and its connection with the construction of particular identities which until the 18th century were based on genders and not on sex, performances in these sites should be read through the articulation of these concepts and self-presentation. Rather than talking about identities, we should talk about practices, experiences and behaviours that can be diverse, therefore questioning the Cartesian divide between mind and body and the Augustinian precept of a truthful identity based on a single body. Although "in Western culture we perceive the self as "naturally" placed in the body, as a kind of precultural given" (Scheper-Hughes and Lock in Low, 2003), technologies favour the display of multiple bodies that crosscut multiple identities, which are performed through techniques of re-presentation where the subjects's agency stresses one or another performance, depending on the sites. As Bryan Turner stated in 1984, we 'have bodies' and 'are bodies'. It is quite common, as informants have told us, to enter gay online sites to "identify" who's there: itineraries are then drawn, and interesting subjects are followed up through other online representations (messenger, blogs, facebook, etc). Thus, though parting from a specific platform, our research should aim to reach particular practices. Summing up, as Tucker says,

> the body has become a pivotal theoretical and empirical puzzle for the social sciences, one that has been approached with renewed vigour in the last twenty years... For Chris shilling, the problem is understanding and theorizing the multidimensionality of the body; for Nick Crossley, the primary issue is attempting to overcome mind-body dualism; and in Coupland and Gwyn they try to understand how theory and method interrelate in articulating the body in social theory (Tucker, 2006:433).

If bodies are the physical places where class, gender, race relations, sexuality and age meet and are embodied and practised, and the body is a ubiquitous signifier of class (Skeggs, 1997), then we should draw our attention to the signifiers through which they are displayed.

Cyberspace can be seen as a a space for the re-presentations and transformation of the body. It breaks the conception that body equals identity. A user can have different representations in different webs, and has the control on how to represent, how to use his identity, what physical and sociocultural traits to show and explain. Free from spatial and time limitations, interactions can become more fluid and fragmented. But what our study reveal is that we cannot consider cyberspace as an homogeneous space. There are different social contexts "in the cyberspace", and in dating websites, for example, the relationship between body and identity is not questioned, but reinforced.

Analysing online settings: bearwww.com

For Giddens (1994), the body becomes part of a project of personal identity development related to consumption. Assuming his ideas, how do we connect our body's image and self-description in these sites? What decisions do we make in order to self-represent us? Choice, control, agency seem to have occupied the place left by "limited repertoire" or "closed script" in relation to identity and body. But indeed, in the case we will overview, images and body descriptions exist in relation to cultural patterns of consumption. We refer to what Whittle calls "the control of the uncontrollable", what seems "to have missed the point for many internet users –that is the amazing ability it gives people to publish to others. You can do this as often or as little as you want, at very low capital cost, with even less ongoing cost" (Whittle, 2001:162).

In the Internet, identity play is not only related to anonymity and gender-crossing, in the case we are going to study, the individual identity signs (age, sex, race, etc.) are not simulated or hidden but they are in many times, openly exposed and shared. Gomez et al. argue that this presentation of the self through images of the own body, far from being a simple practice where the individuals show themselves "just like they are", is a complex playful practice that entails the use of body images not only as a representation of the self, but as cultural objects for self expression, social interaction and identity performance.

> ... in some sites where the body becomes "the" object to be photographed, written about,pictured or video recorded. The body representation becomes the raw material for performing the identity online; and the ego, the success and the fame of embodied people on the net, turn into a the key feature of this new game instead of the anonymity of the players. (Gómez et al., 2007).

Body remains a matter of priority in the gay male culture and there exist images of hyperidealized male bodies. But among them,

> Bears comprise a subculture of gay men who valorize the larger, hirsute body. This research interrogates Bear culture as a gendered strategy for repudiating effeminacy that simultaneously challenges and reproduces norms of hegemonic masculinity... Bear culture yields a number of sexually innovative practices that disperse pleasure across the body and disrupt genitally centered, phallus-and-receptacle interpretations of sex. However, the subversive potential of these practices is significantly undermined by an attendant set of practices that reflect heteronormative and hegemonically masculine interpretations of sex (Hennen, 2005:25).

Bears "have divorced effeminacy from same-sex desire and created a culture that looks like a bunch of 'regular guys'.. they are subversive but the subversion has to do with reorganizing sexuality and very little to do with challenging gendered assumptions" (Hennen, 2005:41).

The self-identification of gay men as Bears originated in San Francisco in the 1980s as an outgrowth of the gay biker and later on the leather and "girth and mirth" communities. It was created by men who felt that the mainstream gay culture was unwelcoming to men who did not fit into a particular twink body norm (hairless and young). Also, many gay men in rural America never identified themselves with the stereotypical urban gay lifestyle (see Wikipedia, "Bear" (gay slang). As Connell (1995) says, "the choice of a man as a sexual object is not just the choice of a body-with-a-penis, it is the choice of embodied masculinity".

Bear Magazine, their first publication, started as a photocopied flyer in 1987. Nowadays, Bears are organized in Associations (in our country we can speak of Madbear in Madrid, Bearcelona, in Barcelona, or Guadalkibear in Seville as the most prominent examples), meet in specific clubs that can or cannot be opened to others (women included) and organize gatherings all over the world (cruises, parties...), apart from launching periodical publications. And of course, they have web pages and one of the most visited gay dating sites in Spain, which is the webpage in which we have centered our analysis: www.bearwww.com.[3]

In this site, the assumption that "cyberspace presents a safe area where body image and presentation are not amongst the initial aspects of personal judgement and social hierarchy within the transgendered community" (Whittle, 2001) cannot be assumed.

Interactions cross-cut any division between virtual and physical, as do discussions on barebacking (practices and debates), as commented by Wakeford. Although Tsang and Bornstein state that 'electronic cruising has replaced bar hopping' (Tsang in Wakeford, 2004:117), strategies used by our informants part from the unspecific visit to sexually marked webs, such as the one mentioned above. Once there, one of the first steps is to carry on a search based on place of residence. In our case, as most of our fieldwork has been carried out in Valencia, our informants live there and search for other users in that city. Once in the gallery, they start interacting within the platform, but they will eventually end up interacting in other sites, as we already explained. If interactions are what they call "successful", they drive users to a personal meeting, normally conducive to sexual interaction.

In this site, identity is performed through combining different formats, mainly a photo (which is compulsory) and text. The use of other devices such as audio and video, flickr, blogs, photologs, SMS messages and so on, self-produced or grabbed from the Net, take place in a second step.

Through visual and textual representation, users play with the "script", the "audience", their routine, and place the body in a central position.

The website combines a free-of-charge and a subscription mode. Users may register free of charge to post their own pages, although to communicate with another "gallery" user, they will have to pay a fee that ranges from $9.95 for 30 days to $69 for one year. The system allows a free possibility of contact.

Bearwww.com is organized as a database, a gallery with a series of fields that the user has to fill in, known as the user profile. In it, the players can write a text to define the image they will project to capture the attention of the potential partner. When we log onto www.bearwww.com, we may choose to have a look at the free gallery, which is open to everybody, or enter through our user's name and decide whether we are looking for love, sex or chat. There is a search engine and a post on Aids and the need for protection. Parental control is also possible and all users are required to be above 18. The labels "Bear" and "Aids" function as sexualizers for initiates, as there is nothing on the first page that indicates the gay bias of the site explicitly. An equal significance have the brownish colours used in the page, which remind initiates of the bear flag, and the bear claws, the iconic symbol of the 'community'. The number of connected people is also shown on the home page (at the moment of writing, 6149).[4]

In the menu line on top of the page, we can find different labels: offline, home, gallery, who's online, Guide, Manage my profile and Create my profile.

The upper part of the page is devoted to banners advertising clubs and special events. The connection between online and offline activities is also displayed under the 'Guide' button: we can choose the events, bars, etc that we want to see by selecting all countries or specifying a particular country.

Past events offer the possibility to comment on the experience and have a link to the website, whereas future events offer the possibility to advertise one's attendance and the link to the website. When clicking "who's going", the limited profile is shown (profile which corresponds to the profiles visible offline through the 'gallery' and 'search' buttons and also shown under the 'who's connected' button). Once again, online and offline correspondence is explicit and encouraged, as this resource can be used for actual meetings.

As we have already stated, this site offers two ways of identity representation: through the script of the profile and through a compulsory photograph. For creating one's profile, which is free of charge, one needs a nickname and a password, an email and what is called 'main' photograph: "The main pic must be a pic of you, whether it shows your face or not. For the main pics, nudity is allowed only if genitals are not shown. Landscapes or other subjects are not allowed. The webmaster has the right to accept pics or not." Under these data, we find requirements for body characterization of one's self and of the desired ones. Body becomes clearly central at this point.

In this site, there is no place for gender play or sex identification games: users are not offered the possibility to label themselves, as it is assumed that they are all males. Nevertheless, difference is created by self-identification with a body type such as "bear", "cub", "chubby", "daddy" or "musclebear" and through the inclusion of otherwise intimate and delicate information, such as sexual roles (top, bottom, versatile, not said) and sexual practices.[5] After this information, we find the agreement button and the users IP address, which can function as a means to make users aware that they are being controlled.

The search engine on the home page, linked to the gallery, also requires explicitation of corporeal characteristics, such as body size, age and muscle complexion, body and facial hair (which is a defining trait of this group) and sexual role and practice. The events that the users are attending and the fact of being online can also be used as search criteria.

Some users take advantage of the possibilities of the system to enunciate the requirements photographs of their contacts must meet: thus, many users require photographs which show the person's face to initiate contacts. Here lays the main differentiation in representation: although most photographs show the person's face, making him identifiable by others, especially in a regional based site as this (the search engine allows to limit search to one city), there are different strategies that users adopt to show and hide at the same time. For example, some photographs show the user's body but not their face (6). Other strategies are the use of hats or caps, sunglasses, veiling one's eyes or using cameras or their flash lights to hide one's face.

Most of the pictures have been shot indoors in order to transmit the intimacy of the moment, and with some exceptions such as the case of a couple from Barcelona who represent themselves in disguise, they usually show the man on his own. Use of black and white photographs, although minor, is a differentiating element as well.

Another matter to have into account is the use/absence of costumes. When a man is dressed, he is wearing casual clothes. Alternatives of representation are limited: people are either dressed, nude or in underwear. Some men pose for the photograph in a sensual or interesting manner, some just pretend to be "natural" as their bodies, in contrast to the disciplined and muscled hyperidealized gay bodies.

It is curious that in many cases, one of the man's arms is elevated, thus indicating that the picture has been taken by the own person. This reinforces the intimacy of the moment and protects the man from unwanted questions. Curiously enough, this does not necessarily correspond to nudes, but to dressed people.

At this point, and without being registered, we can click on the image, and have access to a short profile. In some cases, they just provide information about the sexual role, the classification and the body complexion (as "admirers" do not necessarily have a big body), place, height and weight. Sex role, body type (as typified in the script of the profile) and place of residence are present in all the profiles, but all the other information is discretional.

As Gómez et al. (2007) state, going beyond the stage metaphor of the "presentation of the self" (Goffman), the corporeal images, the bodies themselves, take their prominence in online self-representation in a new relationship with the self identity projecting it as a never ending game (Giddens, 1994; Bauman, 2003; 2005) in what they have called the "playful embodiment" of online identity performance. That is to say, a combination of the performative and theatrical possibilities of the interaction and communication, already studied on the CMC systems, with the media displays of the digital images, specifically those of the own personae body. As we have seen in the case of the profile photographs of this site, the users explore these performative and theatrical possibilities that include hiding, simulation, scenification and play.

Body and online identity performance

In this particular site, the combination of a closed classification system based on corporeal criteria –size and shape-, facial and body hair and age (daddy, chubby, cub...) and compulsory photographs that are required to be "real" (under the terms and conditions of use) limit the possibility of playful strategies of identity and self-representation. A definite erotic and sexual desire expressed through particular practice is also restrictive. There is no possibility for gender or sex play. Having into account city-based search, publicity of events and their relevance in the site, and the itineraries of users that nearly always derive them to offline meetings with previous online contacts, the correspondence of textual and visual representation and the person can even be read as a guarantee of authenticity. This is even more important if we consider that from our experience we can say that interactions on the net and interactions on gay bars and discos are not exclusive, but reinforce (or inhibit, if the person does not turn out to be as expected) each other. Both contexts function as contexts of homosocial exchange. In bars, you meet people thay you can see online, and viceversa, and, therefore, the correspondence of these spheres inhibits stronger manipulation or simulation play.

We have drawn our analysis on males who are identified by their corporeal constitution and their desire not only for other men, but for men with specific traits. Men whose body is central to their self-definition. Men who interact in sites whose aim is the establishment of personal relationship and the consumption of bodies. Thus, they can be a perfect context to illustrate the existing multiplicity of bodies: the 'two bodies" of the social and physical (Douglas, 1978); the "three bodies" of the individual body, social body and body politic (Scheper-Hughes and Lock, 1987) or the "five bodies" with the addition of the consumer body and the medical body to the three (O'Neill, 1985) (in Low, 2003:11). Online dating sites draw upon body as a commodity and self identity is mainly performed through body representation. Cyberspace is not a disembodied "place" but it is constitued by different social contexts where social categorizations are displayed and negotiated. Bodies construct the main basis for online social interaction, and in particular that interaction related to emotional relationships. Embodying practices of gender and sexual identity are intertwined with digital technologies creating meaningful social contexts of interaction.

Notes

1. Ardévol's research on www.match.com, *Dream Gallery* (2005), is a clear precedent for the research we now present.
2. Gómez et al.'s research is part of an ongoing research on identity and body on the Internet conducted by a research group, mediacciones, to which the authors belong.
3. Other popular gay dating sites among Spanish users are www.bakala.org and www.gaydar.es.
4. On Friday 30th May, 2008, 20.01 h.
5. In the Spanish version the terms used for roles are "active" and "passive", whose emotional and stigmatizing potential is more evident than in the English terms.
6. All the pictures analized are from the gallery whose access is open to public.

Bibliography

Ardevol E. (2005), *Dream Gallery: Online dating as a commodity*. Paper to the 6th EASA media Anthropology Network e-Seminar. [http://www.media-anthropology.net/ardevol_dreamgallery.pdf]

Balsamo A. (1998), *On the Cutting edge. Cosmetic surgery and the technological production of the gendered body*. In: Mirzoeff N. (ed.) *The Visual Culture Reader*, London, Routledge, Pg. 685–695.

Bell D. and Kennedy B. M. (2000), *The Cybercultures Reader*. London, Routledge.

Bel D. (2001), *Meat and Metal*. In: Holliday, Ruth and John Hassard (eds.), *Contested Bodies*, London, Routledge, Pg. 168–178.

Berry K. (2007), *Embracing the Catastrophe: Gay Body Seeks Acceptance. Qualitative Inquiry*, Vol. 13, Pg. 259–281.

Butler J. (1990), *Gender Trouble*. London, Routledge.

Butler J. (1993), *Bodies that Matter. On the Discursive Limits of Sex*. London, Routledge.

Callen A. (1998), *Ideal masculinities: an anatomy of power*. In: Mirzoeff N. (ed.), *The Visual Culture Reader*, London, Routledge, Pg. 603–617.

Connell R. W. (1995), *Masculinities*. Cambridge, Polity Press.

Douglas M. (1978), *Símbolos naturales*. Madrid, Alianza Editorial.

Enguix B. (1996), *Poser y Deseo. La Homosexualidad masculina en Valencia.* Valencia: Alfons el Magnànim.
Giddens A. (1994), *Modernidad e Identidad del yo. El yo y la sociedad en la época contemporánea.* Barcelona, Península.
Goffman E. (1968), *Estigma. La Identidad Deteriorada.* Buenos Aires, Amorrortu.
Gomez E., Ardévol E. and Estalella A. (2007), *Playful embodiment and identity performance on the Internet.* Paper to the AoIR Conference, Canada, Vancouver.
Greco M. and Fraser M. (2005), *The Body: A Reader.* London, Routledge.
Heim M. (1993), *The Metaphysics of Virtual Reality.* Oxford, Oxford University Press.
Hennen P. (2005), *Bear Bodies, Bear Masculinity: Recuperation, Resistance, or Retreat?* Gender Society, Vol. 19, Pg. 25–43.
Herrmann A. F. (2007), *People Get emocional About their Money: performing masculinity in a Financial Discusión board.* Journal of Computer-mediated Communication, Vol. 12 (2).
Hetherington K. (1998), *Expressions of Identity. Space, Performance, Politics.* London, Sage.
Hine C. (2004), *Etnografía virtual.* Barcelona, editorial UOC.
Holliday R. and Hassard J. (2001), *Contested Bodies.* London, Routledge.
Howson A. (2005), *Embodying Gender.* London, Sage.
Kuntsman A. (2004), *Cyberethnography as home-work.* Anthropology Matters Journa, Vol. 6, 2, Pg. 1–10.
Le Breton D. (1990), *Antropología del cuerpo y modernidad.* Buenos Aires, Editorial Nueva Visión.
Low S. M. (2003), *Embodied Space(s): anthropological theories of body*, Space and Culture, Vol. 6, Pg. 9–18.
Martinez Oliva J. (2005), *El Desaliento del Guerrero. Representaciones de la masculinidad en el arte de las décadas de los 80 y los 90.* Murcia, Ad Hoc.
Mirzoeff N. (ed.) (1998), *The Visual Culture Reader.* London, Routledge.
Nyboe L. (2004), *You said I was not a man: Performing Gender and Sexuality on the Internet.* Convergence. The International Journal of Research into New Media Technologies, Vol. 10, Pg. 62–80.
O'Riordan K. and Phillips D. J. (2006), *Queer Online. Media, Technology and Sexuality.* New York, Peter Lang Publishing.
Richardson D. and Seidman S. (eds.) (2004), *Handbook of Lesbian and Gay Studies.* London, Sage.
Shilling C. (2005), *The Body in Culture, Technology and Society.* London, Sage 2005.
Skeggs B. (1997), *Formations of Class and Gender.* London, Sage.
Tseelon E. (2001), *Masquerade and Identities.* London, Routledge.
Tucker I. (2006), *Towards the Multiple Body.* Theory Psychology; Vol. 16 (3), Pg. 433–440.
Turner B. S. (2006), *Body.* Theory, Culture, Society, Vol. 23 (2-3), Pg. 223–229.
Van Zoonen L. (2002), *Gendering the Internet. Claims, Controversies and Cultures.* European Journal of Communication, Vol. 17 (1), Pg. 5–23.
Wajcman J. (2006), *El Tecnofeminismo.* Valencia, Cátedra.
Wakeford N. (2004), *New technologies and 'cyber-queer' research.* In: Richardson, Diane and Steven Seidman (eds.) *Handbook of Lesbian and Gay Studies*, London, Sage, Pg. 115–144.
Weinstein E. (1965), *Reviewe of 'Stigma'.* The American Journal of Sociology, Vol. 70, n. 5 (Mar., 1965), Pg. 636–636.
Whittle S. (2001), *The trans-Cyberian Mail Way.* In: Holliday, Ruth and John Hassard (eds.) *Contested Bodies* London, Routledge, Pg. 153–167.
Wikipedia, "Bear (gay slang)" (13 May 2008).
Zafra R. (2005), *Netianas. N(h)acer mujer en Internet.* Madrid, Lengua de trapo.

Chapter 12
From the developing mind to developing technology

A. Frisiello, M.C. Piccirillo

Abstract

Humans use the body to know and to experience the world. The body is the metaphor with which we build artefacts and creations. Technology reveals–in its shape and functions–the human action behind it. Anthropomorphism could contribute to bridging the (seemingly wide) gap between the body and technology, by showing that in everyday life the human being adopts as his explicatory principle: his body, his mental models, his experiences, his history. The imaginative function is one of the main characteristics of our mind, which is strictly related to the bodily dimension in conveying sensations, impulses, and contributing to identity. In new technologies this function has found a new role, at the boundary between reality, imagination and virtual space. New technologies create spaces of new presence, based on new social rules. People can assume different identities, modifying their behaviour, languages, and the mind; by relating to the mind, technology becomes an agent of social change. This paper looks at the human being as the result of the interaction between the biological substratum (brain and body) and all of our relations and emotional experiences, or as Siegel (1999) describes it, the developing mind, in relation to the environment and direct experiences. Moreover, it presents the joint experience of a psychotherapist and a psychologist working on cognitive ergonomics. The authors strongly believe that from the observation of clinical cases and the evolution of ICT technologies, devices, services, virtual environments, etc., it is possible to infer the unpredictable implications of technologies on everyday life, in order to comprehend how the imagination, individual evolution, and ultimately our minds are changing.

What neurophysiological changes do technological experiences cause in humans? Do these changes, in their turn, have feedback effects on technologies? If it is true that humans create artefacts on the basis of their bodily and functional experiences, according to the principle of anthropomorphism, in what ways do technologies reflect the minds that create and use them? This paper addresses the relationship between new technologies and the human being in modern-day society, observing such technologies as interacting systems, to propose an integrated view of this relationship. The considerations presented in this paper come from the work and experience of a multidisciplinary work group at the Istituto Superiore Mario Boella (ISMB), where one of the authors is a researcher, and her encounters with a clinical psychotherapist.

For several years now, the ISMB has investigated technologies to support the independent living, with particular emphasis on the engineering of systems and models for homecare services for elderly people. The engineers, computer scientists and psychologists who work together at the Services and Applications

Laboratory have designed and developed a tele-homecare system, which has been trialled in various contexts with real users. The trials conducted with elderly people, caregivers and the professional workers who assist them, have enabled us to observe certain aspects related to the introduction of the technology within the sphere of personal care. The research and development methodology used, focused fundamentally on the person, has been enriched with case studies taken from the clinical practice of psychotherapy, which has enabled us to reflect in greater depth, and share our considerations on the role that technology plays in individual and social change.

Introduction

The body and the mind have always been the means through which humans know and experience the world. They are the metaphor with which we build our artefacts and creations, including the new technologies, which have driven humans today into new fields of action, in which to act, experience and learn.

Our relationship with new technologies has developed in various ways over time, to the point that today it might be said that humans are no longer the subject, and technology no longer a mere tool at our disposal. At some point, technological criteria surpassed humans in their objectives, capacities and speed. As Galimberti explains, "*technology has come to constitute the human environment, and in measuring ourselves against the criteria of rationality and efficiency, we have no hesitation in subordinating the needs of humans to the needs to the device*" (Galimberti, 1999:36).

The relationship between humans and machines has intensified increasingly, as technologies become ever more immersed and widespread in all spheres of daily life. This heightened association, this intensified interaction between humans and technology has changed both. The change witnessed in humans is therefore also a consequence of our use of technologies. As we continue to observe how technology is evolving, we ask ourselves where the new frontiers of technology are headed–from social networks to virtual worlds. Another question we might ask is how humans are changing in relation to new technologies.

Anthropomorphism

Evolutionary theory explains human beings as the product of the animal evolution. Nevertheless, humans are differentiated from other animal species by the fact that we do not have a genetically pre-established habitat. As humans, in fact, we seek out our environment, we determine it, we change it, and we build it, guided primarily by our needs. Piaget (1958) defines this ability to adapt oneself to the environment as a fundamental expression of the intelligence. Technology is one of the main means through which humans have established a certain control over the environment: humans have been able to prevail over other species and over nature precisely because of our ability to use and create tools and artefacts that have become increasingly more complex.

The tools humans create, including technologies, take on the role of an extension of the body and of the physical and mental capacities of humans, expanding on them. Of all our artefacts, those belonging to the digital age are vested in a very special way with anthropomorphic qualities. Personal computers in particular, which today are so massively present in all spheres of daily life (home, work, school, mobility, etc.,), have been rapidly "anthropomorphized" in the imagination of users, so much so that they are attributed qualities which until their invention were reserved exclusively to humans (Grazia, 1999).

Nevertheless, technological anthropomorphism should not only be understood in a formal sense, that is, insofar as it reflects human shapes and functions; rather, the concept needs to be broadened to include more intrinsic aspects, such as our emotivity and relational nature, pertaining to the sphere of human needs.

Given that a need is a state of tension produced by the lack of something which would otherwise fulfil our physiological, psychological and social requirements, it can be said that new technologies represent one of the possible ways in which we seek to satisfy our needs.

In this search, humans often make use of *imaginative transposition*, that is the capacity to move from one object to another in avoidance. Imaginative transposition is a cognitive mechanism that helps us avoid the disappointment that comes from the impossibility of satisfying a need. Moving beyond known and existing approaches in the real world, this capacity for abstraction suggests new ways (which are also symbolic) of satisfying to some extent the underlying need.

> "Taking A for B is a symbolic act which seeks a connection between diverse things, taking as its starting point a shared characteristic which the need points out. If the act is successful, the symbolic connection between A and B becomes a stable perspective which can be repeated at pleasure, independently of effective need. The set of these symbolic connections which are acquired through successful acts constitutes our experience. However, experience cannot be had by simply limiting oneself to reproducing in the future what we have acquired in the past: memory would suffice for this. The future can be faced via the transposition of one's imagination" (Galimberti, 1999:210).

Digital environments constitute the materialization of imaginative transposition, to which individuals are drawn because they are driven by an intrinsic relational need. They are an alternative and innovative approach to satisfying needs which otherwise would remain unfulfilled and frustrated. "*There is no interiority which does not grow on the fabric of our actions in the world. And vice versa, there are no actions which are not connected with our interior world*" (Galimberti 1999, p.215). Within the scope of these actions, our interactions with digital artefacts should also be considered.

The relational mind

Anthropomorphism permeates the ways in which humans establish a relationship with the world we experience, through the qualities of the objects we use. People, according to Siegel (2001), are shaped fundamentally by two elements: the genome and experience. On today's experiential horizon, new technologies play an important role, contributing to change in humans. The theatre in which these processes unfold is the *relational mind* (Siegel, 2001). In defining the mind *relational*, Siegel explains how the mind, in all its biological and cognitive dimensions, is formed through the interaction of neurophysiological and experiential processes. Lying halfway between the neurosciences and psychotherapy, Siegel scientifically merges the interpersonal and neurobiological bases for the development of human beings. He recognises and emphasises the organic substratum, affirming that the mind emerges from brain activity. The mind is generally defined as a "*subjectively perceived, functional entity, based ultimately upon physical processes but with complex processes of its own: it governs the total organism and its interaction with the environment*". Some definitions place greater emphasis on the biological sphere, stressing how *the neuronal activity of the brain creates the mind;* others identify the mind with cognitive functions, likening it to a *flow of information.* Siegel introduces and adds an extra element: "*the human mind emerges from patterns in the flow of energy and information within the brain and between brains*". This implies that social interaction and interpersonal experience influence genetically-determined growth and the way in which we mentally reconstruct reality.

When we are born, the brain is the least differentiated of all our organs. It has been shown how the development of the brain is a process that depends on experience. The experiences we go through stimulate the formation of synapses and activate neuronal circuits, creating new circuits and consolidating existing ones. These circuits are the expression of our genes modified by experience. The counterpart to this process is called *pruning*, where the absence of stimuli leads to the death of cells.

According to this approach, relationships shape the development of nerve connections, as the human brain is an open and dynamic, living system–that is, the brain is a system subject to change in response to the ongoing changes in our environment, able to receive specific stimuli insofar as it is functionally tied to other systems, especially to other brains. This interpersonal neurobiology rereads individual differences in relation to the interaction between hereditary traits, environmental influences and causal events. It also shows, with increasingly greater evidence, that the mental functions that lie at the basis of human experience–memory, attachment, emotion, representation–are not in fact functionally separated, from a biological point of view, from cognitive functions (Grazia, 1999).

In what way can the mind of an individual influence the development and functions of the mind of another person? The key lies in the exchange of information. An analogy can help us understand what Siegel means by this. Just as we exchange (send and receive) communications and messages with other people on a daily basis, the fundamental components of our minds exchange information between themselves. While the voice, e-mail, and fax are the channels for interpersonal communication, at the neurophysiological level the channels are slightly different, namely neurotransmitters, hormones and complex neural profiles. Both these levels of exchange are active in connections between two minds: information transfer may occur via different means, but it happens simultaneously. For example, physical vicinity conveys information through the transmission of heat, tactile stimulation, and visual contact–all fundamental aspects of the parent-child relationship in the early years of a child's life (Hofer, 1994; Supolski, 1997: in Siegel, 2001). Physical contact directly influences cerebral electrical activity in the people involved (McCraty et al., 1998: in Siegel, 2001); however, in the absence of such contact, information flow is conveyed by other, verbal and non-verbal, behavioural responses which activate processing and representation processes in the mind of the person who receives the messages. In particular, non-verbal messages such as facial expressions, gestural expressiveness and posture, communication times and prosody activate the emotional centres of the brain, influencing significantly the way meaning is attributed. In this way, apparently external phenomena act at a very profound level on mental states (Siegel, 2001:269–270).

The relational mind in humans is more than just a state of being; it is a need, which drives us to build relationships with the external world, so as to maintain a suitable psychological equilibrium. The effects of isolation, which is at the same time social solitude and deprivation of sensorial stimuli, are a demonstration of this. Being in isolation causes humans to lose their sense of self, of the boundary between the self and the world, and the sense of existence. Sensorial and affective stimuli, in terms of presence and quality, strongly influence the attachment style that a person subsequently develops.

Research by Spitz (1962) has shown how premature separation experiences (of a child from its mother) which are not supported by affective stimuli lead to passiveness, weakness, insomnia and physical aggressiveness, also against oneself. Other studies on the same topic have further shown that the presence of negative physical stimuli is actually more bearable than the total absence of stimuli. These studies explain how the need for recognition emerges and acts–that is, the need that other people show signs that we exist for them. This need is important also because the image we build of ourselves and the quality of our existence in the world very much depends on the recognition we receive from others (Berne, 1971, 1967; Magrograssi, 2003). These aspects help explain the intrinsic capacity of humans to relate to the external world.

At the basis of this relational predisposition we find neurophysiological entities. Mirror neurons are the cells responsible for our capacity to recognise ourselves as autonomous beings and make empathetic and affective contact with others (Iacoboni, 2008). As Gallese, one of the Italian scientists who discovered the mirror neurons, explains, a pre-linguistic process would seem to be at the basis of empathy, a process of *embodied simulation* (Gallese, 2005), whereby some motor neurons act immediately before cognitive processing sets in. Empathy is based on an internal simulation of an action, which enables us to identify it in our experiences in an automatic or non-conscious way. The activation of mirror neurons is possible due to what Rizzolatti and Gallese call our *action vocabulary* or *motor repertoire,* or, alternatively, *behavioural repertoire.* In other words, the reaction of the neuronal mechanism would seem to be closely tied to the capacity of the observer to reproduce in his turn the action that is observed (Bracco, 2005:37). *Intentional consonance* is what Gallese calls the opposite state of feeling alienated from others, which, in normal conditions, leads to the implicit and inevitable creation of an alter-ego (Gallese, 2007).

Through imitation processes and empathy–the result of the activation of mirror neurons–the individual is able to relate to other people, also by means of a machine: it is no longer necessary that people see each other to be able to relate to each other. All this explains how it is possible, through the simple use of technology, to feel bodily sensations of pleasure, excitement, or malaise, and the consequent cognitive-affective processing of these processes. Machines thus become a vehicle for human relations, by reason of which they also take on all the relative characteristics. The physiological, sensorial and cognitive mechanisms that are activated in relational contexts mediated by digital technologies seems the same mechanisms that are activated in "traditional" relationships.

Our minds, therefore, are able to accept new media as something "natural", as the possibility of interaction activates mental processes that are genetically present in all people be they young or old, regardless of their age.

With the introduction of new technologies, humans have changed genetically, so much so that some authors speak of *homo tecnologicus.* What we can say is that a *digital revolution* has occurred, capable of changing our mental and sensorial system and shaping a *new culture*, made of modes and forms that are quite different from the past (Cantelmi p.21 in: Fabbroni, 2006).

The relational side of technology

New technologies have become part of the genetic heritage of all of industrialised society, simply because they have access to the very basis of the human organism. We have made these tools such an integral part of our genetic heritage that they have become an essential part of us. New technologies are a sphere of experience, relational objects and subjects just like other human beings, as they relate to humans even though they are Other to us. Although dependent on us, they enjoy an autonomy which leads them to assume a fundamentally human characteristic: they are relational and social.

This perspective broadens our view of technology, assigning it the role of subject in relation to humans, with its own life and characteristics. If we consider technology merely an object, its role is limited to that of a tool at the service of humans. On the other hand, if we see technology as a subject, capable of conveying relations with other individuals, it takes on the human characteristics and rules valid for interpersonal relationships. The result is that technologies are *relational.*

Digital technologies are human artefacts, though they are the only artefacts with which it has been possible to establish a dialogic relationship based on the use of language. The possibility of applying the rules of human communication to interactions with machines opens the way for the same dialogic dynamics as those that take place between people. In this way, technological objects become *relational artefacts*, as Turkle (2006) defines them, as they present, for all intents and purposes, real internal states that can influence the people with whom they interact. Fusion with new technologies is today potentially total. This is the age of embodied interactions, of the natural-born cyborg, in which the human body and the hardware of technology are merged into a new system sharing the one body. Previously the body represented the boundary and limits of subjective action, which could be extended through technological 'prostheses'; today the body is no longer a limit, but rather a possibility, which continues to interact with the dimension of identity and of the self, in interactions, relations, and affective life.

We have seen how our demands and expectations when it comes to technologies are strictly tied to the fundamental needs of humans. The importance of the neurophysiological sphere, the role of cognitive functions, and the involvement of emotive reactions are prevailing concepts in human-machine interactions, and at the same time the founding concepts of human psychology.

Cognitive ergonomics and clinical psychotherapy are fields which are usually seen to be disconnected, though in reality they are united by a common objective: the search for personal well-being. Ergonomics, in contrast with other human and social sciences which study humans, does not act on individuals but on the world that surrounds them, designing artefacts that take fully into account cognitive, sensorial, bodily and emotive characteristics. In clinical practice, the well-being of humans is conceived in the same terms, within the scope of therapy.

Below we report some of our experiences from ergonomic and clinical practice, which contributed to our considerations on the relational use of new technologies.

Affective human factors

Ergonomics or Human Factors (HF)–which, according to the Ergonomics Society, are interchangeable terms–represents a field of study and practice focused on understanding how the relationship between humans, machines and the environment works, with a view ultimately to correcting or redesigning the relationship on the basis of personal needs, according to the criteria of efficiency, satisfaction, safety and comfort (Novara et al. 1996). The contributions made to this branch of study are many, and have come from a wide range of different disciplines: from cognitive psychology and psychophysiology, to sociology, engineering and computer science. In studying the interactions of humans and machines, the focus traditionally has been placed on cognition, representing the point of encounter (and confrontation) between the two entities. Cognitivism understands cognitive processes of humans in computational terms, with the relationship between biology and the mind likened to the relationship between hardware and software, or where "*mental states are considered largely independent of the mental machine which forms the substratum of those states*" (Grazia, 1999). However, many people have worked, and still work today, to confute this straightforward analogy between humans and machines, and expand on it. The psychology of embodied cognition considers the mind not as a set of brain circuits, but rather as the result of a more widely distributed phenomenon, which does not lie solely in the brain. Human Computer Interaction (HCI) studies stress the importance of embodiment and the rules that characterise mediated interaction. Here we also find the *direct manipulation* approach, which is based on the idea that people interact with digital objects in the same way that they interact with physical artefacts (Mantovani, 2000). This approach has long been predominant in the field of Human Factors: in the study of interaction modes, in design issues, in guidelines, and in the way usability is evaluated. The underlying basis of this approach is the concept that we think and act using our bodies as a reference, and hence that the most effective metaphors are those tied to embodied experience. Therefore, an artefact or interface is more effective the less it demands of people in terms of learning requirements, and the more it adapts to our perceptive-motor schema (Morganti et al., 2005) –and, we would add, to our goals, our motivations and our needs.

All human activities are always directed upon an object, be it physical or ideal. There exists no activity without an underlying need. Guiding the complex structure of human activity is expectation, the anticipation of the action, which at the level of the activity takes the form of motivation (Riva, 2004). Taking this concept as their starting point, some authors have recently begun to shift their attention to the emotive aspects that come into play in interaction between humans and machines. Here they speak of *New Human Factors* or *Affective Human Factors* to refer to the field of ergonomic research which makes emotional responses and aspirations, affective needs and social relations between people central to design criteria (Talbot, 2005). Affective Human Factors privilege the emotive aspects of individual experience over functional aspects (though the latter are nevertheless taken as fundamental in designing effective interactions). In this way, project goals tied to usability and the utility of applications and artefacts are extended to include criteria directly tied to the quality of the individual's experience, along with his or her motivations and relational contexts.

An example of a practical application of this approach is given by the design activities for a new technological system at the Istituto Superiore Mario Boella, where a methodology focused on the person was adopted. The User Centred Design (Rubin, 1994) was applied, with special emphasis placed on those aspects of the user experience tied to the sphere of emotion and well-being. In designing DREAM, an evolved tele-homecare system for elderly people, focus was placed on these aspects throughout the project–in the preliminary definition of user requirements, in the design and engineering of the object and the service, and successively during field trials, conducted in various real-life environments, with real users (in own homes and nursing homes).

The results mentioned come from qualitative and quantitative studies conducted within the DANTE project[1], in Italy and France, on a sample of 350 people divided into 'future elderly people' (from 50 to 65 years old), 'elderly people' (from 65 to 75), and 'senior elderly people'(over 80). The surveys enabled a range of key issues to be focused on, concerning ageing and the quality of life. It was found that the fundamental needs of elderly people have to do with their autonomy and safety. The ability to count on oneself, to control one's environment (the home mainly), and to play a useful role in society were felt by senior citizens to be of fundamental importance. This was associated with a strong sense of fragility, and hence the overriding need for protection and contact. In short, these were the needs and motivations which guided the design of flexible tele home-care service models, based on similarly flexible technology. The main component of the DREAM system is a watch, an object that can be worn, featuring sensors and integrated functions designed to monitor the activities and state of the person in the environment in which he or she is located. Field trials, using laboratory prototypes, confirmed that the introduction of the watch and the more complex monitoring system was easily acquired and accepted by the involved elderly people. They proved to be interested users, curious about new technologies and highly motivated about progress. Actively involved in a context in which, first of all, a relationship was built, and objectives and information shared, the trial participants provided considerable, and often detailed feedback on how to improve the usability and wearability of the system.[2] The wearable prototypes facilitated relationships between the elderly people involved and their closest friends and relatives (while the trials themselves proved an occasion to establish new relationships).

From the point of view of the researchers, and in particular the technologists involved, the adopted Affective Human Factors approach was of great importance. In their experience, an awareness of the deeper needs of the final user in design projects has proved a strong and effective guide to making decisions of an architectural and technological nature. Knowledge of the deeper needs of final users enabled use-cases and technological solutions to be evaluated much more effectively.

Clinical practice

Everyday clinical practice and ongoing work with individuals and groups enables us to observe the interconnections between the psychic, mental, cognitive, physical, biological and social aspects of individuals. In encountering people directly, clinical work draws out their peculiarities, characteristics and histories. As it has to do with the subjectivity and uniqueness of the individual, clinical practice cannot avoid encountering the social aspects that are reflected in individual people. Clinical study provides a theatre in which the social aspects of people's lives take on a special dimension, where they can be observed in depth and connected with personal meanings, and personal life histories. The perspective offered by clinical study on people has the potential to enrich the way we conceive the relationship between humans and new technologies.

In everyday clinical practice, it can be seen how new media have gradually begun to enter the life histories of patients. Initially they were received with suspicion, and then later with curiosity. Clinical perspectives are generally focused on the study of pathologies. Nevertheless, where the therapist is open to the topic, dialogue can be promoted in therapy on topics which would normally not arise in such settings–topics on which psychotherapists are still often poorly informed. Hence we have an occasion to reflect on the effects that technologies have on people who, through the patient-therapist relationship, seek help in dealing as best they can with the world.

From a clinical point of view, it would seem that new technologies, and the internet in particular, enter easily into synch with the internal world of people, and with their needs–in particular with their need for relations and belonging. In the clinical setting, it can be observed how imaginative transposition enables virtual reality to take the place of real life in satisfying needs that are otherwise frustrated. An empirical analysis of some case studies reveals how people who somatize their malaise in the bodily dimension, with symptoms arising on the sexual and emotive planes, often use digital technologies to relate to others.

From a clinical point of view, technology can make an appearance in adaptive solutions. Let us mention some examples:

A. *a woman who suffers from panic attacks. She has constructed a manner of living which is limited in real time and space, but enriched in the virtual dimension: on the one hand she has difficulty driving her car, but on the other, she has elected the internet as her preferred means to encounter others, so as to remain seated in a place she perceives to be safe.*
B. *a patient disturbed by panic attacks. He restricts his social relations to work relationships and the relationship with his partner. To meet other people without having to deal with what he feels to be a physical limit–the fear and shame of a possible panic attack–he uses the internet.*
C. *a man who constrains his sexual curiosity in real life, to uphold an image of himself which is consistent with his personal and spiritual values, and satisfies that curiosity by secretly visiting websites of a pornographic or erotic nature.*

For these people, virtual reality represents a real opportunity to satisfy their needs without having to address the obstacles that inhibit them from living positively as they would like. In these cases, digital contents and technologies represent an initial response strategy which they otherwise would not have been able to activate. For example, in the case of patients suffering from panic attacks, satisfying their relational needs in the virtual world means avoiding that which for them is most painful: emotional failure in romantic relationships, their difficulty in establishing and maintaining affective relations, addressing the angst associated with real encounters where the person is faced with his/her own physical nature and, in particular, the sensations of malaise that the body conveys (tachycardia, excessive perspiration, trembling, etc. which is caused by the feeling of panic or terror).

The utility of this tool lies in the fact that people can experience the freedom of exploring aspects of the self which are present but inhibited due to the discomfort felt. People who freely use the internet experience *other places where they can explore themselves.*

The provided examples illustrate just how often people experience aspects of themselves as limitations. Technological experience, on the other hand, enables the possibility of freely exploring what for them, in real life, is a source of pain and suffering. When reality is experienced as a limitation and the individual seeks to override and control his impulses for fear of them, the person becomes blocked. On the internet, in contrast, they feel they are free, they experience openness and independence, and they explore without fear. The approach towards communication technologies is generally gradual, and the key to the success of these experiences lies in the intimate sharing of a sense of freedom. The experience of "us", of belonging to a virtual community, represents the possibility of contact with others, which may then become the backdrop and support for further experiences, giving these people the possibility to renew themselves continually. Various authors in the clinical and social field are beginning to focus attention on the experience of belonging and its relationship to the sense of identity. In our contemporary world, belonging has become a difficult and problematic issue that cannot be taken for granted, cutting across social, anthropological, political and economic disciplines along with psychology (Franceschetti, 2005). Talk of belonging is indicative of a change in perspective: after the narcissistic accentuation of individualism over belonging (Lash, 1999), the concept of identity has increasingly returned to be understood in terms of "we", with its roots in the socio-cultural setting. In our post-modern world, the crisis in ties of belonging has made the search for identity more uncertain and tragic, highlighting the need for new ties, and for a new community (Franceschetti, 2005; Bauman, 2001). The internet and new technologies have quite effectively inserted themselves into this new context, enabling people to uphold their individuality while being part of a community. The ties that are built between people on the internet are psychologically anomalous and interesting, as they are based on the situation of two intimate strangers united in their search for new ways to build their identities (Fabbroni, 2006). This *technological magic* enables people to approach the Other without having to encounter it in its entirety, and hence without having to reveal their fragility, because the Other stands beyond the physical dimension of reality, though it is still real and not imaginary.

Such explorations of the virtual world, where shared with the therapist, become an occasion to transform one's limits into resources, by shifting relational skills and experiences from the digital to the real. This fresh awareness of oneself and the world, where shared and transformed into new abilities through therapy, helps patients overcome their block and face real life.

New technologies can also represent a benevolent outlet through which certain dissatisfactions in the real world can be overcome. Precautions are needed against inappropriate usage to prevent the risk that the individual may become alienated in the illusion of having satisfied his or her real needs in the virtual world, leading to a *loss of consciousness* that experiences mediated by machines

can provoke in human beings, as prophesied by Weizenbaum. Understanding technologies in their relational sense, and never as a surrogate for real relations and contexts, enables their positive and healthy potential to be realized. Even in cases where they fill real shortfalls, it is always necessary to focus on the real satisfaction of the need so as to prevent a vicious circle from being created.
The relational nature of technology further suggests the importance of an assisted approach: these tools should be presented and explained, though certainly not demonized, and their constructive and informed use promoted, so as to open up new possibilities for success in the real world.

Conclusions

In this article we sought to present a holistic view of humans in relation to new technologies. Taking as our starting point a psychological point of view in the fields of neurosciences, ergonomics, and psychotherapy, we arrived at the relational characteristic as the explicatory principle of the relationship between humans and technologies.

In this paper, we highlighted the aspects upon which the human need for relations and belonging is based. What emerged is how the experience of technologies and interaction with them influence the human brain and mind, modifying them as far down as the neurophysiological substratum. Such changes make it increasingly easier for humans to relate to digital technologies, regardless of their age: these technologies shape our physiology and cognitive functions, adopting a language which is becoming progressively more natural for us. For this reason, although they may have their own rhythms and embodiments and present their own specific characteristics, we tend to perceive technologies as resembling humans.

Such mimesis risks underpinning a fundamental cognitive error, as although technology is the child of humanity, it is Other to us. In relating to technology, humans see themselves reflected back, thus falling prey to the illusion of being like the tool, dematerialising our humanity and neglecting the characteristics and needs of our bodies and minds.

In this process of 'dematerialization', which today affects many different spheres of everyday life (work, money, etc.), humans in reality seek concrete realization in the material of technology. The real world does not disappear, but remains beyond the digital space; it does not dematerialize. If we keep a firm grip on this concept, what we find is that real life is enriched with new possibilities, new spaces, times, tools and relations. Technologies are part and parcel of the real world and of our society.

The experiences mentioned, relating to the fields of ergonomics and clinical practice, led each of us to arrive at rather similar considerations. What we have witnessed is the enormously positive potential of new technologies, which can be activated most effectively when new media are used within relationships. In experiences with elderly people and clinical experiences with patients, technology played quite a natural role because it was never taken as a surrogate or substitute for either therapeutic or social relations. Rather, technology was more of a *bridge to the future*, opening the eyes of people to brand new perspectives.

Preserving the differences between the real world and its virtual counterpart will, therefore, be useful, distinguishing between two types of interactions which, although emerging from the same root, evolve in two distinct directions: relations mediated by relational technologies, and interpersonal relations. In doing so, the beauty will lie in beginning or continuing a relationship while accepting this fundamental difference.

Recognising the independence of the two entities could help prevent addiction and the misuse of the relationship between humans and new technologies, which would seem to be amongst the strongest ever witnessed in human history.

Notes

1. DANTE Project - *Domani Anziani oggi Nuove Tecnologie/Demain bel Age aujurd'hui Nouvelles Technologies* [INTERREG III A Alcotra Italy-France], 2004–2006.
2. For further details see: Bestente et al. "*Dream: emergency monitoring system for the elderly*". 6th Conference of the International Society for Gerontechnology, May 2008.

Bibliography

Berne E. (1964), *Games people play*. Random House, Inc. Copyright renewed 1992 by Ellen Berne, Eric Berne, Peter Berne and Terence Berne.

Berne E. (1961), *Transacional Analysis a systematic individual and social psychiatry*. New York: Grove Press.

Bracco M. (2005), *Empatia e neuroni specchio. Una riflessione fenomenologica ed etica*. In: *Comprendere*, Pg. 33–53.

Bauman Z. (2001), *Voglia di comunità*. Bari: Laterza.

Cantelmi T. (2006), *Saggio introduttivo*. In: Fabbroni B. *La solitudine in un click*, Roma: Edizioni Universitarie Romane.

Fabbroni B. (2006), *La solitudine in un click*. Roma: Edizioni Universitarie Romane.

Franceschetti G. (a cura di) (2005), *Attacchi di panico e postmodernità*. Milano: Franco Angeli.

Galimberti U. (1999), *Psiche e techne. L'uomo nell'età della tecnica*. Milano: Feltrinelli.

Gallese V. (2005), *Embodied simulation: from neurons to phenomenal experience*. Phenomenology and the Cognitive Sciences, Vol. 4, Pg. 23–48.

Gallese V. (2007), *Dai neuroni specchio alla consonanza intenzionale. Meccanismi neurofisiologici dell'intersoggettività*. Rivista di Psicoanalisi, Vol. LIII, 1, Pg. 197–208.

Grazia A. (1999), *La ricerca in campo informatico e la nuova scienza della mente*. In: Psiconline 26/04/1999, Pg. 151–193.

Iacoboni M. (2008), *I neuroni specchio. Come capiamo ciò che fanno gli altri*. Torino: Bollati Boringhieri.

Lash C. (1999), *La cultura del narcisismo*. Milano: Bompiani.

Magrograssi G. (2003), *Le carezze come nutrimento*. Milano: Baldini Castoldi Dalai.
Mantovani G. (a cura di) (2000), *Ergonomia*. Bologna: Il Mulino.
Morganti F. e Riva G. (2005), *Conoscenza Comunicazione e Tecnologia*. LED Edizioni Universitarie - www.ledonline.it.
Novara F. e Sarchielli (1996), *Fondamenti di psicologia del lavoro*. Bologna: Il Mulino.
Piaget J. (1947), *Le Jugement et le raisonnement chez l'enfant*. Neuchatel et Paris, Delachaux & Niestl.
Riva G. (2004), *Psicologia dei nuovi media*. Bologna: Il Mulino.
Rubin J. (1994), *Handbook of Usability Testing: How to Plan, Design, and Conduct Effective Tests*. New York: John Wiley & Sons, Inc.
Siegel D. J. (2001), *The developing mind*. Tr. it. *La mente relazionale. Neurobiologia dell'esperienza interpersonale*, Milano: Cortina.
Spitz R. (1962), *Il primo anno di vita del bambino. Genesi delle prime relazioni oggettuali*. Firenze: Giunti Barbera.
Talbot J. (2005), *New interactions? The role of 'Affective Human Factors' in design*. Ergonomic Australia, Vol. 19, no. 1, Pg. 21–25.
Turkle S., Taggart W., Kidds C. D. and Dast O. (2006), *Relational artifacts with children and elders: the complexities of cyber-companionship*. In: *Connection Science*, Vol. 18, No, 4, December, Pg. 347–361.

Postscript

Thomas Spence Smith

Social scientists over the years have contributed their own specialized analysis to knowledge of the body. Gutted though such a body may seem—it is not the body of the physiologists—it has nonetheless proven itself to be a subject on which social life leaves its mark. Yet what we have come to know about the body, in all of its socially constructed, publicly controlled, religiously segregated, artistically idealized, and medically managed forms, is only a small fraction of what this rich subject promises to reveal. There are still great gaps in our understanding, and there is no consensus on the direction of future research. In the human sciences, this is all too apparent in the cacophony of voices attempting to bring theoretical discipline to the study of this subject. For the last few years especially, professional interest in the body has mushroomed, after decades in which bodily considerations were seldom among the topics discussed at annual meetings or given proper hearing in professional journals. Now "The Body in the Social Sciences" has become a Research Committee of the International Sociological Association. And the hunt is underway for a framework of concepts and theoretical insights that will convert the cacophony into a meeting of minds.

In these two volumes, compiled and edited by Biancamaria Pirani and Ivan Varga, appear fresh studies of the body, undertaken by sociologists, anthropologists, philosophers, architects, artists, communications specialists, health scientists, and others from kindred academic and professional communities, all of whom have affiliated themselves, formally or informally, with the ISA's new Research Committee. At the center of their work is not, with several exceptions, the body as it is studied by the health sciences or by biological scientists. For want of a better characterization, it is instead the body-in-society—the body as it is socially constructed, publically scripted, religiously controlled, variously idealized in different historical settings and across different social groups, and as it is coupled interactively to technology in increasingly complex relationships. When the body is studied in any of these ways, it is seldom understood alone and nearly always becomes a screen on which we can discern the forces of society.

Alongside this body-in-society perspective stands other work that has taken a biological turn. Here we are asked to recognize how social life itself is sometimes best grasped as an extension of our physiology—how social interaction, attachment behavior, and social networks themselves cannot be understood without grasping how their structure and dynamics are extrapolations of the same powerful physiological forces unifying a mother and her newborn. Society itself, in this view, is a system of networked bodies, bodies whose social links are strongly constrained by the physiology of bonding, care, and parenting.

Where do these several themes, ancient and modern, come from? The "body" has figured in modern social thought for nearly two centuries. Seldom, however, has it had a leading voice. Fascinating to observers, yes; but drawn to the center of social thought, never. The exceptions, notably, were in the writings of Mauss and Durkheim, whose singular focus on the Social *sui generis* left a legacy that has worked its way into modern social constructivism, phenomenology, and various forms of symbolic interactionism, from Mead to Blumer and beyond. Alongside Durkheim and Mauss, sometimes following, sometimes contemporaneously, came the profoundly interesting investigations of anthropologists—ethnographers from anthropology's classical period, from Morgan to Malinowski and Radcliffe-Brown, each of whom found himself, if not actually living with their subjects, studying societies where bodies weren't hidden from view, clothed and fashioned as they had been in the West, but occasionally naked and often ornamented in symbolic ways. Bodies, in these non-western settings, were understood conceptually in the local categories of myth, history, and origins. To anthropologists, unsurprisingly, the body often appeared as an emblem of society.

This had been Durkheim's conclusion, too. Religions everywhere were a fertile source of body commentary and doctrine. Studying religion led naturally to analyzing the social and cultural origins of attitudes toward the body. In particular, the functions of the body, its place in a world shaped by immaterial forces, and its potential to pollute sacred space made the control of the body a religious duty. Something of the same normative pressure, of course, had also been a theme in public philosophy from antiquity onwards. Images of the body, and the training of the body, figured in normative reasoning about how to build the best society: about forms of political organization and the means to best educate citizens to fulfill civic obligations. All of these themes have entered into our modern interest in the body, as witnessed by the chapters in these two volumes.

Some of these chapters are empirical—studies of fitness centers and internet postings, of the clustering of artists into communities, of the body as an extension of technology used in the piloting of planes, of socio-geographical mapping, of locative technologies such as GPS devices, of the quarantining of architectural sites, of the commercialization of women's breasts, of performative gender, of the dissociative issues in acting, of aboriginal identity-making, and other subjects. Still others are analyses of utopian bodies, studies that seek to provide us with conclusions about shifts in bodily ideals that are correlated to changes in political organization. One chapter takes Obama as a charismatic figure, and highlights the social conditions in which the hunger for charisma develops. Interspersed with these empirical studies are examples of theoretical research and formal modeling—commentaries on memory, evolution, and the theoretical integration of biological and cultural knowledge; formal models of innate mechanisms active in interaction that have formed the basis for computational studies, simulations that explore the neurosociological foundations of social networks.

I have learned a good deal from reading these two volumes. Yet reading them has also highlighted for me where weaknesses exist in our control of this subject. What we are witnessing in the agenda of the Research Committee on

The Body and the Social Sciences are developments that the physicist Richard Feynman (2001) saw in the history of physics. In an essay on the relationship of physics to mathematics, Feynman contrasted two traditions in the teaching of mathematics; one he called the Babylonian tradition, the other the Greek tradition. In Babylonian schools, students learned by working various problems, eventually grasping general rules. They also learned some geometry—how to calculate the area of circles or the volume of a cube, the Pythagorean Theorem, and other matters. Compared to the Greek tradition, the Babylonians taught mathematics on a problem-by-problem basis, and although they were able to reach solutions nearly as exact as the Greeks, they were unable to deduce consequences or axioms from postulates or theorems. This the Greeks could do, following the discoveries of Euclid. Greek mathematics was top-down and fully controlled by deductive rules. The Babylonians, by contrast, had little islands of mathematical knowledge, not all connected. Without bridges, getting from one island to another was not always a matter of certainty. This, according to Feynman, is the way work in physics proceeded for a long time—by the slow accumulation of rich descriptions, both empirical and theoretical, that arose on small, sometimes isolated, islands of research, many of which were not linked to others by bridges. It is the eventual development of these bridges that makes all the difference to the unity of a discipline. In physics, the low temperature world could then be seen as continuous with the study of galaxies or the modeling of turbulence.

This kind of disciplinary unity is missing in the human sciences. We have a few islands that have been well-described empirically—occupational stratification, the structure of social networks, the diffusion of innovation, complex organizations, and others—but it nearly impossible to get from one of these islands to another using theoretical tools that do real work explaining observations on both. You can't get far in understanding occupational prestige with theories that explain the diffusion of innovation. And there are few tools that link any of these subjects to studies of the body. Generally speaking, of course, the study of the body merely reflects the same theoretical and conceptual fragmentation found elsewhere in the human sciences. This isn't because specialists have no interest in adjacent subjects. It is because social theory everywhere is missing something.

What is missing? Post-disciplinary dialogue is one thing, as Donald Levine has suggested (see Levine 1995, Camic and Joas 2004). The divisions among academic disciplines are nowadays historical relics of ancient boundary disputes—markers of old turf wars, not markers of unbridgeable intellectual divisions in the study of the natural world. Indeed, research that proceeds unaware of work in adjacent fields risks, if not being eclipsed by that work, succumbing to the fate of intellectual insularity wherever we find it. It ends up eventually reaching and then limited by an asymptote, the limit on its capacity to understand its own subject, a limit formed by narrowly focusing on its subject solely in the terms of its own traditions. Sociology has this problem. Consider a simple (to many sociologists heretical) question. Has sociology been helped or hindered by harkening back over the years to Durkheim's turf wars—to his struggle to define sociology as a separate discipline? Against the backdrop formed by the intellectual hegemonies of the 19th century university, Durkheim had tried to show that there was a class of phenomena he called social facts. These were subjects, like the social suicide rate, that were social *sui generis*; they exhibited regularities that were resistant to explanation in view of non-social facts. This every sociologist knows, pointing to Durkheim's work as a model of good science. But has such a methodological rule-of-thumb helped or hindered us in developing the sociological study of the body? Too many of our subjects are automatically given a Durkheimian *imprimatur*, as if their variation, like that of the social suicide rate, could only be understood in terms of purely sociological variables. So the answer to my question requires us to know whether the body as studied, say, by physiologists or evolutionary biologists, indicates ways to extend our understanding of the body-in-society. And here, I think, the answer has been obvious for a long time, even if we have been blinded to it by our insularity.

There are core subjects in sociology that reappear in the research agendas of other disciplines, from neuroscience to psychology. One of these is the body. Others include the study of interaction, attachment, bonding, dominance, and the social dyad. For example, dyads—mothers and newborns, mating partners, interaction pairs—have been described by research in physiology, evolutionary biology, neuroscience, psychology, and sociology. Dyads are, to use the language of the philosophy of science, "robust" natural entities, and the focus on them from different disciplinary starting points has produced convergent observations and generalizations. What this indicates is that dyads have properties that are invariant under different theoretical descriptions. Dyads are real, their properties incorrigible, not reified concepts. The same can also be said of the body, at least as it is studied in the natural sciences, and this has far-reaching implications for the sociological study of the body. One thing that follows, for example, is that sociological indifference to the robust properties of the body opens sociological theory to endless specification error—to proposing superfluous explanations of bodily matters, explanations of properties or dynamics that have already been well-described empirically and theoretically, in non-sociological terms, at lower levels of analysis. This amounts, at times, to spuriously attributing causal significance to social variables when in fact they themselves are merely covariates of unrecognized processes at more atomic levels.

This conclusion, as Willliam Wimsatt (2007) might say, argues for a kind of science that is both reductionist and emergentist (Callebaut 1993). And no real scientist these days, as Wimsatt has so cogently argued, can afford not to be both. But this raises another question. When is one entitled to treat a social phenomenon as an emergent—as, that is, a phenomenon that cannot be reduced to variation at lower levels of analysis? One theory of emergence would say that an entity is an emergent when it is not subject to linear decomposition—that is, when it can't be shown to result merely from adding together lower-level or more atomic entities, like so many snowflakes in a drift. A snowdrift or a sandpile is a linear composition of snowflakes or grains of sand. But few phenomena in the natural world are really linear in this sense. Instead, they result from combining entities or processes at lower levels of analysis in ways that entail feedback—

complex patterns of reciprocal causation that reinforce and stabilize the coupling of simpler entities into more complex structures. The result is then a nonlinear composition, an emergent entity that resists linear decomposition.

Follow me a step further in this reasoning, because it leads to some interesting conclusions about the body and its study. Well-described examples of emergent structures from the natural world include coupled cyclic reactions that sometimes emerge in chemical systems, emergent entities that Manfred Eigen (Eigen and Schuster 1979) has called hypercycles. Hypercycles, or coupled autocatalytic and cross-catalytic reaction sets, enable more complex emergent chemical structures to be built out of simpler components. This happens, for example, if the reaction sets work more efficiently when they are coupled than when they aren't—when, by virtue of their coupling, the interactions among their components are marked by higher frequency dynamics or stronger molecular interactions. High frequency dynamics then stabilize the emergent entities. Eigen argued that it is only on the basis of such couplings that nature could build more complex systems out of simpler components—life itself, for example. Needless to say, hypercycles are complex nonlinear structures.

All such complex emergent systems are marked by self-organization, sometimes called autopoeisis. Self-organization, as it occurs in the natural world, depends on introducing energy into a system of interacting elements such that, under certain parameter conditions, their interaction exhibits patterns, organization, or stable structural features. Prigogine's classic example (1978, 1980) is convection in a fluid system, the torroidal flow of molecules that emerges when water is heated and its temperature is maintained between an upper and lower threshold. Below the lower threshold, we see random or Brownian motion of the water molecules; above the upper threshold, the water boils and the self-organized torroidal flow dissolves into chaos.

What does this have to do with the study of the body? You might think that it suggests only that the body is a complex system. There is no real question about that. By itself, such a conclusion would lead us to claim that sociological analysis of the body, in purely sociological terms, is warranted only when what sociologists want to explain is emergent—not subject to linear decomposition. Second, it means that describing and explaining bodily properties or dynamics will always entail unraveling complex systems of feedback. Third, it means that self-organization is likely to be at work in producing the patterns or features in question. But something even more fundamental is implied by this analysis. We get to the heart of the matter by asking where the body leaves off and social life begins?

Here the answer, at least the answer I am proposing, will seem both more and less intuitive to social scientists. In a sense, I am asking if methodological individualism ought to be the epistemological framework of our analysis. And my answer is that it ought never to blight our view of social life again. Social life, like the natural systems I have mentioned, is a complex, nonlinear system—many such systems. It rests on social interaction, and social interaction is a nonlinear process entailing reciprocal dependencies and feedback of many kinds. And though individual persons are entities that enter into our understanding of social life, there are more fundamental units of analysis. As you might suppose from what I have already said, the core entity of social life—the minimal social system—is the dyad. Dyads are easier to explain than individuals because they are more fundamental—the building blocks out of which more (or less) complex systems of the social world emerge. Dyads are innate in social life—the elementary structures to which all social life has a tendency to return. They are the social life's ground form. Perturb an individual person, and he or she will seek to form a dyad—to couple into a system with another person. Pass a stressor through a community, and the community will tend to disaggregate into dyads. Perturb more complex social systems such as formal organizations, and again the dyad will emerge as a fallback structure. There are parameter conditions for these claims, but they aren't that difficult to establish empirically. By contrast, individuals are far less robust entities. They do not function well when isolated. Indeed, isolation eventually produces pathology in individuals—illness, madness, collapse, death. But bring individuals back into dyads, and they will stabilize and regain the strength to function socially. Indeed, incorporation into social life is a condition of individual functioning.

Much of what is sociologically interesting about the body arises from the tendency of persons to form these small social systems. This is really a physiologically motivated predisposition. It arises from the fact that our physiology does not end at the body, but completes itself through interpersonal channels. Indeed, evolution has privileged interpersonal over intrapersonal management of comfort—of how we feel. Cognition, by contrast, is slow relative to the speed with which physiological processes operate. So for children who are stressed or frightened, and for the rest of us as well, it is much better to get a hug than to think about getting hug. For this reason, some of the robust patterns in our social life can be understood only as extensions of physiological processes. These include the natural tendency to modulate anxiety through attachment behavior—an innate predisposition. We also benefit as a species from our inability to be indifferent to others' suffering. It draws us together. It supports social integration. It fuels normative behavior. This mutual caregiving is akin to what Shelly Taylor (2002) calls the "tending instinct." Women, under conditions of threat and arousal, group together with their children and with others close to them. Under the same conditions, men band together and engage in classic fight/flight behavior and threat management—aggressive defense of the community, attacking predators, or warding off enemies. All of these patterns are evidence of something we might call social physiology.

Social physiology depends on feedback between internal physiological systems and psychological, social and cultural systems. Cognitive, normative, and cultural controls all figure in how social physiology works out. In some senses, physiology itself is a kind of memory—learned responses of the species that have been conserved evolutionarily. Yet much of this memory is not in our heads but locked instead into collective and social behavior, immanent in the patterns that appear in our interaction and attachment. This appears most clearly (though certainly not only) in the attachment of an infant to its mother. To say this is to say that infant-caregiver attachment and interaction recapitulates evolutionarily conserved memory. These recapitulations regulate or modulate physiological activity, which in turn fuels the recapitulations themselves. As children undergo developmental changes, thinking and normative pressure, cognition and autobiographical memory, thereafter mix into the interpersonal management of comfort and feelings, sometimes controlling them, sometimes driven by them. Figuring out how they all work together is the ultimate problem for the human sciences. Right now, many of these subjects are like the small islands Feynman described in physics and in Babylonian mathematics. Building the bridges that link them ought to shape the research agenda of the future.

The fact that robust natural features in social life are studied at many levels of analysis, in disciplines as seemingly far apart as neuroscience and political science, physiology and sociology, actually indicates where some of the most important islands are and, hence, where the bridges ought to be constructed. At the present, theory that works on one of these islands mostly develops apart from what happens on others. You can't reach studies of clustering in networks from studies of the social construction of personal identity or of the performance of gender. There is too little theoretical consensus. We need far more developed empirical and theoretical description of the separate islands on which our research proceeds, and we need bridges that will span the gulf between research on social construction and research on the neurosociology of attach-ment, interaction, and the growth of social networks. But bridges among these subjects will first require that sociologists become less insular—more interested in what happens on other islands. Only then will the missing bridges begin to appear—when sociologists recognize that getting from one island to others can unify and deepen their discipline. This kind of development will depend on showing how work in one place helps to solve problems appearing in another, and it hints at the outline of future research and future professional development. Down with insularity, up with bridges.

Rochester, New York
June, 2009

Bibliography

Callebaut W. (ed.) (2003), *Taking the Naturalistic Turn, or How Real Philosophy of Science is Done*. Chicago: University of Chicago Press.

Camic C. and Hans J. (2004), *The Dialogical Turn: New Roles for Sociology in the Post-disciplinary Age*. London: Rowman and Littlefield.

Eigen M. and P. Schuster (1979), *The Hypercycle: A Principle of Natural Self-Organization*. Part C. "The Realistic Hypercycle." Naturwissenschaften, Vol. 65, Pg. 341–369.

Feynman R. (2001), *The Character of Physical Law*. Messenger Lectures, 1964, Boston: MIT Press.

Levine D. (1995), *Visions of the Sociological Tradition*. Chicago: University of Chicago Press.

Prigogine I. (1978), *Time, Structure, Fluctuation*. Science, Vol. 201, Pg. 777–85.
Prigogine I. (1980), *From Being to Becoming: Time and Complexity in the Physical Sciences*. New York: Freeman.
Taylor S. (2002), *The Tending Instinct: Men, Women, and the Biology of our Relationships*. New York: Holt, Winston and Company.
Wimsatt W. (2007), *Emergence as Non-Aggregativity and the Biases of Reductionisms*. In: *ReEngineering Philosophy for Limited Beings: Piecewise Approximations to Reality*. Cambridge: Harvard University Press.

About the contributors

Roberto Cipriani

Roberto Cipriani is full professor of Sociology, and chair of the Department of Education at University Rome 3. He has been president of the Italian Sociological Association, and Editor-in-Chief of *International Sociology*, journal of the International Sociological Association. He his author of articles and books translated into 8 languages: "Sociology of Religion. A Historical Introduction", in particular, is available in English, Italian, French, Spanish, Portuguese and Chinese.

Bianca Maria Pirani

Bianca Maria Pirani is Professor of Sociology of Cultural and Communicative Processes at the Faculty of Sociology of Sapienza, University of Rome, and President of the Research Committee 54 "The Body in the Social Sciences" of International Sociological Association. She is Member of the Editorial Board of SSIS [Sage Studies in International Sociology].

She has published widely on the ever increasing insights into the relations between human bodies and social knowledge specifically as follows:

She is editor of thr monograph *Bodily Order. Mind, Emotion and Social Memory*, "Current Sociology", vol.53, n.2, monogr aph 1, March 2005, Sage publication, ISSN 0011-3921, and co-editor with Ivan Varga of the book *The New Boundaries between Bodies and Technologies*, Cambridge Scholars Publishing, Newcastle, NE5 2JA, UK: 2008, ISBN 1-84718-499-5.

Combining both evolutionary and social perspectives, her forthcoming book *The Dance of the Bodies: the Social Places of Technology* introduces the notion of "acting body" in the wider system of the Western Social Sciences. The book places equal emphasis on data and theory, on laboratory experiments and real world applications and ultimately promotes the idea of the body as "active agent" in the processes of self-formation and social invention.

Ivan Varga

Ivan Varga, Professor Emeritus of Sociology, Queen's University, Kingston, Ontario, Canada, Honorary President, Research Committee 22 (Sociology of Religion) of International Sociological Association. Member, Editorial Board, *International Sociology.*

Volume One

Mark Featherstone
Mark Featherstone is Senior Lecturer in Sociology at Keele University, UK. His areas of specialism are social and political thought and psychoanalysis. His current research focuses on notions of utopia and dystopia in social and political thought and he recently published a monograph on this topic entitled *Tocqueville's Virus* (Routledge).

Jeffrey A. Halley
Jeffrey A. Halley teaches sociology at the University of Texas at San Antonio, where he is Director of the CNRS Laboratory for the Sociology of the Arts, Culture, and Communications. A past President of the Research Committee on the Sociology of the Arts, International Sociological Association, he is currently editor of two volumes of the journal, *Sociologie de l'art* (Sociology of Art), on Theory/Epistemology/Literature, and Rationalization and Resistance/Postmodernism. Recently he was Guest Professor, École des hautes études en sciences sociales, Paris and Marseille, France.

James P. Ordner
James P. Ordner received his Master's degree in sociology from The University of Texas at San Antonio in 2008, and is currently enrolled in the Ph. D program at the University of Kansas. His academic interests include critical theory, the problem of representation, and environmental sociology.

Pierre Bouvier
Pierre Bouvier, docteur d'Etat, is Professor emeritus of socioanthropology at the University of Paris X – Nanterre and member of the LAOIS research lab.

His most recently book is entitled *Le lien social* (Gallimard 2005, 2nd ed. 2007), An upcoming book on Frantz Fanon and Aimé Césaire, a "biographie croisée", will be published shortly.

Gaspare Mura
Gaspare Mura, professor emeritus of the Pontifical Urbaniana University of Rome, is the professor of "philosophical hermeneutics" at the Pontifical Lateran University and of the Holy Cross University of Rome. He was the Director: of the "International Institute for the Study of Atheism" at the Pontifical Urbaniana University; of the Urbaniana University Press and of the journal of philosophy and theology "*Euntes Docete*". Prof. Mura was also literary editor of the "Citt Nuova" editors of Rome, for whom he edited important works of philosophy and directed numerous collections and series of philosophical works. His brilliant editorial carrier performed was awarded in 2007 by the papal honor of "Commentator of the order of St. Sylvester". He is currently President of the "Academy of Humanities and Social Sciences"(A. S. U. S.) in Rome, which is founded by several professors with

the objective to promote interdisciplinary dialogue between the humanities, social sciences, empirical sciences and the science of religious convictions. He is also a consulter to the Pontifical Council for Culture. He asset numerous scientific publications of philosophical and theological and his own philosophical thought known internationally as the "veritative hermeneutics". Among his writings: *Hermeneutics and truth, history and problems of philosophy of interpretation* (1990, 1992), (this work received the Culture Award of the Presidency of the Council of Ministers); *Think the word, for a philosophy of the encounter* (2001); *Introduction to hermeneutics of the truth* (2005); *The religious phenomenon today: tradition, change, denial* (2001); *Holy Text and religions, compared Hermeneutic* (2006).

Umberto Melotti

Umberto Melotti is professor of Political Sociology and Sociology of Ethnic Relations at the Faculty of Sociology of Sapienza, University of Rome. His works in English include *Marx and the Third World* (Macmillan, London, and Humanities Press, Atlantic Highlands, N.J.; also available in Italian, Spanish and Chinese) and his contributions to *Sociobiology of Ethnocentrism* (ed. by V. Reynolds et al., Croom Helm, London, and Georgia U.P., Atlanta), *The Politics of Multiculturalism in the New Europe* (ed. by T. Modood and P. Werbner, Zed Books, London and New York) and *The New Boundaries between Bodies and Technologies* (ed. by B.M. Pirani and I. Varga, Cambridge Scholar Publishing, Newcastle). He was one of the main contributors of the *Dictionary of Race, Ethnicity and Culture* (Sage, London).

Chan Langaret

Chan Langaret is lecturer on socioanthropology at Paris X University. After researches on the comic book writers and players of role playing games, he currently works on a socioanthropological approach of the play actors, especially the links between the person and the characters.

Ana Lúcia de Castro

Ana Lúcia de Castro has a Ph.D in Social Sciences. She is a lecturer in the Department of Anthropology at Paulista State University (UNESP-Araraquara). She has experience in Urban Anthropology, and works with the following subjects: body cult, body and society, mass media, contemporary culture, identity and the consumer.

Salvatore Polito

Salvatore Polito is a sociologist and Gender Studies Scientific Coordinator of the Centre for Media and Cultural Studies CMCS-CRISC. His main field of interest and research is the analysis on contemporary development of cultural, political and social models of masculinity. He is also researcher in sociology of consumption. About these fields of study he has published several books and articles.

Natalie Haber-Barker
Natalie Haber-Barker, Loyola University Chicago is currently working on her dissertation project, which investigates the work and commodification of the Hooter waitress. Additionally, she is an adjunct professor at Loyola University Chicago and Indiana University.

Raewyn Connell
Raewyn Connell is University Professor at the University of Sydney. One of Australia's leading social scientists, she has worked in sociology, education, political science, history and gender studies. Her books include "Gender: in World Perspective", "Southern Theory", "Masculinities", and "Schools and Social Justice".

Claudia Malacrida
Claudia Malacrida is a University Scholar in the Social Sciences and Associate Professor in Sociology at the University of Lethbridge in Alberta, Canada. She has published widely on motherhood and loss (Mourning the Dreams, Left Coast Press) motherhood and ADHD (Cold Comfort: University of Toronto Press) and the body (Sociology of the Body: A Reader, Oxford University Press). Her current research focuses on mothers and disability in Canada and the UK, and on oral histories of institutional survivors.

Volume Two

Alan Dunning
Alan Dunning is an interdisciplinary, media artist, exploring the spectral and technologized body. He is the Head of Media Art + Digital Technologies at the Alberta College of Art + Design in Calgary.

Paul Woodrow
Paul Woodrow is Professor of Fine Arts at the University of Calgary. He has been collaborating on the art and science work the Einstein's Brain Project for the last thirteen years.

Tatiana Mazali
Tatiana Mazali has a Phd in Communication Sciences. She is researcher in the context of Sociology of Communication, Knowledge and Culture. She is researcher and lecturer at Polytechnic of Turin. Among his most recent publications: *Giovani, media e consumi digitali*, Liguori Editore, 2008; *Ex-Peau-Sition: nomadic bodies and new technologies*, in M.B. Pirani e I. Varga (eds.), *The new boundaries between bodies and technologies*, Cambridge Scholars Publishing, 2008; *Ex-Peau-Sition, il corpo riscritto nelle performance artistiche*, Guerini & Associati, 2007.

Marina Maestrutti

Marina Maestrutti, PhD in Epistemology, history of sciences and techniques, assistant professor in sociology at Paris I Panthéon-Sorbonne University. She works on the imaginaries accompanying nanotechnology (NST), and analyzes their cultural background and their potential impact on the cultural integration of technological innovations.

Chris Speed

Dr. Chris Speed is Reader in Digital Spaces across the Schools of Architecture and Landscape Architecture at the Edinburgh College of Art, where he teaches undergraduate, masters and supervises PhD students.

Caroline Moricot

Caroline Moricot is Maître de Conférences (Assistant Professor) in Sociology at University Paris 1, Panthéon-Sorbonne and member of the CETCOPRA (Centre d'Etude des Techniques, de COnaissances et des PRAtiques). She works on the stakes of appropriation of the automated devices in the fieldwork of civil and military aviation. She is Co-responsible for the Working Group "Body, techniques and society" of the French Association of Sociology.

Gianni Corino

Gianni Corino is a lecturer and artist at the Institute of Digital Art and Technology, University of Plymouth, working on interactive media.

Thomas S. Smith

Thomas Spence Smith is Professor of Sociology at the University of Rochester. He is the author of numerous scholarly articles and books, including the widely praised theoretical monograph *Strong Interaction* (1992). In recent years he has undertaken formal and computational research aimed at exploring sophisticated new theoretical models of social interaction that are based on important discoveries in social neuroscience.

Sidney M. Greenfield

Sidney M. Greenfield is Professor of Anthropology Emeritus at the University of Wisconsin-Milwaukee and co-chair of the Columbia University Seminars on Religion and Brazil. He has conducted ethnographic research in the West Indies, New Bedford, Massachusetts, various parts of Brazil and ethnohistorical and historical research in Portugal and the Atlantic Islands on problems ranging from family and kinship, patronage and politics, the history of plantations and plantation slavery and entrepreneurship to Spiritist surgery and healing and syncretized religions in Brazil. Author and/or editor of eight books, producer, director and author of five video documentaries, he has published some 130 articles and reviews in books and professional journals.

Diane Hafner

Diane Hafner lectures in the School of Social Work and Human Services at the University of Queensland, Australia. Her research focuses on interactions between Australian Indigenous people and public institutions such as museums; the social contexts of memory and emotion; and questions of social marginalisation and personal experience.

Deborah Brian

Deborah Brian is a lecturer in the School of Social Work and Human Services at the University of Queensland, Australia, where she lectures on working with Indigenous people in the context of professional practice. Her research interests include understanding the cultural contexts of memory production, and archaeological interpretations of the relationships between people and place.

André Suchet

Andre Suchet finishes a Ph. D. at the Institut de Géographie Alpine, University of Grenoble, France. His work talks about the history and the social geography of the outdoor sporting cultures. In particular, his findings discuss the influence of conflicts between groups in the structuring of the practices and their evolution.

Begonya Enguix

Begonya Enguix, Ph. D. in Social and Cultural Anthropology, is a lecturer at the Arts and Humanities Department (Universitat Oberta de Catalunya, Barcelona, Spain). Her research is focused in Gender and Sexuality, and Urban and Media Anthropology.

Elisenda Ardévol

Elisenda Ardévol, Ph. D., is Senior Lecturer at the Arts and Humanities Department (Universitat Oberta de Catalunya, Barcelona, Spain). Her research is focused in Media Anthropology and Ethnography of TIC mediated social interaction.

Antonella Frisiello

Antonella Frisiello has a degree in Clinical Psychology and a Master Degree in Cognitive Ergonomics. She works as Human Factor Specialist at ISMB, research centre on new technologies.

Cristina Piccirillo

Cristina Piccirillo is Psychologist and Clinical Trasnsactional Analyst. She works as Psychoterapist in Turin.